CAMBRIDGE LIBRARY COLLECTION

Books of enduring scholarly value

Earth Sciences

In the nineteenth century, geology emerged as a distinct academic discipline. It pointed the way towards the theory of evolution, as scientists including Gideon Mantell, Adam Sedgwick, Charles Lyell and Roderick Murchison began to use the evidence of minerals, rock formations and fossils to demonstrate that the earth was older by millions of years than the conventional, Bible-based wisdom had supposed. They argued convincingly that the climate, flora and fauna of the distant past could be deduced from geological evidence. Volcanic activity, the formation of mountains, and the action of glaciers and rivers, tides and ocean currents also became better understood. This series includes landmark publications by pioneers of the modern earth sciences, who advanced the scientific understanding of our planet and the processes by which it is constantly re-shaped.

A System of Mineralogy

Robert Jameson (1774–1854) was a renowned geologist who held the chair of natural history at Edinburgh from 1804 until his death. A pupil of Gottlob Werner at Freiberg, he was in turn one of Charles Darwin's teachers. Originally a follower of Werner's influential theory of Neptunism to explain the formation of the earth's crust, and an opponent of Hutton and Playfair, he was later won over by the idea that the earth was formed by natural processes over geological time. He was a controversial writer, accused of bias towards those who shared his Wernerian sympathies such as Cuvier, while attacking Playfair, Hutton and Lyell. He built up an enormous collection of geological specimens, which provided the evidence for his *System of Mineralogy*, first published in 1808 and here reprinted from the second edition of 1816. Volume 1 deals with what Jameson terms 'earthy minerals', including diamonds, rubies and feldspar.

Cambridge University Press has long been a pioneer in the reissuing of out-of-print titles from its own backlist, producing digital reprints of books that are still sought after by scholars and students but could not be reprinted economically using traditional technology. The Cambridge Library Collection extends this activity to a wider range of books which are still of importance to researchers and professionals, either for the source material they contain, or as landmarks in the history of their academic discipline.

Drawing from the world-renowned collections in the Cambridge University Library, and guided by the advice of experts in each subject area, Cambridge University Press is using state-of-the-art scanning machines in its own Printing House to capture the content of each book selected for inclusion. The files are processed to give a consistently clear, crisp image, and the books finished to the high quality standard for which the Press is recognised around the world. The latest print-on-demand technology ensures that the books will remain available indefinitely, and that orders for single or multiple copies can quickly be supplied.

The Cambridge Library Collection will bring back to life books of enduring scholarly value (including out-of-copyright works originally issued by other publishers) across a wide range of disciplines in the humanities and social sciences and in science and technology.

A System of Mineralogy

VOLUME 1

ROBERT JAMESON

CAMBRIDGE UNIVERSITY PRESS

Cambridge, New York, Melbourne, Madrid, Cape Town,
Singapore, São Paolo, Delhi, Tokyo, Mexico City

Published in the United States of America by Cambridge University Press, New York

www.cambridge.org
Information on this title: www.cambridge.org/9781108029735

This edition first published 1816
This digitally printed version 2011

ISBN 978-1-108-02973-5 Paperback

SYSTEM

OF

MINERALOGY.

A SYSTEM OF MINERALOGY.

BY

ROBERT JAMESON,

REGIUS PROFESSOR OF NATURAL HISTORY, LECTURER ON MINERALOGY, AND KEEPER OF THE MUSEUM IN THE UNIVERSITY OF EDINBURGH; Fellow of the Royal and Antiquarian Societies of Edinburgh; President of the Wernerian Natural History Society; Honorary Member of the Royal Irish Academy, and of the Honourable Dublin Society; Fellow of the Linnean and Geological Societies of London, and of the Royal Geological Society of Cornwall; Member of the Physical and Mineralogical Societies of Jena; of the Society of Natural History of Wetterau,

&c. &c. &c.

SECOND EDITION.

VOL. I.

EDINBURGH:

Printed by Neill & Company,

FOR ARCHIBALD CONSTABLE AND COMPANY, EDINBURGH; AND LONGMAN, HURST, REES, ORME & BROWN, LONDON.

1816,

PREFACE.

Mineralogy, although a science of comparatively modern date, has, within a short period of time, made rapid advances. It was first successfully cultivated in Germany. In Great Britain, so distinguished in all the other sciences and arts of life, it was, until lately, almost entirely neglected. Now, however, it has become with us a subject of general interest and attention, and, like Chemistry, is considered as a necessary branch of education. The establishment of Lectureships and Societies, having Mineralogy as one of their principal objects, is a strong proof of the public feeling of the importance and utility of this science. Within a few years, several of the Universities have founded Professorships of Mineralogy; and that munificent and patriotic association, the Honourable Dublin Society, have lately added to their establishment a Lecturer on this science. This example has been followed by other public bodies, and also by private associations.

The establishment of the Wernerian Natural History Society of Edinburgh in 1808, directed, in this part of the empire, the particular attention of naturalists to Mineralogy. Three years afterwards, an expression of the same feeling was manifested in England, by the establishment of the Geological Society of London,—a Society which has attracted to the study of Mineralogy a number of naturalists, distinguished for talent, enterprise and activity. And even in the remote county of Cornwall, a Geological Society has been lately founded, under royal patronage. But the present enthusiasm displayed throughout this country in the study of Mineralogy, is not entirely owing to the exertions of teachers, and the spirit excited by Societies: it has been also fostered, encouraged, and directed, by the writings of individuals. Of these, the most eminent is KIRWAN, whose System of Mineralogy excited very general attention, was long the standard book on this subject, and has been of infinite benefit to Mineralogy. Since the publication of the second and most valuable edition of that work, and which contained the first English account of the Wernerian System, several other authors have, by their writings, directed the studies, and assisted the labours of mineralogists. Among these, Dr KID of Oxford has distinguished himself, as the author of a treatise, entitled, "Outlines of Mineralogy." Dr THOMSON, in his System of Chemistry, has dedicated a volume to the science of Mineralogy, in which that distinguished

stinguished chemist has proposed several judicious improvements in the prevailing mineralogical systems; and Dr MURRAY, in his System of Chemistry, gives a view of the natural characters and chemical properties of the different species, adopting the chemical arrangement. ARTHUR AIKIN, Esq. Secretary to the Geological Society of London, has published a useful "Manual of Mineralogy." And the work now presented to the public attention, professes to have the same claims and intentions. It is an enlarged and improved edition of my *System of Mineralogy*, published several years ago. It contains, as far as my knowledge extends, a full statement of the principal facts known in regard to the Oryctognostic Characters and relations of Simple Minerals; and I believe it will not be found deficient in what respects the Chemical Characters and Composition. The Geognostic relations, and Geographical distributions, are given as fully as is consistent with the plan of the treatise; and the Economical Uses, and History of the Species, are also considered. The arrangement followed is that of WERNER, somewhat modified. I have not attempted to enter into the minuter details of Crystallography, because these are not consistent with the nature of this work, which is to be considered as a *popular view* of the Natural History of Simple Minerals.

The Plates which accompany the descriptions, are intended to represent only the more obvious forms of crystallised minerals, and in general such as are most fre-

 quently

quently met with, and are therefore most likely to engage the attention of the student.

Each description is preceded by a List of Synonymes from the best authors, and these are arranged chronologically; the first being the name as it occurs in the earliest writer; the last, that of the most modern author. Mr THOMAS ALLAN's treatise, entitled, "Mineralogical Synonymes," will be found useful to the Student of Mineralogy. The descriptive language used throughout the work, is explained in my "Treatise on the Characters of Minerals.

Since this work was put to the press, I have been favoured with communications from different quarters, which have contributed materially to its usefulness. I feel particularly indebted to my friends G. B. GREENOUGH, Esq. Dr MACCULLOCH, and Dr FITTON of Northampton. Nor can I allow to pass unacknowledged the polite attentions and communications of Dr CLARKE of Cambridge, Dr MURRAY, Dr THOMSON, Dr MACKNIGHT, Mr KÖNIG of the British Museum, Mr LOWRY of London, and Mr VIVIAN.

EDINBURGH, *April* 1816.

LIST

OF

WORKS

Quoted in this Treatise, and of the Abbreviated Titles employed.

Names of Authors.	Titles of the Works.	Abbreviated titles.
Theophrastus.	Περι λιθων, History of Stones, with an English version, and notes, including the modern history of the gems described by that author. By Sir John Hill. London, 1774.	Theophrastus.
Plinius.	C. Plinii sec. Historiæ Naturalis libri xxxvii.; quos interpretatione et notis illustravit J. Harduinus, Paris 1732, t. iii. fol. C. Plinii sec Hist. Nat. libri xxxvii. ex recenssione J. Harduini, Studius Societ. Bipont. 1783, 1784. Basil, 1546, fol.	Plin.
Agricola.	G. Agricola de Re Metallica l. xii. et de natura Fossilium, l. x. &c. Basil, 1546, fol.	Agricola.
Bacci.	De Gemmis et Lapidibus pretiosis, 1611.	
Boetius de Boot.	Gemmarum et Lapidum Historia, 1647.	
Boyle.	An Essay about the original Virtues of Gems, wherein are proposed and historically illustrated some conjectures about the consistence of the matter of precious stones, and the subjects wherein their chiefest virtues reside. By the Honourable Robert Boyle, Esq. Fellow of the Royal Society, London, 1672.	

Names of Authors.	Titles of the Works.	Abbreviated titles.
Wallerius.	John Gottschalk Wallerii Systema mineralogicum, 8vo.	Wall.
Linnæus.	Linne System. Naturæ, t. iii. cura J. F. Gmelin, Leipsiæ, 1793.	Lin.
Brückmann.	Urban Friederick Benedict Brückmann's Abhandlung von Edelsteinen, 2te Ausgabe. Braunschweig, 1773.	
	U. F. B. Brückmann's Beitræge zu seiner Abhandlung von Edelsteinen Braunschweig, 1778.	
	U. F. B. Brückmann's Gesammlete und eigene Beiträge zu seiner Abhandlung von Edelsteinen. Braunschweig, 1783.	
Dutens.	Dutens des Pierres precieuses, et des Pierres fines, avec les moyens de les connoitre. A Paris & Bale, 1778, 8vo.	
Rome de Lisle.	Cristallographie ou Description des formes propres a tous les corps du regne mineral. Avec figures et tableaux synoptiques de tous les cristaux connu. Par M. De Rome de Lisle, 2de edit. Paris 1783, 4. t. 8vo.	R. de L.
Bergmann.	Torb. Bergmann's Sciagraphia regni mineralis, secundum principia proxima digesti. Manuel du Mineralogiste, ou Sciagraphie du regne mineral, distribue d'apres l'analyse chimique, par M. Torb. Bergmann, traduite et augmentee des notes par Monge. A Paris, 1784.	Bergmann.
La Metherie	Theorie de la Terre, t. i. et ii. 1797.	Lam.
Kirwan.	Elements of Mineralogy, by Richard Kirwan, Esq. President of the Royal Irish Academy, &c. 2 vols. 8vo. 1794—96.	Kirw.
Napione.	Elementi di Mineralogia, 1 vol. 8vo. Turin, 1797.	Nap.
Werner.	Cronstedt's Versuch einer Mineralogie übersetzt, 1 vol. 8vo. Leipsic, 1780.	Wern. Cronst.
Idem.	Ausführliches und systematisches Verzeichniss des Mineralien-kabinet des Pabst von Ohain, 2 vols. 8vo. Freyberg, 1791, 1792.	Wern. Pabst.

Names of Authors.	Titles of the Works.	Abbreviated titles.
De Born.	Catalogue methodique et raisonne de la collection des Fossiles, de Mademoiselle Eleonore de Raab. Vienne, 1790.	De Born.
Saussure.	Voyages dans les Alpes, par Horace Benedict Saussure. 4 vols. 4to, 1779, 1796.	Saussure.
Wiedenman.	Wiedenman's Handbuch der Oryctognostischen Theils der Mineralogie. Leipzig, 1794, 8vo.	Wid.
Reuss.	Lehrbuch der Mineralogie nach des H. O. B. R. Karsten Tabellen, von F. Ambros Reuss. 4 vols. 1803, 1808.	Reuss.
Hauy.	Traite de Mineralogie par Hauv, 4 vols. Paris, 1801.	Hauy.
Brochant.	Traite elementaire de Mineralogie, suivant les principes du Professor Werner, 2 vols. Paris, an ix.	Broch.
Ludwig.	Handbuch der Mineralogie nach Werner, von C. F. Ludwig, Professor in Leipzig, 2 vols. 1803, 1804.	Lud.
Suckow.	Suckow's Anfangsgründe der Mineralogie nach der neuesten Endeckungen. Leipzig, 1803.	Suck.
Bertele.	Bertele's Handbuch der Minerographie. Landshut, 1804.	Bert.
Mohs.	Der Herrn von der Null mineralien-kabinet nach einem, durchaus auf aussere Kennzeichen gegrundeten systeme geordenet, beschrieben, &c. von F. Mohs, 3 bünde. Vien. 1804.	Mohs.
Haberle.	Characterisende Darstellung der Mineralien mit hinsichit auf Werner et Hauy's beobachtungen, von Dr Carl Const. Haberle. Weimer, 1806, 8vo.	Hab.
Lucas.	Tableau Methodique des Especes Minerales, par J. A. Lucas. Premiere Partie, Paris 1806. Seconde Partie, Paris 1813.	Lucas.
Leonhard.	Systematisch-tabellarische, Übersicht et Characteristik der Mineral Körper, von C. C. Leonhard, K. F. Merz, et Dr J. H. Kopp. Frankfurt am Maine, 1806, fol.	Leonhard, Tabel.
Brongniart.	Traite Elementaire de Mineralogie, par Alexandre Brongniart. 2 vols. Paris 1806.	Brong.

Names of Authors.	Titles of the Works.	Abbreviated titles.
Brard.	Manuel du Voyager et du Geologue Voyager, par C. P. Brard. 8vo, Paris 1808.	Brard.
	Traité des Pierres Précieuses, &c, 8vo, Paris, 1808.	
Karsten.	Mineralogische Tabellen met Ruchsicht auf die neuesten Entdeckungen aufgestellt und mit erlauternden anmerkungen versehen, von D. L. G. Karsten, Berlin, 1808, fol.	Karsten, Tabel.
Hausmann.	Entwurf eines Systems der unorganischen Natur Körper, von J. F. L. Hausmann. Cassel, 1809, 8vo.	Haus.
Kidd.	Outlines of Mineralogy, by J. Kidd, M. D. Professor of Chemistry in the University of Oxford, 2 vols. 8vo. Oxford, 1809.	Kid.
Hauy.	Tableau Comparatif des Resultats de la Cristallographie et de l'analyse chimique relativement à la Classification des Mineraux, par M. l'Abbé Hauy, 8vo. Paris, 1809.	Hauy, Tabl.
Lenz.	Erkenntnisslehre der anorgischen Natur Körper, von Dr J. G. Lenz. Giessen, 1813.	Lenz.
Hoffmann.	Handbuch der Mineralogie, von C. A. S. Hoffmann. 1811, 1812, Freyberg.	Hoff.
Oken.	Oken's Lehrbuch der Naturgeschichte 1ter Theil Mineralogie, 8vo, Leipzig, 1813.	Oken.
Bournon.	Catalogue de la Collection Mineralogique du Comte de Bournon. Londres, 8vo, 1813.	Bournon.
Hausmann.	Handbuch der Mineralogie von Joh. Friedr. Ludw. Haussmann, 3 bänd, 8vo. Göttingen, 1813.	Haus. Handb.
Aikin.	A Manual of Mineralogy. By Arthur Aikin, Secretary to the Geological Society, London, 1814.	Aikin.
Steffens.	Vollstandiges Handbuch du Oryktognosie, Erster Theil. Halle, 1811.	Steff.

Names of Authors.	Titles of the Works.	Abbreviated titles.
	JOURNALS, &c. referred to.	
Gehlen.	A. F. Gehlen's Newes allgemeines Journal der Chemie, Berlin, 1803, 1806, 6 bände 8vo.	Gehlen.
Schweigger.	J. C. Schweigger's Journal für Chemie und Physik. Nürenberg, since 1811, 8vo.	Schweigger.
Journal de Physique.	J. C. de Lamethrie, Journal de Physique, de Chimie, et d'Historie Naturelle. A Paris.	Journ. d. Phys.
Bergmännishe's Journal.	Bergmännishe's Journal, herausgegeben von Köhler et Hoffman, Freyberg.	Bergm. Journ.
Leonhard.	Leonhard's Taschenbuch für die gesammte Mineralogie, 1807 to 1815.	Leonhard, Taschenbuch
Moll's.	Ehrenberg. Freiherrn von Moll's Jahrbücher der Berg und Hüttenkunde, Salzburg, 1797, 1801, 5 bände, 8vo.	
	——— Annalen der Berg und Hüttenkunde, Salzburg, 1801—1805, 3 Bde 8vo.	
	——— Ephemeriden der Berg et Hüttenkunde, Nürnberg, 1805—1809, 5 Bde.	
	——— Neue Jahrbücher der Berg und Hüttenkunde, Nürnberg, since 1808, 8vo.	
Journal des Mines.	Journal des Mines, publié par le Conseil des Mines à Paris. Since 1794, in 8vo. numbers.	Journ. d. Min.
Annales du Museum.	Annales du Museum d'Histoire Naturelle; à Paris. Since 1802, in 4to numbers.	Annal. d. Mus.
Weber & Mohr.	Friedr. Weber's und D. M. H. Mohr's Archiv für die Systematische Naturkunde. Leipzig, 1804, 8vo. As continuation of this work,	Weber & Mohr.
	——— Beiträge zur Naturkunde, Kiel, 1805—1810, 2 Bde. 8vo.	

Names of Authors.	Titles of the Works.	Abbreviated titles.
Hisinger and Berzelius.	Afhandlingar i Fysik, Kemi, och Mineralogi, Utgifne af W. Hisinger och J. Berzelius, Stockholm, 1806—1810.	
Magazin Naturforschender Freunde.	Magazin der Gesellschaft Naturforschender Freunde zu Berlin. Since 1807, in 4to numbers.	Magaz. Gesel. Nat. f. Fr.
Klaproth.	Klaproth's Beiträge zur Chemischen Kenntniss der Mineral Körper, Berlin. 1795—1810, 5 Bde. 8vo.	Klap. Beit.
	J. F. John, Zwei Fortsetzungen des Chemischen Laboratoriums, unter d. Titel. Chemische untersuchungen mineralischer, vegetabilischer, und animalischer Substanzen. Berlin, 1810, 1811, 2 Bde. 8vo.	
	Thomson's Annals of Philosophy.	
	Transactions of the Royal Society of London.	
	Transactions of the Geological Society.	
	Transactions of the Royal Society of Edinburgh.	
	Memoirs of the Wernerian Natural History Society.	

In order to preserve as much as possible a uniformity of size in the different Volumes of this Work, I have placed three Appendices, containing the New Species, New Localities, and a Tabular View of the different Systems of Mineralogy, at the end of the *Second* Volume.

TABLE OF CONTENTS

OF

VOLUME FIRST.

SYSTEM OF ORYCTOGNOSY.

CLASS I.—EARTHY MINERALS.

I. Diamond Family.

II. Zircon Family.

III. Ruby Family.

IV. Schorl Family.

V. Garnet

V. Garnet Family.

VI. Quartz Family.

VII. Pitchstone

VII. Pitchstone Family.

VIII. Zeolite Family.

IX. Azurestone Family.

X. Felspar Family.

XI. Clay Family.

XII. Clay-Slate Family.

XIII. Mica Family.

XIV. Lithomarge Family.

XV. Soapstone Family.

XVI. Talc Family.

MINERAL

MINERAL SYSTEM.

CLASS I.

EARTHY MINERALS.

I. DIAMOND FAMILY.

Diamond.

Demant, or Diamant, *Werner.*

Adamas, *Plinius,* Hist. Nat. l. xxxvii. c. 4.—Alumen lapidosum pellucidissimum hyalinum, *Lin.*—Gemma vera colore aqueo, *Cartheus.*—Diamant, *Rome de Lisle,* t. ii. p. 189.—Quartzum nobile, *Born,* t. i. p. 56.—Diamond, *Kirw.* vol. i. p. 393.—Diamant, *Estner,* b. ii. s. 54.—*Id. Emm.* b. i. s. 187.—Le Diamant, *Broch.* t. i. p. 153.—*Id. Hauy,* t. iii. p. 287.—Diamant, *Reuss,* b. iii. s. 198 —*Id. Lud.* b. i. s. 57.—*Id. Suckow,* 1r th. s. 80,–85.—*Id. Bertele,* s. 333, 335.—*Id. Mohs,* b. i. s. 3,–16.—*Id. Haberle,* s. 161.—*Id. Lucas,* p. 91.—*Id. Brong.* t. ii. p 58.—Diamant, *Brard,* p. 205.—Diamond, *Kid,* vol. ii. p. 31—*Steffens,* b. i. s. 3.

External Characters.

The most common colours of the Diamond are white and grey. The varieties of white are, snow-white, grey-

 ish-

ish-white, and yellowish-white; of grey, ash-grey, smoke-grey, bluish-grey, pearl-grey, yellowish and greenish grey.

Besides these two colours, it occurs blue, red, brown, yellow, and green.

Of blue, the only variety is indigo-blue, which appears to pass into red.

Of red, the varieties are rose-red and cherry-red; from the latter it passes into clove-brown and yellowish-brown; from this into ochre-yellow, orange-yellow, wine-yellow, lemon and sulphur yellow; further, into siskin-green, asparagus-green, pistachio-green, leek-green, and, lastly, into mountain-green: which latter passes into greenish-grey, and greenish-white.

The clove-brown passes into blackish-brown, pitch-black, and greyish-black *.

Of all the colours, blue and black are the rarest.

The colours are generally pale and light, seldom deep, and very seldom dark. It exhibits a most beautiful play of colours, in the direct rays of the sun, or in candle-light, particularly when cut.

It occurs in rolled pieces, in indeterminate and spherical grains; also crystallised in the following figures:

1. Perfect equiangular octahedron, or double four-sided pyramid, in which the lateral planes are sometimes straight, sometimes convex †. This is the fundamental figure, or that from which all the others are supposed to be derived. Plate I. fig. 1.
2. Octahedron, with alternate larger and smaller planes.

3. Te-

* Mohs, in his Description of Von der Nüll's cabinet, mentions a greyish-black diamond.

† Diamant primitif, Hauy.

3. Tetrahedron, with slightly truncated angles.
4. Segment of the preceding figure.
5. Two segments, same as N° 4. joining or adhering by their bases, forming a twin-crystal.
6. Perfect tetrahedron.
7. Tetrahedron, in which all thc angles are acuminated by three planes, which are set on the lateral planes.
8. Octahedron, in which all the edges are truncated by cylindrical convex planes.
9. The rhomboidal or garnet dodecahedron, with convex planes.
10. The preceding figure somewhat elongated.
11. Garnet dodecahedron, in which the lateral planes are divided, in the direction of the longest diagonal; and the acuminating planes in the direction of the shorter diagonal.
12. Acute, double, six-sided pyramid, in which the lateral planes of the one are set on the lateral planes of the other, and again flatly acuminated by six planes, which are set on the lateral planes. It is a twin crystal.
13. Flat, double, three-sided pyramid, with convex planes, in which the lateral planes of the one are set on the lateral planes of the other, and each of the angles of the common base acuminated by four planes, which are set on the lateral planes. It is a twin crystal.
14. Flat, double, three-sided pyramid, in which the lateral planes of the one are set on the lateral edges of the other, and the angles on the common base truncated.
15. Octahedron, in which the edges are bevelled.

16. Octahedron, in which each of the planes is divided into three compartments; the dividing edges running from the middle point of the plane to the angles.
17. Octahedron, in which the edges are bevelled, and the bevelment once broken.
18. Octahedron, in which each of the planes is divided into six compartments; three of the dividing edges run from the middle of the plane to the middle of the edges; and three to the angles of the octahedron. Plate I. fig. 2.
19. Six-sided table, with oblique terminal planes *.

The crystals are small, and seldom middle-sized: they occur loose and single; rarely two or three are irregularly aggregated together.

The surface of the octahedrons is smooth or streaked; that of the grains uneven, granulated, sometimes approaching to drusy, and frequently rough.

Externally, its lustre alternates from splendent to glimmering; internally, it is always splendent, often specular-splendent, and the lustre is adamantine.

The fracture is straight and perfect foliated, with a fourfold equiangular cleavage; and the cleavages are parallel with the sides of the octahedron.

The fragments are octahedral or tetrahedral.

It rarely occurs in distinct concretions; and these are small, and fine granular.

It is seldom completely transparent; more generally it rather

* Some authors mention the cube as one of the forms of the diamond: Weiss assures us he saw a cubic diamond in the King of Prussia's collection in Berlin, and Mr Koenig of the British Museum informs me, that Mr Lowry has in his collection a cubic diamond truncated on its edges.

rather inclines to semitransparent; but the black variety is nearly opaque: and it refracts single.

Hard in the highest degree; it scratches all other known minerals.

Rather easily frangible.

Rather heavy, approaching heavy.

Specific gravity 3.518, *Cronstedt.*—3.521, *Muschenbroeck.*—3.521, *Wallerius.*—3.500, *Brisson.*—3.600, *Werner.*—3 5185 to 3.55, *Haüy.*—3.51 to 3.53, *Brong.*—3.488, *Lowry.*

Constituent Parts.

Boetius de Boot, in his History of Gems, published in 1609, conjectured that the diamond was an inflammable substance. In 1673, Boyle discovered, that when exposed to a high temperature, part of it was dissipated in acrid vapours. In 1694 and 1695, experiments were made in the presence of the Grand Duke of Tuscany, which confirmed those of Mr Boyle, and shewed that the diamond, although the hardest of minerals, agrees with combustible bodies, in being combustible. In 1704, Sir Isaac Newton, in his great work on Optics, hinted, that from its very great refracting power, it might be an unctuous substance coagulated *.

Since that period, the diamond has been very often examined by chemists, and they find, that when heated to the temperature of 14° of Wedgwood's pyrometer, or not so high as the melting point of silver, it gradually

 dissipates

* Newton does not appear to have been acquainted with the experiments made in Tuscany; and, besides, a considerable part of his work on Optics was written in 1675.

dissipates and burns, and combines with nearly the same quantity of oxygen, and forms the same proportion of carbonic acid, as charcoal. Hence it consists principally of carbon.

Physical Characters.

When rubbed, whether rough or polished, it shews positive electricity; whereas quartz, and the other precious stones, if rough, afford negative electricity, but when polished, positive electricity. It becomes phosphorescent when exposed to the rays of the sun. Many diamonds, however, do not become phosphorescent, although agreeing in colour, form, transparency, &c. with those which readily become luminous. The smaller acquire this property by a much shorter exposure to the light than the larger ones; sometimes a diamond that is not phosphorescent, by the mere action of the solar rays may be made so, by previously immersing it for some time in melted borax. Vid. Grosser in *Journ. de Physique*, vol. xx. p. 270.

Geognostic Situation.

In Golconda, and other districts in Hindostan, the diamond occurs imbedded in an ochry earth, of a yellow or red colour; at the foot of high mountains, which are said to contain quartz. This ochry earth is frequently mixed with quartz, sand, and rolled masses of different kinds, and the whole is sometimes so firmly conglutinated, as to form a pretty solid conglomerate. In the Brazils,

zils, it occurs also in alluvial deposites. According to Dandrada, the Brazilian diamonds are found partly loose in the sand and mud of rivers, and partly in a superficial crust, on the surface of hills, which consists of layers of sand with rolled masses of different kinds, and is denominated *Cascalho*. In this crust it occurs in octahedrons; but in roundish grains, and garnet dodecahedrons, in the sand of rivers.

The original repository of the diamond is still unknown. Werner suspects that it occurs, like zircon, pyrope, and spinel, imbedded in rocks belonging to the newest flœtz-trap formation.

Geographic Situation.

Asia.—The diamond was first found in this quarter of the globe, and is still collected there, although not in such quantity as formerly. It occurs principally in the provinces of Golconda and Visapour, Bengal, and the island of Borneo.

America.—Diamonds were first found in America in the district of Serro Dofrio in Brazil, towards the beginning of the last century.

Lord Anson, who performed his voyage round the world in the years 1740–1–2–3 and 4, gives the following account of its first discovery: " I have already mentioned, that besides gold, this country does likewise produce diamonds. The discovery of these valuable stones is much more recent than that of gold, it being as yet scarce twenty years since the first were brought to Europe. They are found in the same manner as the gold, in the gullies of torrents, and beds of rivers; but only in particular places, and not so universally spread through the country. They were often found in washing the gold, before they were known to be diamonds, and were consequently thrown away with the sand and gravel separated from it. And it is very well remembered, that numbers of very large stones, which would have made the fortunes of the possessors, have passed unregarded through the hands of those, who now with impatience support the mortifying reflection. However, about twenty years since, a person acquainted with the appearance of rough diamonds, conceived that these pebbles, as they were then esteemed, were of the same kind: But it is said, that there was a considerable interval between the first starting this opinion, and the confirmation of it by proper trials and examination, it proving difficult to persuade the inhabitants, that what they had been long accustomed to despise, could be of the importance represented by this discovery; and I have been informed, that in this interval, the governor of one of these places procured a good number of these stones, which he pretended to make use of at cards, to mark with, instead of counters. But it was at last confirmed by skilful jewellers in Europe, consulted on this occasion, that the stones thus found in Brazil were truly diamonds, many of which were not inferior, either in lustre or any other quality, to those of the East Indies *."

But Serro Dofro is not the only district in Brazil where this gem is found; it is also collected in the rivers Giquitignogna, Riacho Fundo, and Rio de Peixe: and Dandrada informs us, that there are unopened diamond mines

* Anson's Voyage, 4to, p. 51.

mines at Ceyaba, and in the vicinity of Guara Para, in the province of St Paul.

Uses.

1. The diamond, on account of the splendour of its lustre, its peculiar play of colour, its hardness, and, lastly, its rarity, is considered as the most precious substance in the mineral kingdom, and is particularly valued by jewellers. The diamonds purchased by jewellers are generally in grains or crystals, and sometimes coarsely polished.

The ancients were unacquainted with the art of cutting the diamond, and hence they used it in its natural granular or crystallised state *. Even in the middle ages, this art still remained unknown; for the four large diamonds that ornament the clasp of the imperial mantle of Charlemagne, and which is still preserved in Paris, are uncut octahedral crystals.

The art of cutting and polishing diamonds was probably known to the artists of Hindostan and China at a very early period. European artists, until the fifteenth century, were of opinion that it was impossible to cut the diamond. Robert de Berghen relates, that Louis Berghen, a native of Bruges, in the year 1456, endeavoured to polish two diamonds, by rubbing them against each

* Some antiquaries pretend, that the ancients cut figures on the diamond itself. Govi, for example, cites an antique head of this kind, in the possession of the Duke of Bedford. Lessing, a very acute and skilful antiquary, is of opinion, that these pretended antique cut diamonds are amethysts or sapphires.

each other: he found that by this means a facet was produced on the surface of the diamonds; and in consequence of this hint, constructed a polishing wheel, on which, by means of diamond powder, he was enabled to cut and polish this substance, in the same way as other gems are wrought by emery. James of Trezzo appears to have been one of the first artists who cut figures on the diamond itself. Clement of Biragues, in the year 1564, cut figures on the diamond; and even so early as year 1500, Charadossa cut the figure of one of the Fathers of the Church on a diamond for Pope Julius II. The artists Natter and Costanzi were also famous for cutting figures on the diamond.

Diamonds, according to Jeffries *, are cut and manufactured by jewellers into *brilliants* and *rose diamonds* the former being for the most part made out of the octahedral crystals, and the latter from the spheroidal varieties.

To fashion a rough diamond into a *brilliant*, the first step is to modify the faces of the original octahedron, so that the plane formed by the junction of the two pyramids shall be an exact square, and the axis of the crystal precisely twice the length of one of the sides of the square. The octahedron being thus rectified, a section is to be made, parallel to the common base, or *girdle*, so as to cut off $\frac{5}{18}$ths of the whole height from the upper pyramid, and $\frac{1}{18}$th from the lower. The superior and larger plane thus produced, is called the *table*, and the inferior and smaller one is named the *collet*: in this state it is called a *complete square table diamond.* To convert it into a brilliant, two triangular facets are placed on each

* A Treatise on Diamonds and Pearls, &c. by David Jeffries, jeweller, 2d edit. 1751.

each side of the table, thus changing it from a square into an octagon; a lozenge-shaped facet is also placed at each of the four corners of the table, and another lozenge extending lengthwise along the whole of each side of the original square of the table, which, with two triangular facets set on the base of each lozenge, complete the whole number of facets on the table side of the diamond, viz. eight lozenges, and twenty-four triangles. On the collet side are formed four irregular pentagons, alternating with as many irregular lozenges, radiating from the collet as a centre, and bordered by sixteen triangular facets adjoining the girdle. The brilliant being thus completed, is set with the table side upwards, and the collet side implanted in the cavity made to receive the diamond

The regular *rose diamond* is formed by inscribing a regular octagon in the centre of the table side of the stone, and bordering it by eight right-angled triangles, the bases of which correspond with the sides of the octagon; beyond these is a chain of eight trapeziums, and another of sixteen triangles. The collet side also consists of a minute central octagon, from every angle of which proceeds a ray to the edge of the girdle, forming the whole surface into eight trapeziums, each of which is again subdivided by a salient angle (the apex of which touches the girdle), into one irregular pentagon, and two triangles.

In the formation either of a brilliant or a rose diamond of regular proportions, so much is cut away, that the weight of the polished gem is no more than half that of the rough crystal out of which it was formed; whence the value of a cut diamond is esteemed equal to that of a similar rough diamond of twice its weight, exclusive of the cost of workmanship. The weight, and consequently ly the value, of diamonds, is estimated in *carats*, one of which is equal to four grains, and the difference between the price of one diamond and another, *cæteris paribus*, is as the squares of their respective weights. Thus, the value of three diamonds of one, two, and three carats weight respectively, is as one, four, and nine. The average price of rough diamonds that are worth working, is about £2 Sterling for the first carat; and consequently in wrought diamonds, exclusive of the cost of workmanship, the cost of the first carat is £8. In other words, in order to estimate the value of a wrought diamond, ascertain its weight in carats, and fractions of a carat, multiply this by two, then multiply this product into itself, and finally multiply this latter sum by £2. Hence a wrought diamond of

1 carat, is worth	£8
2	32
3	72
4	128
5	200
6	288
7	392
8	512
9	612
10	800
20	3,200
30	7,200
40	12,800
50	20,000
60	28,800
70	39,200
80	51,200
90	64,800
100	80,000

This

This rule, however, actually holds good only in the smaller diamonds of 20 carats and under; the larger ones, in consequence of the scarcity of purchasers, being disposed of at prices greatly inferior to their estimated worth. The value of some of the most perfect diamonds exceeds that given in the table; but for a stone that is flawed, cloudy, or of a bad colour, sometimes three quarters of the whole tabular value must be deducted.

2. The transparent snow white variety is considered to be the most valuable; the green and yellow varieties are also much esteemed; the green and blue varieties were formerly more valued than at present; and the least valuable are the grey and brownish varieties. Black diamonds are much prized by collectors.

3. The principal use of the diamond is in jewellery; it is also used by lapidaries for cutting and engraving upon the hardest gems, and by clock-makers in the finer kinds of clock-work; in the glass-trade, for squaring large pieces or plates of glass, and among glaziers for cutting their glass.

4. Zircon is sometimes substituted for diamond, but may be distinguished from it by the muddiness of its colours, its very feeble play of colour, inferior lustre and hardness.

5. We may here give a short account of some of the most remarkable diamonds noticed by authors.

The great Brazilian diamond, in the possession of the Queen of Portugal, is the largest that has been hitherto discovered, its weight being stated at 1680 carats. Many are of opinion, that this remarkable stone is a fine white-coloured topaz; and therefore the largest undoubted diamond, is that mentioned by Tavernier, which was in the possession of the Great Mogul, and which that traveller found to weigh 279- carats. Its form and size are equal to that of half a hen's egg, and it is cut in the

the rose form. Before cutting, it weighed 900 carats. It was found in the mine of Colore, to the east of Golconda, about the year 1550.

The magnificent diamond in the crown of the sceptre of the Emperor of Russia, deserves next to be noticed. It is perfectly pure; weighs 195 carats; and is the size of a pigeon's egg. It was one of the eyes of a Brahminical idol; and was stolen by a French grenadier, who disposed of it at a very low price; and lastly, after passing through three other hands, it was offered to the Empress Catharine of Russia, who purchased it for about L. 90,000 ready money, and an annuity of about L. 4000 more.

The late Grand-Duke of Tuscany had in his possession a diamond, of a pale yellowish colour, but beautifully formed, and which weighed 139½ carats.

The last diamond we shall mention, is the Pitt or Regent Diamond, which was brought from India by an English gentleman of the name of Pitt, and was sold by him to the Regent Duke of Orleans, by whom it was placed among the Crown jewels of France. It is cut in the form of a brilliant, and is said to be the most beautiful diamond hitherto found. It weighs $136\frac{4}{16}$ carats, and was purchased for L. 100,000, although it is now valued at double that sum.

II. ZIRCON

II. ZIRCON FAMILY.

Zircon.

This species is divided into two subspecies, Common Zircon, and Hyacinth.

First Subspecies.

Common Zircon*.

Zirkon, *Werner*.

Topazius clarus hyalinus jargon, *Wall.* t. i. p. 252.—Jargon de Ceylan, *Romé de Lisle,* t. ii. p. 229. *Id. Born.* t. i. p. 77.—Zirkon, *Wid.* s. 233. *Id. Kirwan,* vol. i. p. 257. *Id. Estner,* b. ii. s. 35. *Id. Emm.* b. i. s. 3.—Giargone, *Nap.* p. 105.—Zircon, *Lam.* t. ii. p. 204. *Id. Broch.* t. i. p. 159. *Id. Hauy,* t. ii. p. 465.—Gemeiner Zirkon, *Reuss,* b. i. s. 56.—Zircon, *Lud.* b. i. s. 58. *Id. Suck.* 1r th. s. 166. *Id. Bert.* s. 304. *Id. Mohs,* b. i. s. 16. *Id. Lucas,* p. 89.—Gemeiner Zircon, *Hab.* s. 1.—Zircon Jargon, *Brong.* t. i. p. 269.—Zircon, *Kid,* vol. i. p. 125. *Id. Brard,* p. 106. *Id. Steffens,* b. i. s. 7.

External Characters.

The principal colour is grey: it also occurs white, green, and brown; and rarely yellow, blue, and red. White and brown are the extremes of its colour-suite, and the intermediate colours are grey, yellow, green, blue, and red.

The colours are generally pale, seldom dark.

It

* The word Zircon, is by some authors considered to be of Indian origin: others derive it from the French word *jargon,* which was applied to all those gems, which, on being cut and polished, had somewhat of the appearance of diamond.

It occurs in angular, or roundish original grains; and crystallised in the following figures:

1. Rectangular four-sided prism, rather flatly acuminated on the extremities with four planes, which are set on the lateral planes under equal angles *, fig. 3. This is the fundamental figure †.
2. The fundamental figure truncated on the lateral edges.
3. The fundamental figure bevelled on the angles between the acumination and the prism, and the bevelling planes set on the edges between the acumination and the prism ‡, fig. 4. When these bevelling planes become larger, so that they meet and intersect each other, there is formed
4. A four-sided prism, acutely acuminated on the extremities by eight planes, of which two and two meet under very obtuse angles, and are set on the lateral planes of the prism. This acumination is frequently rather flatly acuminated by four planes, which are set on the obtuse edges of the first acumination.
5. N° 3. in which the edges between the acumination and the prism are truncated ||, fig. 5.
6. When the prism of N° 1. disappears, there is formed an octahedron or double four-sided pyramid.

The crystals are generally small and very small, seldom middle-sized, and occur loose or imbedded.

The

* Zircon prisme, Hauy.

† The primitive form of Zircon, according to Hauy, is an octahedron, composed of two four-sided pyramids, applied base to base, whose sides are isosceles triangles. The inclination of the sides of the same pyramid to each other, is 124° 12 : the inclination of the sides of the one pyramid to to those of another, 82° 50 ; the angle of the summit is 73° 44.

‡ Zircon plagiedre, Hauy.

|| Zircon soustractif, Hauy.

The surface of the crystals is sometimes rough, sometimes smooth and shining, and that of the grains uneven and glistening.

The grains are glistening, the crystals shining.

Internally it is splendent, passing into shining, and the lustre is intermediate between adamantine and resinous, but rather more inclining to the first.

The fracture is perfect flat conchoidal: sometimes an imperfect foliated fracture may be observed, in which the folia are parallel with the lateral planes of the prism.

The fragments are sharp-edged.

It is sometimes transparent, but more commonly it is only semi-transparent, and translucent. It refracts double.

It is much harder than quartz, but softer than diamond.

It is rather easily frangible.

Specific gravity, from 4.416 to 4.700, *Werner*. 4.557, to 4.721, *Lowry*.

Chemical Character.

It is infusible, without addition, before the blowpipe.

Constituent Parts.

	Zircon of Ceylon.	Zircon of Norway.
Zirconia, -	69.00	63
Silica, - -	26.50	33
Oxide of Iron, -	0.50	1
	96.00	99
	Klaproth, Beit. b. i. s. 222.	Id. *Klaproth*, b. iii. s. 271.

The Geognostic and Geographic Situations are the same with those of the second subspecies.

Observations.

1. This species is characterised by its colour-suite, the principal members of which are grey, green, and brown,

generally of a muddy aspect, its suite of crystals, adamantine lustre, conchoidal fracture, considerable hardness, and weight.

2. It is distinguished from *Hyacinth*, by colour, crystallization, kind of lustre, and perfect conchoidal fracture: from *Diamond*, by its crystallization, greater weight, inferior hardness, conchoidal fracture, and its grey muddy colours: from *Topaz*, by its crystallization, smooth lateral planes, kind of lustre, fracture, greater hardness, and weight: from *Vesuvian*, by lustre, perfect acumination, strong double refracting power, greater hardness, and weight: from *Chrysolite*, by crystallization, lustre, greater hardness, and weight; and from all other cut and polished *gems*, by its exhibiting a stronger double refracting power.

Second Subspecies.

Hyacinth *

Hiacinth, *Werner.*

Topazius flavo-rubens, Hyacinthus, *Wall.* t. i. p. 252.—Hiacinth, *Wid.* s. 254. *Id. Kirw.* vol. i. p. 257. *Id. Estner,* b. ii. s. 141. *Id. Emm.* b. i. s. 205.—Giacinto, *Nap.* p. 109.—L'Hyacinthe, *Broch.* t. i. p. 163.—Hiacinth, *Reuss,* b. i. s. 62. *Id. Lud.* b. i. s. 59. *Id. Suck.* 1r th. s. 172. *Id. Bert.* s. 308. *Id. Mohs,* b. i. s. 23. *Id. Hab.* s. 2.—Zircon Hyacinthe, *Brong.* t. i. p. 270.—Hyacinth, *Kid,* vol. i. p. 126. *Id. Steffens,* b. i. s. 7.

External Characters.

The most frequent colours are red and brown, more rarely yellow, grey, and green; and the rarest is white. The principal colour is hyacinth-red, which passes on the one

* The Hyacinth of the ancients appears to have been either amethyst or sapphire. The name hyacinth, is derived from that of the plant denominated *hyacinthus* by the ancients, which is supposed to be the Hyacinthus orientalis, *Lin.*

one side into orange-yellow; on the other into reddish-brown, brownish-red, and flesh red.

It occurs sometimes in angular grains; more frequently crystallised in the following figures:

1. Rectangular four-sided prism, rather acutely acuminated on both extremities by four planes, which are set on the lateral edges. This is the fundamental figure, fig 6
2. The preceding crystal, slightly truncated on the lateral edges, fig. 7.

When these truncating planes become broader, and the lateral planes smaller, there is formed

3 A four-sided prism, differing from that of zircon, in the acumination being rather more acute
4. N° 2. in which the edges between the lateral and acuminating planes are also truncated, fig. 8.

When the prism of N° 1. becomes so low that the two acuminations touch each other, there is formed

5. An irregular garnet dodecahedron; and when the prism entirely disappears,
6 A flat octahedron is formed.

The crystals are small and very small, seldom middle-sized. They are all around crystallised

The surface of the crystals is smooth and splendent.

Internally it is specular-splendent, and the lustre is intermediate between resinous and vitreous.

The fracture is perfect straight foliated, with a double rectangular cleavage, and the folia are parallel with the diagonals of the prism N° 1 *

* Mohs observed a sixfold cleavage, of which two of the cleavages are parallel with the lateral planes, and four with the acuminating planes; and Haberle conjectures that this mineral has twenty-one cleavages, but of all these two only are very distinct, and are those mentioned above.

The fragments in general are sharp-edged, and sometimes prismatic, with three sides.

It is sometimes transparent, sometimes semi-transparent, or only translucent: it refracts double.

It is hard in a high degree; scratches quartz with ease.

It is rather easily frangible.

Specific gravity, 4.200, 4.300, *Guyton.* 4.545, 4.620, *Klaproth.* 4.525, 4.780, *Mohs.*

Chemical Characters.

Before the blowpipe it loses its colour, but not its transparency, and is infusible without addition.

Constituent Parts.

	Hyacinth of Ceylon.	Hyacinth of Expailly.	
Zirconia,	70.00	64.00	66.0
Silica,	25.00	32.00	31.0
Oxide of Iron,	0.50	2.00	2.00
Loss,	4.50	1.50	1.00
	100.00	100	100
	Klaproth, Beit. b. i. s. 231.	*Vauquelin*, Jour. d. Mines, N. 26. p. 106.	

Observations.

1. It is characterised by its colour-suite, the central colour of which is hyacinth-red, its crystallizations, resinous lustre, foliated fracture, and great specific gravity.

2. It is distinguished from *Common Zircon* by its colours, crystallization, external and internal lustre, and fracture: from *Precious Garnet*, by its crystallization, resinous lustre, foliated fracture, greater hardness, greater weight, and infusibility. Even the garnet dodecahedron of hyacinth cannot be confounded with that of precious garnet:

garnet; for if the dodecahedral garnet be viewed as a six-sided prism, the dodecahedral hyacinth will appear as a four-sided prism; and in the garnet, the adjacent planes meet under angles of 120°, but in the hyacinth under angles of 124° 12′ and 117° 54′.

3. The common zircon has been frequently confounded with the Diamond, and the hyacinth with several mine rals, as appears from the following enumeration. 1. The oriental hyacinth of Rome de Lisle (t. ii. p. 282.) is orange-coloured sapphire. 2. The occidental hyacinth is yellow-coloured topaz (Dutens, *Des pierres prec.* p. 62.) 3. Cruciform hyacinth is cross-stone. 4. Brown volcanic hyacinth is vesuvian. 5. White hyacinth of Somma is meionite. 6. Hyacinth of Compostella is iron-shot quartz. 7. Hyacinth of Dissentis (Saussure, *Voyages dans les Alpes,* n. 1902.) is a variety of garnet.

Geognostic Situation of the Zircon species, including Common Zircon and Hyacinth.

It occurs in grains and crystals, imbedded in transition sienite; also imbedded in basalt and lava, and dispersed through alluvial soil, along with sapphire, spinel, ceylanite, pyrope, tourmaline, augite, olivine, iron-sand, iron-pyrites, and gold.

Geographic Situation of the Zircon Species, including Common Zircon and Hyacinth.

Europe.—It has been found in several places in this quarter of the globe, not only loose in the sand of rivers, but also in its original repository. Thus it occurs in considerable quantity, along with sapphire and iron-sand, in what is called *volcanic sand,* in the rivulet of Rioupezzouliou, near Expailly in Auvergne; also near to Pisa, and in the supposed volcanic sand of the Vicentine. In

 the

the vicinity of Trziblitz and Podsedlitz in Bohemia, it occurs in a clayey alluvial deposite, near rocks of the newest trap formation, along with pyrope, sapphire, and iron; also, in very small grains, in auriferous sand, in Silesia; and in the trap rocks around Lisbon, and in those of Spain.

It was first found in its original repository at Friedrickschwärn, in the district of Christiania in Norway, where it occurs in considerable abundance in transition sienite. Faujas St Fond found it imbedded in basalt near Expailly: Cordier in a similar rock in the mountain of Anise, also in Auvergne; and Weiss found it imbedded in a volcanic scoria in the same country. In the year 1812, I found it imbedded in a rolled mass of sienite in the shire of Galloway.

As a —In the island of Ceylon, where this mineral was first found, it occurs in the sand of rivers, along with spinel, sapphire, tourmaline, and iron-sand. It has been observed in a similar situation in the district of Ellore in Hindostan: and it is mentioned by Reuss as a production of Asiatic Russia.

America.—A Spanish mineralogist, M. Henri Amana, presented Hauy with some small crystals of zircon, which had been collected in the province of Antioquia, in the kingdom of Santa Fe de Bogota: it is mentioned as a mineral of Brazil; and Mr Solomon W Conrad found it imbedded in a primitive rock, composed of felspar and quartz, near Trenton in New Jersey, in the United States .

Africa.—It is said to occur in Teneriffe.

Uses,

* Bruce's Mineralogical Journal.

Uses.

As common zircon is considered by jewellers one of the gems, it is frequently cut and polished, and used for ornamental purposes. The greyish-white and yellowish-white varieties are the most highly valued, on account of their resemblance to the diamond. The darker coloured varieties can be deprived of their colour by exposure to heat: hence artists generally employ this method, when they intend to employ zircon in place of diamond. Like the diamond, it is cut into the table, rose and brilliant forms, and is used for jewelling watches, ear-pendents, necklaces, and, on account of the intermixture of grey in the colour, it is particularly valued in some countries as an ornament in mourning-dress. When cut, it exhibits in a faint degree the play of colours of the diamond; and hence it is not unfrequently sold as an inferior kind of diamond. The Hyacinth is also esteemed by jewellers, and, when pure, and of considerable size, is employed in various kinds of ornamental work. But it seldom occurs large, and in trade, other minerals, as cinnamon-stone, pale garnets, and rock-crystals, are frequently substituted for it.

III. RUBY FAMILY.

This Family contains the following species: Automalite, Ceylanite, Spinel, Sapphire, Emery, Corundum, and Chrysoberyl.

1. Automalite.

Automolith, *Werner*.

Automalit, *Eckeberg*, in N. Allgem. Journal der Chemie, 5. B. s. 422,—455.—Automalit, Corindon zincifere, Jour. de Phys. an 14, p. 270.—Automalit & Fahlunit, *Karst.* Tabel. p 102. —Gahnite, *Von Moll.*—Spinelle zincifere, *Hauy*, Tabl. p. 67. Automalith, *Steffens*, b. i. s. 32.

External Characters.

Its colour is muddy duck-blue, which inclines very much to mountain-green.

It has been hitherto found only crystallised, and in the following figures:

1. Perfect octahedron.
2. Octahedron, with alternate larger and smaller planes.
3. Tetrahedron, truncated on the angles.
4. Segment of the tetrahedron.
5. Two segments, N° 4. joined together, so that re-entering angles are formed on the three corners of the figure.

The crystals are small and middle-sized; all around crystallised: and the planes are smooth.

Externally it is glistening; and the lustre is pearly, inclining to semi-metallic. Internally it is shining on the principal fracture, but glistening on the cross fracture, and the lustre is resinous.

The fracture is foliated, and exhibits a fourfold cleavage, parallel with the planes of the octahedron: a flat conchoidal fracture is also to be observed.

The fragments are splintery, or angular, and not very sharp-edged.

It is opaque, or faintly translucent on the edges.

It is so hard as to scratch quartz.

It

It is brittle.
It is rather easily frangible.
It is heavy.
Specific gravity, 4.261, *Hisinger*. 4.696, *Hauy*.

Chemical Character.

It is infusible before the blowpipe.

Constituent Parts.

Alumina,	60	- -	42
Silica,	4	- -	4
Oxide of Zinc,	24	- -	28
Iron,	9	- -	5
Sulphur,	0	- -	17
Loss,	3	Undecomposed,	4
	100		100
	Eckeberg, J. de Phys, an 14, p. 270.		*Vauquelin*, Annales du Mus. t. vi. p. 33.

Geognostic and Geographic Situations.

It occurs imbedded in talc-slate, along with galena, and has been hitherto found only at Fahlun at Sweden.

Observations.

In its crystallizations it nearly resembles both Ceylanite and Spinel: it is distinguished from the former by its more distinct green colour, foliated fracture, inferior hardness, superior specific gravity, and chemical composition: from the latter by colour, inferior lustre, perfect foliated fracture, low degree of transparency, inferior hardness, greater specific gravity, and chemical composition.

2. Ceylanite.

2. Ceylanite.

Zeylanit, *Werner*.

Schorl ou Grenat brun, *Romé de Lisle*, t. iii. p. 180. Note 21. —Ceylanit, *La Metherie*, Journ. de Phys. 1793, p. 23.—Pleonaste, *Hauy*, t. iii. p. 17. *Id. Broch.* t. ii. p. 525.—Ceylanite, *Reuss*, b. ii. th. ii. s. 38. *Id. Lud.* b. ii. s. 148. *Id. Suckow*, 1r th. s. 148.—Pleonast, *Bertele*, s. 284. *Id. Mohs*, b. i. s. 100. *Id. Lucas*, p 52. 263.—Spinelle pleonaste, *Brong.* t. i. p. 438. *Id. Steffens*, b. i. s. 27.

External Characters.

Its colour is muddy duck-blue, and greyish-black, which approaches to iron-black.

It occurs in blunt angular pieces, and grains; and crystallised in the following figures:

1. Octahedron, either perfect, or truncated on the edges. Figs. 9, and 10.
2. Octahedron, having each of its angles acuminated by four planes, which are set on the lateral planes. Fig. 11.
3. Garnet or rhomboidal dodecahedron. Fig. 12.

The crystals are small, and very small, seldom middle-sized; and sometimes imbedded, sometimes superimposed.

Externally the angular pieces and grains are rough and glimmering, or glistening, but the crystals are smooth and splendent.

Internally it is splendent, and the lustre is resinous, inclining to semi-metallic.

The fracture is perfect, and rather flat conchoidal.

The fragments are angular, and very sharp-edged.

It is translucent on the edges.

It is hard: scratches quartz, but not so readily as spinel.

It is rather easily frangible.

It

It is rather heavy; approaching to heavy.

Specific gravity, 3.7647, or 3.7931, *Hauy.*

Chemical Character.

It is infusible before the blowpipe.

Constituent Parts.

Alumina, - -	68
Magnesia, - -	12
Silica, - - -	2
Oxide of Iron, - -	16
Loss, - - -	2
	100

Collet Descotil, Ann. de Chem. xxxiii.

Geognostic and Geographic Situations.

This mineral was first found in the Island of Ceylon, where it occurs in the sand of rivers, along with tourmaline, zircon, sapphire, and iron sand. It also occurs in the ejected unaltered rocks at Monte Somma. These rocks are sometimes calcareous, sometimes composed of leucite, felspar, mica, quartz, and olivine, and contain in their cavities octahedral crystals of ceylanite. It occurs also in the trap rocks near Andernach on the Rhine, and in the supposed volcanic rocks of Valmaargue, Montferrier, and at Lestz near Montpellier.

It thus appears to be an inmate of the flœtz-trap formation; probably also of volcanic rocks; and if the rocks of Somma are primitive, of primitive rocks.

Observations.

1. This mineral is distinguished from the other species of the Ruby family, by its dark colours: from *Spinel*, by its

its resinous lustre, inferior hardness, greater weight, and inferior transparency: When it occurs in grains, it is apt to be confounded with *Tourmaline*, but its resinous lustre, greater weight, and its not becoming electric by heating, distinguish it from that mineral.

2. It was first established as a distinct species by La Metherie, and afterwards acknowledged by Werner and Hauy. Hauy has lately changed his opinion in regard to it, and now arranges it in the system as a variety of Spinel.

3. It is not certain that the blue-coloured octahedral crystals, in the primitive limestone of Acker, are varieties of ceylanite.

3. Spinel.

Spinell, *Werner*.

Rubinus balassus, Rubinus spinellus, *Wall.* t. i. p. 247.—Rubis spinelle octaedre, *Romé de Lisle,* t. ii. p. 224.—Spinel, & Balass Rubies, *Kirw.* vol. i. p. 253.—Spinel, *Estner.* b. ii. s. 73. *Id. Emm.* b. i. s. 56, & b. iii. s. 252.—Rubino Spinello, *Nap.* p. 118.—Rubis, *Lam.* t. ii. p 224.—Spinel, *Hauy,* t. ii. p. 496. *Id. Broch.* t. i. p. 202. *Id. Bournon,* Phil. Trans. 1792, part ii. p. 305. *Id. Reuss,* b. ii. th. 2 s. 31. *Id. Lud.* b. i. s. 67. *Id. Suck.* 1r th. s. 449. *Id. Bert.* s. 281. *Id. Mohs,* b. i. s. 101. *Id. Hab.* 36. *Id. Lucas,* p. 42.—Spinelle rubis, *Brong.* t. i. p. 436.—Spinell, *Brard,* t. i. p. 113.—Spinel Ruby, *Kid,* vol. i. p. 143.—Spinell, *Steffens,* b. i. s. 23.

External Characters.

The principal colour is red, from which there is a transition on the one side into blue, and almost into green, on the other side into yellow and brown, and even into white. Thus it passes on the one side from carmine-red into cochineal-red, crimson-red, and cherry-red, into plum-blue, violet-blue, and indigo-blue; the indigo-blue sometimes

times inclines to duck-blue, which is nearly allied to green: on the other side it passes from crimson-red into blood-red, and hyacinth-red, into a colour intermediate between orange and ochre yellow, into yellowish-brown, and reddish-brown. From the cochineal-red it passes through rose-red into reddish-white. The colours are seldom pure, being generally somewhat muddy. The blue and white varieties are rare, and the green variety is very rare.

It occurs, sometimes in grains, more frequently crystallised. The grains are usually rolled crystals.

The following are its crystallizations:

1. Perfect octahedron, which is the fundamental figure. Fig. 13.
2. Octahedron, with alternate larger and smaller planes.
3. Tetrahedron, slightly truncated on the angles. Fig. 14.
4. Perfect tetrahedron. Fig. 15.
5. Tetrahedron, deeply truncated on the apex.
6. Segment of figure 3.
7. Two segments of the tetrahedron, truncated on the angles, as in figure 3. joined together in a conformable manner by their bases, forming a *twin-crystal* with three re-entering angles.
8. Two segments of the tetrahedron, truncated on the angles, (as in figure 3.), joined together by their bases in an unconformable manner, so that the extremities of the segments project. *Twin-crystal.*
9. Two crystals, N° 5. attached by their bases. *Twin-crystal.* Fig. 16.
10. A crystal of N° 6. attached by its base to the lateral plane of a crystal N° 5. *Twin-crystal.*
11. A crystal of N° 10. attached to one of N° 13. *Triple-crystal.*
12. Octahedron, in which two opposite planes are much larger than the others.

11. A

13. Thick cquiangular six-sided table, in which the terminal planes are set alternately oblique on the lateral planes. Sometimes the table is elongated, when it assumes more the appearance of a
14. Very oblique four-sided table, which is truncated on both the acute angles.
15. Acute rhomboid, in which the two acute angles are truncated. It is formed by the lessening of two opposite planes of the octahedron.
16. Octahedron, truncated on the edges. Fig 17.
17. Garnet dodecahedron. Fig. 18.
18. Octahedron, in which the axis is oblique, the edge of the common bases is truncated, and the apices sometimes rounded off.
19. Rectangular four-sided prism, acuminated by four planes, which are set on the lateral planes. Fig. 19.
20. Lengthened or cuneiform octahedron. Fig 20.

All the planes of the crystals that originate from the fundamental figure are smooth; whereas those that are derived from truncations on the edges are streaked.

The crystals are generally small and very small; seldom middle-sized .

Externally and internally the spinel is splendent, and the lustre vitreous.

The most frequent fracture is flat conchoidal, sometimes also concealed foliated; and the folia are parallel with the sides of the octahedron.

The fragments are angular and sharp-edged, or splintery.

It alternates from translucent to transparent, and refracts single.

It is hard in a high degree. It scratches quartz very readily; but is scratched by sapphire.

It

* Brard mentions a fine spinel, weighing 215 grains, which was intended for Josephine, the wife of Buonaparte.

It is brittle.

Rather heavy, approaching to heavy.

Specific gravity, 3.500, 3.789, *Werner*. 3.645, *Hauy*. 3.570, 3.590, *Klaproth*. 3.705, *Lowry*. 3.523, *Mohs*.

Chemical Characters.

Before the blowpipe it is unalterable without addition; but is fusible with borax.

Constituent Parts.

Alumina, - -	82.47
Magnesia, - -	8.78
Chromic acid, - -	6.18
Loss, - - -	2.57
	100

Vauquelin, J. M. N° 38. p. 89.

Geognostic and Geographic Situations.

It is found in the kingdom of Pegu, and at Cananor in the Mysore country; and in the island of Ceylon, accompanied with zircon, tourmaline, and ceylanite. It also occurs in drusy cavities along with vesuvian and ceylanite, &c. in the granular ejected limestone of Vesuvius.

We are still ignorant of the class of rocks in which it occurs. Werner conjectures that it may, like zircon and pyrope, be an inmate of flœtz trap rocks *.

Uses.

* In the magnificent collection of the late Honourable Mr Greville, now in the British Museum, there are two interesting specimens, which, although they do not enable us to ascertain the repository or kind of rock in which the spinel occurs, make us acquainted with some of its accompanying minerals. In one of the specimens, crystals of spinel are imbedded in calcareous-spar, and accompanied with crystals of mica, magnetic pyrites, and a substance which Count de Bournon believes to be asparagus-stone; and in the other specimen, the spinel is imbedded in adularia, and is accompanied with magnetic pyrites.

Uses.

It is used as a precious stone, being cut for various ornamental purposes; but it has neither the hardness nor fire of the red sapphire or oriental ruby. When it weighs four carats, (about sixteen grains), it is considered of equal value with a diamond of half the weight. Figures are sometimes cut upon it. It does not appear that the ancients ever cut figures on this mineral; for there is no mention made of antique engraved gems of this kind, by any of their writers; and in the vast collections of engraved gems, preserved in different parts of Lurope, there are none of spinel.

Observations.

1. *Distinctive Characters.*—*a.* Between Spinel and octahedral *Zircon:* In zircon, the octahedron is more obtuse than in spinel; and the specific gravity of zircon is higher, it being 4.4, whereas spinel is only 3.7.—*b.* Between Spinel and *Oriental Ruby* or *Red Sapphire:* Red sapphire is not only harder, but heavier than spinel.

2. The carmine-red variety is the *Spinel-ruby* of the jeweller: the cochineal-red variety is the *Balais-ruby* of jewellers, so named from Balacchan, the Indian name of Pegu, where this variety is found: the violet blue spinel is the *Almandine of Pliny;* is so named from Alabanda, a town in Lesser Asia, near which it was found; and the orange-yellow variety is the *Rubicelle-ruby* of jewellers.

3. M. Verina informs me. that Werner has formed a new subspecies of spinel, under the name *Salamstone,* which is the Indian name of that mineral. The following are its characters: " Colours are red and blue. It occurs in grains; and crystallised in the following figures.

(1.) Six-sided prism, in which the lateral planes are ribbed, and the alternate angles are truncated.

2. These

(2.) These truncations sometimes increase so much that they form a three-planed acumination, the apex of which is frequently truncated.

(3.) When the prism of the preceding figure becomes shorter, and at length disappears, (if the acumination is obtuse,) there is formed a flat, double, three-sided pyramid, in which the lateral planes of the one are set on the lateral planes of the other, and the remains of the six-sided prism form truncations on the common base. If the acumination is acute, there is formed a nearly cubical figure, having two diagonally opposite angles truncated.

(4.) When the apices of the preceding crystals are deeply truncated, a table is formed.

Internally, it is shining.

The fracture is conchoidal, and also concealed foliated, parallel with the terminal planes of the prism.

It is generally only translucent, and exhibits a pearly light on its surface.

It is somewhat heavier than true spinel; but in other characters agrees with it.

It occurs principally in the peninsula of India.

M. Verina is of opinion, that many of the rubies of commerce are varieties of this mineral.

5. Spinel was first established as a species by Romé de Lisle and Werner, and separated from sapphire, with which it had been confounded.

4. Sapphire.

Sapphir.—*Werner.*

Saphirus, *Wall.* t. 1. p. 248.—Rubinus orientalis, *Id.* p. 247. —Topazius orientalis, *Id.* p. 251.—Rubis d'orient, *R. de L.* t. 2. p. 212.—Oriental ruby, sapphire, and topaz, *Kirw.* t. 1. p. 250.—Sapphir, *Estner,* b. 2. s. 86.—*Id. Emm.* b. 1. s. 67. & b. 1. s. 251.—Zaffiro et rubin-zaffiro, *Nap.* p. 113. & 121.—Saphir, *Broch.* t. 1. p. 207.—Telesie, *Hauy,* t. 2. p. 480. —Perfect corundum, *Greville* and *Bournon,* Lond. Phil. Trans. 1798 & 1802.—Saphir, *Reuss,* b. 2. th. 2. s. 24.— Rubin, *Reuss,* b. 2. th. 2. s. 20.—*Id. Lud.* b.1. s. 67.—*Id. Suck.* 1. th. s. 446.—*Id. Bert.* s. 280.—*Id. Mohs,* b. 1. s. 128.— *Id. Hab.* s. 36.—Telesie, *Lucas,* p. 40.—Corindon telesie, *Brong.* t. 1. p. 427.—Corindon hyalin, *Brard.* p. 110.—Sapphire, *Kid,* vol. i. p. 137.—Corindon hyalin, *Hauy,* Tabl. p. 30.—Sapphire, *Steffens,* b. i. s. 14.

External Characters.

Blue and red are its principal colours; it occurs also grey, white, and yellow. From indigo-blue it passes through smalt-blue, Berlin-blue, azure-blue, lavender-blue, into a kind of flesh-red, rose-red, crimson-red, peach-blossom red, and cochineal red. It occurs also pearl-grey, bluish-grey, milk-white, reddish-white, yellowish-white; which latter inclines strongly to lemon-yellow *.

It frequently shews two colours at once, particularly blue and white, and sometimes blue and red. Three colours, as white, blue and grey, are sometimes seen in the same crystal.

It

* A deep-green variety is mentioned by authors

It occurs in small rolled pieces, and crystallised. Its crystallizations are as follow:

1. Very acute, equiangular, simple, six-sided pyramid. Fig. 21.*.
2. Preceding figure truncated on the summit. Fig. 22.
3. Perfect six-sided prism, Fig. 23.; sometimes truncated on the alternate angles, Fig. 24.
4. Acute, double, six-sided pyramid, in which the lateral planes of the one are set on the lateral planes of the other. Fig. 25.
5. The preceding figure acuminated on the extremities by six planes, which are set on the lateral planes.
6. The preceding figure truncated on the extremities.
7. N° 4. truncated on the extremities. Fig. 26.
8. N° 3. acutely acuminated with six planes, which are set on the lateral planes.
9. The preceding crystal truncated on the summit. Fig. 27.
10. N° 1. acutely acuminated by six planes, which are set on the lateral planes.
11. The preceding figure truncated on the summit.

The crystals are small, and middle-sized, and all around crystallised. The planes of the crystals are generally transversely streaked, and, when fresh, are usually splendent.

Internally, its lustre is splendent and vitreous, sometimes inclining to adamantine.

The fracture is conchoidal or concealed foliated, with a fourfold cleavage. Three of the cleavages are parallel with the truncating planes on the alternate angles of the six-sided prism; the fourth parallel with the terminal planes of the prism.

 The

* Formerly Haüy viewed the primitive form of sapphire as a regular hexahedral prism; now he considers it as a slightly acute rhomboid.

The fragments are angular, and sharp-edged.

It alternates, from transparent to translucent; and the translucent varieties frequently exhibit a six-rayed opalescence.

It refracts double.

It is, after diamond, the hardest substance in nature.

It is brittle, and easily frangible.

Intermediate, between heavy and rather heavy.

Specific gravity 4,320, 4.000, *Werner*.—4.283, 3.999, *Hauy*.—4 000, *Hatchet* and *Greville*.—4.161, 3.907, *Bournon*. Yellow Sapphire, 3.916; Blue sapphire, 3.985; Red sapphire, 3.975, *Lowry*.

Constituent Parts.

Blue sapphire.			Blue sapphire,		Red sapphire, or oriental ruby.
Alumina,	98.5	-	92,0	-	90.0
Lime,	0.5	Silica,	5.25	-	7.0
Oxide of iron,	1.0	-	1.0	-	1.2
		Loss,	1.75		1.8
	100		100		100
Klap. Beit. b. 1, s. 88.		*Chenevix*, Phil. Trans. 1802.			*Chenevix*, Phil. Trans. 1802

Chemical Characters.

It is infusible before the blow-pipe.

Geognostic Situation.

It occurs in alluvial soil, in the vicinity of rocks belonging to the newest flœtz-trap formation.

Geographic Situation.

Europe.—It occurs in alluvial soil, along with pyrope, zircon, and iron-sand, at Podsedlitz and Trziblitz in Bohemia. In France, on the banks of the stream Riou Pezzouliou,

Pezzouliou, near Expailly; also at Brendola in the Vicentine, and in Portugal.

Asia.—It is found particularly beautiful in the Capelan Mountains, twelve days journey from Serian, a city of Pegu, in Ceylon; and, it is said, also in Persia.

Use.

This mineral is, next to diamond, the most valuable of the precious stones. It is cut with diamond-powder, and polished by means of emery. The white and pale-blue varieties, by exposure to heat become snow-white, and when cut, exhibit so high a degree of lustre, that they are used in place of diamond. The most highly prized varieties are the crimson and carmine red; these are the Oriental Ruby of the jeweller, and, next to the diamond, are the most valuable minerals hitherto discovered. The blue varieties, the Sapphire of the jeweller, are next in value to the red. The yellow varieties, the Oriental Topaz of the jeweller, are of less value than the blue or true sapphire.

It does not appear that the ancients ever engraved figures upon this mineral All the engraved sapphires preserved in collections, are of modern date; and of these, one of the most beautiful is a red sapphire, or oriental ruby, on which is cut the figure of Henry the Fourth of France. This gem was engraved by the celebrated artist Coldere, and was in the collection of the late Duke of Orleans.

Observations.

1. Sapphire was first established as a distinct species, and separated from spinel, with which it had been confounded, by Rome de Lisle and Werner.

2. It would appear, from the observations of Werner, that spinel is very nearly allied to sapphire; but it may be questioned, whether or not the varieties of spinel which are considered as pointing out this connection, are not themselves varieties of sapphire. The varieties are, the acute rhomboid, cube truncated on two diagonally opposite angles, and the six-sided prism having its three alternate angles truncated.

3. The following are the names given to the different varieties of this species:

a. Blue sapphire. *True* or *oriental sapphire.*
b. Red sapphire. *Oriental ruby.*
c. Yellow sapphire. *Oriental topaz.*
d. Violet-blue sapphire. *Oriental amethyst.*
e. Pearl-grey or bluish-grey sapphire. *Vermeille* or *Calcedonic ruby.*
f. Green sapphire. *Oriental Emerald.*

4. Certain varieties of sapphire exhibit particular kinds of opalescence, and these have received the following denominations:

(1) *Girasol sapphire.* This variety exhibits a pale-reddish and bluish reflection upon a transparent ground.

(2) *Opalescent sapphire.* This variety shews a very bright pearly opalescence.

(3) *Asteria sapphire.* When this variety is cut *en cabochon,* it exhibits, in a direction perpendicular to the axis of the crystal, a silvery star of six rays.

5. We may here remark, that the epithet *Oriental,* frequently applied to the finer kinds of gems, was adopted, in consequence of its having been observed, that the hardest precious stones came from the East.

5. Emery.

5. Emery.

Schmiergel.—*Werner*.

Smirgel, *Reuss*, b. 2. th. 2. s. 156.—*Broch.* t. 2. p. 292.—Smirgel, *Lud.* b. 2. s. 183.—*Id. Suck.* 2. th. s. 298.—*Id. Bert.* s. 427.—*Id. Mohs*, b. 1. s. 136.—Corindon granuleux, *Lucas*, p. 260.—*Id. Brard.* p. 111. Emeril, *Brong.* t. 1. p. 431.—Corindon granulaire, *Hauy*, Tabl p. 30.—Schmirgel, *Steffens*, b. i. s. 21.

External Characters.

Its colour is intermediate between greyish-black and bluish-grey.

It occurs massive and disseminated: and the massive is sometimes intermixed with other minerals.

Its lustre is glistening, passing into glimmering, and is adamantine.

The fracture is fine and small grained uneven; sometimes splintery.

The fragments are angular, and rather blunt edged.

It sometimes occurs in fine granular distinct concretions.

Is slightly translucent on the edges.

It is hard in a high degree; scratches quartz.

Is rather difficultly frangible.

Heavy.

Specific gravity 4.0.

Constituent Parts.

Alumina,	-	86.0
Silica,	-	3.0
Iron,	- -	4.0
Loss,	- -	7.0
		100

According to *Tennant*, Phil. Trans. for 1802.

Geognostic Situation.

We know only of the geognostic situation of the Saxon emery, which occurs in beds of talc and steatite, along with blende and calcareous spar, in primitive clayslate.

Geographic Situation.

Europe.—It is found at Ochsenkopf near Schwartzenberg, and Eibenstock in Saxony. Jersey and Guernsey are mentioned as localities of this species; but Dr Macculloch, who examined these islands, could neither find it, nor learn that it ever had been discovered there *

It occurs abundantly in the island of Naxos, in large loose masses, at the foot of primitive mountains, and also at Smyrna. It is mentioned as a production of Parma in Italy, and of Ronda, in the kingdom Granada, in Spain.

Asia.—Near the town of Charlowa in the Altain Mountains.

America.—Mexico and Peru.

Use

* Macculloch, Geological Transactions, vol. i. p. 12.

Use.

It is used for polishing hard minerals and metals, and hence is an important article in the arts. Before using, it is ground into powder of various degrees of fineness, according to the use that is intended to be made of it. The different kinds of powder are obtained, by repeatedly diffusing the ground emery in water, and allowing the water to settle a longer or shorter time, according as we wish a fine or coarse powder. It is used with water for polishing stones; but with oil for polishing metals.

Observations.

Magnetic iron-stone, and specular iron-ore, from their being frequently used in place of emery, in polishing hard bodies, have been confounded with it.

6. Corundum.

Korund & Demant-Spath, *Werner*.

Korund, *Wid.* s. 237.—Adamantine spar, *Kirw.* vol. i. p. 335. Demant-Spath, *Emm.* b. 1. s. 9. & b. 3. s. 229.—Spato adamantino, *Nap.* p. 223.—Corindon, *Lam.* t. 2. p. 356.—Le spath adamantine, *Broch.* t. 1. p. 356.—Corindon, *Hauy,* t. 3. p. 1.—Imperfect corundum, *Greville* & *Bournon, Phil. Trans.* 1798 and 1802.—Korund & Demant-spath, *Reuss,* b. 2. th. 2. s. 16, & 12.—*Id. Lud.* b. 1. p. 103.—*Id. Suck.* 1. th. s. 439.—*Id. Bert.* s. 290.—*Id. Mohs,* b. 1. s. 112, & 120.—Corindon harmophane translucide & Corindon harmophane opaque, *Lucas,* p. 259, & 260.—Corindon adamantine & Corindon adamantine noiratre, *Brong.* t. 1. p. 429, 430.—Corindon harmophane, *Brard,* p. 110.—Corindon harmophane opaque, *Hauy,* Tabl. Comparat. p. 30.—Korund & Demant-spath, *Steffens,* b. i. s. 17, & 19.

External

External Characters.

Its colour is greenish-white, of various degrees of intensity, which passes into light greenish-grey, and even into mountain-green, asparagus-green, Berlin-blue, and azure-blue; it is sometimes also pearl-grey, which passes into flesh-red, cochineal-red, crimson-red, and hair-brown.

When cut in a semicircular form, it often presents an opalescent star of six rays.

It occurs massive, disseminated, in rolled pieces, and crystallised. Its principal crystallizations are the following:

1. Equiangular six-sided prism.

2. Same prism, having its alternate angles truncated.

3. Same prism, having its terminal edges and alternate angles truncated.

4. When the truncations on the angles of N° 2. increase very much in magnitude, there is formed a three-planed acumination, in which the acuminating planes are set on the alternate lateral edges of the prism.

5. When the truncations on the edges increase very much, a six-planed acumination is formed; and when the prism becomes very short, or disappears, there is formed a simple six-sided pyramid; and if the prism is acuminated on both extremities, a double six-sided pyramid; and in both cases the summits of the pyramids are truncated.

The crystals are middle sized.

Externally, they are dull and rough.

The lustre of the principal and cross fracture is shining and glistening, and is either vitreous inclining to resinous, or pearly inclining to adamantine.

The

The fracture is perfect foliated, with a four-fold cleavage. Three of the cleavages are parallel with the truncating planes on the alternate angles of the six sided prism, and the fourth with the terminal planes of the prism. The cross fracture is small, and imperfect conchoidal.

The fragments are rhomboidal.

It shews a tendency to straight lamellar concretions.

It alternates from translucent to translucent on the edges, and it refracts double.

It is so hard as to scratch quartz; but is softer than sapphire.

It is rather easily frangible.

Is rather heavy, approaching to heavy.

Specific gravity 3.710, *Klaproth.*--3.873, *Hauy.*—3.875, *Bournon.*

Constituent Parts.

	Corundum of the Carnatic.	Of Malabar.	Diamond spar, or Corundum of China.
Alumina,	91.0	86.5	84.0
Silica,	5.0	7.0	6.50
Iron,	1.5	4.0	7.50
Loss,	2.5	2.5	2.
	100	100	100
		Chenevix.	*Klaproth.*

Geognostic Situation.

In India, it occurs imbedded in a rock composed of felspar, fibrolite, quartz, hornblende, and mica, and is sometimes accompanied with pistacite, talc, garnet, and zircon. In Italy, it is imbedded in mica-slate.

Geographic

Geographic Situation.

It is found in the Carnatic, and on the coast of Malabar; also in China, and in the department of Serio in Italy.

Use.

It is used for cutting and polishing hard minerals; and it is reported, although probably incorrectly, that the Chinese use it as an ingredient in their porcelain.

Observations.

This mineral was known to Dr Woodward as early as 1768. Dr Black ascertained that it differed from all other minerals then known. Its complete history was first given by the Honourable Mr Greville and the Count de Bournon.

7. Chrysoberyl.

Krysoberyll, *Werner*.

Chrysolithus colores reflectens varios; chrysoberyllus, *Wall.* t. i. p. 216.—Krisoberill, *Wid.* s. 246.—*Id. Kirw.* vol. i. p. 261.; *Estner*, b. ii. s. 63.; *Id. Emm.* b. i. s. 19.—Crisoberillo, *Nap.* p. 134.—Chrysopal, *Lam.* t. ii. p. 244.—Le Chrysoberil, *Broch.* t. i. s. 167.—Cymophane, *Hauy*, t. ii. p. 491.—Chrysoberyll, *Reuss*, b. ii. s. 48.—*Id. Lud.* b. i. s. 60.—*Id. Mohs*, b. i. s. 42.—Cymophane, *Lucas*, p. 41.—*Id. Brong.* t. i. p. 425.—*Id. Brard.* p. 111.—*Id. Hauy*, Tabl. p. 30.—Krysoberyll, *Steffens*, b. i. s. 12.

External

External Characters.

Its chief colour is asparagus-green: which passes on the one side into apple-green, mountain-green, and greenish-white; on the other side, through light-olive and oil-green, into light yellowish-grey, which inclines strongly to brown, and even passes to reddish-brown.

It exhibits a milk-white opalescence *, which appears in general to float in the interior of the mineral.

It occurs in blunt angular, rolled pieces, that sometimes approach to the cubic form. It occurs very seldom crystallised, and the following are its crystallizations:

1. Long and thick six-sided table, having longitudinal streaked lateral planes, and sometimes the lateral edges truncated †. Fig. 28.
2. When the truncating planes increase, the table passes into a double six-sided pyramid; and the summits of the acuminations are sometimes truncated. Fig. 29.

The crystals are small.

The surface of the rolled pieces is intermediate between rough and smooth, and is glistening.

Externally the crystals are shining; internally splendent, and the lustre is intermediate between resinous and vitreous, but more inclining to the first.

The fracture is perfect conchoidal, and sometimes foliated in the direction of the axis of the prism.

Fragments

* It is said that the opalescence does not always occur.

† According to Hauy, the primitive form of chrysoberyl is a rectangular parallelopiped.

The fragments are angular, and sharp-edged.

It is semitransparent, sometimes inclining to transparent, and refracts double.

Hard; scratches quartz easily, and spinel sensibly.

Brittle.

Rather easily frangible.

Rather heavy.

Specific gravity 3.600, 3.720, *Werner.*—3.710, *Klaproth.*—3.7961, *Hauy*—3.8, *Lucas,*—3.550, *Lowry.*

Constituent Parts.

Alumina, -	71.5
Silica, - - -	18.0
Lime, - - -	6.0
Oxide of iron, - -	1.5
Loss, - -	3.0
	100

According to *Klaproth*, b. 1. s. 102.

Chemical Characters.

Before the blow-pipe it is infusible without addition, *(Lelievre.)*

Geognostic and Geographic Situations.

It occurs in Brazil along with the topaz, probably in alluvial soil. It has been lately discovered in Connecticut

cut in North America, imbedded in a granular rock composed of quartz, felspar, talc, mica, and garnet *.

In the island of Ceylon, it occurs in the beds of rivers, along with sapphires and tourmalines.

Use.

It is considered as a gem of the second order by the jeweller; it is generally cut in a semi-globular form, and to heighten its lustre, and improve the colour, it is set with a gold foil. It is, however, a rare mineral, and seldom to be met with in the possession of jewellers.

The Brazilian variety is better fitted for the purpose of the jeweller than the North American.

Observations.

1. Rolled pieces or pebbles of chrysoberyl might be confounded with pebbles of *sapphire*; but their green colour, milk or bluish white opalescence, frequent cubic form, rough and glistening surface, and inferior hardness, distinguish them sufficiently from that mineral. The crystallised varieties are distinguished from crystallised sapphire by colour and surface.

2. I have placed it immediately after corundum, on account of its near alliance to it.

3. It was first established as a distinct species by Werner, in the Bergmännisches Journal, 3 Jahrg. 2. B. 54.

IV. SCHORL

* In Mr Ferguson of Raith's interesting and valuable collection of minerals, I saw a good specimen of the North American chrysoberyl, in its granitous rock.

IV. SCHORL FAMILY.

This family contains the following species: Topaz, Schorlite, Pyrophysalite; Euclase, Emerald; Iolite *; Schorl, Epidote, Zoisite; and Axinite.

1. Topaz †.

Topaz, *Werner.*

Topazius octaedricus prismaticus, *Wall.* t. 1. p. 251.—Topaze du Brezil, *R. d. L.* t. 2. p. 230.—Topaze de Saxe, *Id.* p. 260. —Topaz, *Wern. Cronst.* p. 97.—*Id. Wid.* p. 267.—Occidental Topaz, *Kirw.* vol. i. p. 254.—Topaz, *Estner.* b. 2. f. 98. *Id. Emm.* b. 1. f. 374.—Topazio, *Nap.* p. 136.—Topaze du Brezil, de Saxe et de Siberie, *Lam.* t. 2. p. 254.—*Id. Broch.* t. i. p. 212.—*Id. Hauy,* t. ii. p. 504.—Topaz, *Reuss,* b. ii. th. ii. s. 40.—*Id. Lud.* b. i. s. 68.—*Id. Suck.* 1. th. s. 455. —*Id. Bert.* s. 294.—*Id. Mohs,* b. i. s. 27.—*Id. Hab.* s. 54. —*Id. Lucas,* s. 43.—*Id. Brong.* t. i. p. 419.—*Id. Brard,* p. 116.—*Id. Kid,* vol. i. p 145.—*Id. Hauy,* Tabl. p. 17.— *Id. Steffens,* b. i. s. 33.

External Characters.

Its principal colour is wine-yellow, which occurs of all degrees of intensity.

The *pale* wine-yellow passes into yellowish-white, greyish-white, greenish-white, mountain-green, and celandine-green.

The

* Iolite has sometimes been arranged under the Ruby Family, but its true place is in the Schorl Family.

† The name Topaz is derived from Topazos, a small island in the Red Sea, where, it is said, the Romans used to collect their topaz, which is the Chrysolite of the moderns.

The *dark* wine-yellow passes from orange-yellow through cherry-red into violet-blue *.

It occurs seldom massive; disseminated, and in rolled pieces; most frequently crystallised.

Its crystallizations are as follow:

1. Oblique four-sided prism, rather acutely acuminated by four planes, which are set on the lateral planes †.
2. N° 1. in which the acuter lateral edges are bevelled; or it may be viewed as an eight-sided prism, in which two and two lateral planes meet under obtuse angles. Fig. 30.
3. N° 1, & 2., with a double acumination; the planes of the second acumination set on those of the first.
4. N° 2. with a triple acumination; in which the planes of the one always rest on those of the others.
5. N° 2. in which the angles on the acute edges, and the summits of the acuminations, are truncated. Fig. 31.
6. N° 4. in which the angles on the acute and obtuse edges are truncated Fig. 32.
7. N° 1, & 2, in which the summits of the acuminations are truncated.

8. The

* The violet-blue variety is very rare: in proof of this, it may be mentioned, that Mr Von der Nüll of Vienna, the proprietor of one of the most beautiful and instructive cabinets in Europe, and which has been excellently described by Mohs, paid 1500 ducats for a single specimen of violet-blue coloured topaz. Vid. Von Moll's Ephemeriden.

† According to Haüy, the primitive form of topaz is a rectangular octahedron.

8. The preceding figure, in which the angles formed by the obtuse lateral edges and the acuminating planes, are bevelled.
9. The bevelling edges of N° 8. truncated.
10. The preceding figure, in which the edges formed by the truncating planes of the bevelment with the surrounding planes, are truncated.
11. N° 2. in which the terminal planes are bevelled, and the bevelling planes set on the acute lateral edges.
12. The preceding figure, in which the angles formed by the proper edge of the bevelment are bevelled. Fig. 33.
13. A lengthened octahedron; formed by the approximation of the bevelling planes.

The crystals are small and very small; very seldom large; and are generally superimposed.

The lateral planes of the crystals are longitudinally streaked; but the acuminating and bevelling planes are smooth; the terminal planes are rough *.

Externally, it is splendent; internally, splendent and vitreous.

The cross fracture is perfect, and straight foliated; the longitudinal small, and imperfect conchoidal.

The fragments are angular and sharp-edged, and also tabular.

The massive varieties occur in coarse and small granular distinct concretions.

It

* The Brazilian and Siberian topazes are more deeply streaked than the Saxon; further, the Brazilian topaz is generally acuminated, but is without truncations; the Siberian, on the contrary, is usually bevelled.

It alternates, from translucent to transparent; and it refracts double.

It is so hard as to scratch quartz; but it has no effect on spinel, which latter scratches it.

It is easily frangible.

It is rather heavy, approaching heavy.

Specific gravity 3.464 to 3.556, *Werner.*—3.556 to 3.564, *Hauy.*—3.540 to 3.576, *Karsten.*—3.532 to 3.641, *Lowry.*

Chemical Characters.

Saxon topaz in a gentle heat becomes white *, but a strong heat deprives it of lustre and transparency: the Brazilian, on the contrary, by exposure to a high temperature, burns rose-red †, and in a still higher violet-blue. Before the blow-pipe it is infusible, but exposed to a steam of oxygen gas it soon melts into a porcellanous bead. It is fusible with borax, but alkali has little effect on it.

Physical Characters.

The topaz of Brazil, Siberia, and Mucla in Asia Minor, and Saxony, when heated, exhibits at one extremity positive, and at the other negative electricity.

 Constituent

* When thus altered, the Saxon topaz is sometimes imposed on the ignorant for diamond.

† Topaz thus altered, is cut and sold by jewellers under the name of Brazilian ruby and pale spinel

Constituent Parts.

	Saxon.	Brazilian.	Brazilian.
Silica,	35.	44.5	29.
Alumina,	59.	47.5	50.
Fluoric acid,	5.	7.	19.
Oxide of iron,		0.5	
Loss,	1.	0.5	2.
	100.	100.	100.
	Klap. b. 4. s. 160.	*Klap. ib.*	*Vauquelin*, Annales du Mus. t. vi. p. 24.

Geognostic Situation.

It occurs in that particular mountain-rock denominated *Topaz-rock;* which is an aggregate of massive topaz, quartz, and schorl; having in the small a slaty, but in the large a granular texture; and in which there are frequent small cavities, lined with crystals of topaz, quartz and schorl, and portions of lithomarge. It occurs in drusy cavities, in granite, along with beryl, and rock crystal; also in veins, which traverse primitive rocks, accompanied, in certain formations, with beryl, rock-crystal, and iron-ochre, when the veins traverse granite; in others with tinstone, arsenical pyrites, sometimes copper pyrites, apatite, fluor-spar, quartz, steatite, when the veins traverse gneiss and mica-slate. It has been also discovered in nests, in transition clay-slate, along with red-coloured quartz, brown-spar, and selenite or gypseous spar; and it is found in rolled pieces in alluvial soil.

Geographic

Geographic Situation.

Europe.—It occurs in large crystals, and rolled masses, in an alluvial soil, in primitive country, in the upper parts of Aberdeenshire *; and in veins, along with tinstone, in killas, at St Anne's in Cornwall. Upon the continent of Europe, it appears most abundantly in topaz-rock at Schneckenstein; also in veins that traverse gneiss, along with tinstone, fluor-spar, and arsenical pyrites, at Ehrenfriedersdorff; and in rounded or angular pieces, and sometimes in crystals of a mountain-green colour, in alluvial soil, at Eibenstock in Saxony; it also occurs at Zinnwald and Geyer in the same country; at Schlackenwalde and Zinnwald in Bohemia, it occurs in veins that traverse gneiss, along with tinstone, fluor-spar, copper-pyrites, and lithomarge. It has been found at Hirschberg, and other places in Silesia, and at the Höllengraben, at Werfen in Salzburg, in nests in transition clay-slate.

Asia.—It occurs both in the Altain and Uralian mountains. In the Altain range, it occurs on the banks of the river Tom; and in the mountain Adon-Tschelan, along with beryl, quartz, schorl, fluor-spar, and lithomarge. About twenty-five leagues north of Catharinenburg, in the Uralian range, it is found in considerable quantity in a kind of granite, resembling that variety known under the name of Graphic Granite. There, it is said to occur in drusy cavities, along with rock-crystal and beryl. It has

* Vid. Wernerian Transactions, vol. i. p. 445,—452.

has been discovered in loose crystals in Kamschatka; and it is said also along with rock-crystal, common quartz, &c. in the river Poyk in Caucasus. The beautiful rose-red variety was discovered at Mukla, in Asia Minor, by an intelligent traveller, our countryman Mr Hawkins. In Ceylon, Pegu, Hawkesbury river in New Holland, and Cape Barren Island in Bass's Straits, it occurs in alluvial soil.

America.—In Brazil, in the beds of rivers. In the National Museum in Paris, there is a large rock-crystal, containing red-coloured topazes from Brazil.

Use.

1. This gem is much prized by jewellers. They divide it into the following different kinds, according to locality, colour, and other properties:—*a. Oriental*, those varieties which are brought from India, and which have a deep and rich orange-yellow colour. They are very highly valued.—*b. Brazilian*, those which are imported from Brazil, which have not so pure a yellow colour as the oriental.—*c. Saxon*, which are well distinguished by their wine-yellow colour, and, when cut, often exhibit a lustre equal to that of the finest oriental varieties.—*d. Bohemian*, those which are found in the tin mines of that country, and which are of small size, deficient in transparency, and have only grey or muddy-white colours, and hence are not esteemed.—*e. Aqua-marine*. Under this name are included the mountain-green varieties of topaz, which are found in the alluvial rocks of Eibenstock, and in veins and drusy cavities in Siberia, and, we may add, in the alluvial soil, in the upper parts of Aberdeenshire, in this country. When cut, they do not

not shew so much fire or lustre as the oriental, and hence are not so highly valued.—*f. Brazilian Ruby* and *Sapphire ;* which include the red and blue Brazilian topaz.—*g. Taurian Topaz ;* which is of a pale-blue colour.

2. This gem was much prized by the ancients. In proof of this, it may be mentioned, that Cleopatra presented a fine stone of this kind to Antony; and that Ovid adorns the chariot of the sun with it. It is cut in the same manner as other gems. When set, it should have a gold foil *. Figures are sometimes engraved on it; and these, when well executed, are very highly valued. In the National Museum in Paris, there is a superb Indian Bacchus engraved on topaz. In the cabinet of the Emperor of Russia there are several fine engraved topazes; and the King of Spain had in his possession a Brazilian topaz, on which was admirably engraved the portraits of Philip II. and Don Carlos.

3. Other minerals are sometimes sold for topaz, as yellow-coloured Rock-crystal, in this country named *Cairngorm-stone*, from the place where it is found, or *Scots topaz* †. Even very fine varieties of Calcedony and Carnelian, when well-cut and set, have been imposed on the ignorant as topaz.

* It is worthy of remark, that this mineral rather readily absorbs dust, and is liable to crack, particularly in the direction of its cleavage.

† The large mass of yellow transparent stone which was preserved in the collection of the Stadtholder under the name of Topaz, was but a fragment of rock-crystal.

4. Coarse kinds of topaz are broke down and used as a kind of emery in cutting hard minerals. Lastly, it may be mentioned, that topaz was formerly kept in the apothecaries shops, and sold as a powerful antidote against madness.

Observations.

1. Topaz may in general be distinguished from all other minerals, by the rhomboidal base of its crystals, straight foliated cross fracture, and longitudinal streaked lateral planes.

2. It cannot readily be confounded with yellow-coloured *Sapphire*, because sapphire is harder and heavier, and does not, like the greater number of topazes, become electric by heating: nor can we mistake red topaz for *Spinel*, because spinel is harder, refracts only single, whereas topaz refracts double; and spinel does not become electric by heating, as is the case with topaz; and green-coloured topaz is readily distinguished from *Beryl*, by the following characters: It does not exhibit cleavages parallel with their lateral planes, as is the case with beryl; its prism has a rhomboidal base, which is not the case with the prism of beryl; and its specific gravity is 3.5, but that of beryl only 2.7.

3. When colour was considered as affording the most certain means of distinguishing th precious stones from each other, many different minerals were associated with the topaz, and varieties of topaz were described as distinct species; as appears from the following tables:

1.

Minerals which have been confounded with Topaz.

1. Yellowish-white sapphire,	Oriental topaz
2. Zircon,	Hyaline topaz, and yellowish-red topaz *.
3. Chrysolite,	Yellowish-green topaz †.
4. Yellowish beryl.	Siberian topaz.
5. Yellowish rock-crystal,	Bohemian, Scotch, or Occidental topaz.
6. Clove-brown and brownish-black rock-crystal,	Smoke topaz.
7. Yellow fluor-spar,	False topaz.

2.

Names given to particular varieties of Topaz.

1. Mountain green topaz,	Aqua-marine.
2. Blue topaz,	Sapphire.
3. Yellow topaz.	Chrysoprase, *Baillon*, Cat. p. 137.
4. Wine-yellow, inclining to red topaz,	Rubicelle.
5. Red topaz,	Brazilian or Balais ruby.
6. Yellowish-green topaz.	Saxon chrysolite.

In

* Wall. edit. 1778, t. i. p. 252.

† Ibid.

4. In the collection of minerals in the Museum of Natural History at Paris, there is a Brazilian topaz which weighs 4 ounces 2 gros, which is the largest specimen in that great national repository. In the upper parts of Aberdeenshire, much heavier and larger specimens of real topaz have been found. In the first volume of the Wernerian Memoirs, we find mentioned a specimen weighing 1 pound 3 ounces 8 drams and 8½ grains, troy weight, from Aberdeenshire; and we understard that larger masses have been since discovered.

2. Schorlite.

Schörtlartiger Berill, *Werner*.

Weisser Stangenschœrl, *W. Cronst.* p. 169.—Schorl blanc prismatique, *R. d. L.* t. ii. p. 420.—Schörlartiger berill, *Wid.* p. 276.—Shorlite, *Kirwan,* vol. i. p. 286. *Estner,* b. ii. p. 207. —Sorlo bianco, *Nap.* p. 152 —Leucolite, *Lam.* t. ii. p. 274.—Leucolite, et Pycnite, *Hauy,* t. iii. p. 236.—Le Berill schorleorme, *Broch.* t. i. p. 124.—Stangenstein, *Reuss,* b. ii. th. i. s. 110..—Schörlartiger berill, *Lud.* b. i. s. 70.—*Id. Mohs,* b. i. s. 155.—Stangenstein, *Hab.* s. 52.—Pycnite, *Lucas,* p. 78. *Id. Brongt.* t. i. p. 418. *Id. Brard,* p. 191.—Topaz septihexagonalé & Topaz cylindroide, *Hauy,* Tabl. p. 18.—Schorlit, *Steffens,* b. i. s. 37.

External Characters.

Its principal colour is straw-yellow, which passes into yellowish-white, greenish-white, asparagus-green, and sulphur-yellow *.

It

* Some varieties, as those from Altenberg in Saxony, are marked with spots of violet-blue, which inclines to cherry-red.

It occurs almost always massive, and crystallised in long six-sided prisms, which are sometimes truncated on the terminal edges and angles, and are generally imbedded.

The crystals are large and middle sized.

Externally and internally its lustre is shining, approaching to glistening, and is resinous.

The cross fracture is imperfect foliated; the longitudinal fracture small and imperfect conchoidal.

The fragments are angular.

It is composed of parallel, thin, and straight prismatic distinct concretions, which are longitudinally streaked.

It is more or less translucent on the edges.

It is hard in a low degree; scratches glass.

It is brittle.

It is uncommonly easily frangible.

Rather heavy, approaching to heavy.

Specific gravity 3.530, *Klaproth.*—3.514, *Hauy.*—3.535 and 3.503, *Haberle.*

Chemical Characters.

Before the blow-pipe it is infusible without addition; with borax it melts into a pure transparent glass.

Physical Characters.

Like topaz, it becomes electric by heating.

Constituent

Constituent Parts.

Silica,	30.	34.	43.
Alumina,	60.	48.	49.5
Fluoric acid,	6.	17.	4.
Lime,	2.		
Iron and manganese,	-	1.	
Water,	1.		1.
Oxide of iron,	-	-	1.
Loss,	1.		1.5
	100.	100.	100.
	Vauquelin, in Brongniart's Mineralogie, i. 419.	*Bucholz*, in Gehlen's Journ. ii. 38.	*Klaproth*, in Karsten's Tabell.

Geognostic and Geographic Situations.

It occurs in a rock composed of quartz and mica, at Altenberg in Saxony. The rock forms a bed several fathoms thick, and of unknown extent. Werner is of opinion that this schorlite rock is subordinate to mica-slate; but Haberle is inclined to place it in the system between gneiss and mica-slate. Its true position is still undetermined. At Schlackenwald in Bohemia, it is imbedded in an aggregate of quartz, tinstone, wolfram, and molybdena. In Siberia it occurs along with mica and quartz; and is at Mauleon in France imbedded in steatite.

Observations.

1. *Distinctive Characters.—a.* Between schorlite and *beryl.* The colour-suite of beryl is different from that of schorlite;

schorlite; both external and internal lustre of beryl is much higher than that of schorlite; and in beryl the lustre is vitreous, whereas it is resinous in schorlite; beryl is easily frangible, schorlite uncommonly easily frangible; beryl scratches quartz, which schorlite does not; beryl has a specific gravity of 2.7; schorlite is 3.5.— *b.* Between schorlite and *topaz*. The colour-suite of topaz is different from that of schorlite; both external and internal lustre of topaz much exceed that of schorlite, and the lustre is vitreous, not resinous, as is the case with schorlite; schorlite occurs in prismatic concretions, which is never the case with topaz; topaz is easily frangible, schorlite uncommonly easily frangible; lastly, topaz is hard in a high degree, schorlite in a low degree.

2. It received the name *schorlite* from Klaproth, on account of general resemblance to schorl. Its colour-suite, crystallization, lustre, fracture, distinct concretions, and hardness, shew, that although it is nearly allied to topaz, yet it cannot well be considered as belonging to the same species. Should the fact stated by some authors prove correct, namely, that the same mass at one extremity is schorlite, and at the other topaz, then it would be necessary to consider this substance as a sub-species of topaz, and which might then be named *schorlous topaz*.

3. Haüy now considers it as but a variety of topaz.

3. Pyro-

3. Pyrophysalite.

Pyrophysalith, *Hisenger.*

Pyrophysalith, Afhandlingar i Fysik, Kemi ocli Mineralogie, 1. th. p. 111,—118.; Tillagning, p. 239, 240.; Annal. de Chem. 1806, n. 173. p. 113,—124.—*Steffens*, b. i. s. 40.—Topaz prismatoide, blanc-verdatre, translucide ou opaque, *Hauy*, Tabl. p. 18.

External Characters.

Its colour is greenish-white and mountain-green.

It occurs massive, in longish kidneys *(niern)*, that sometimes approach to the form of rhomboidal prisms, having angles of 62° and 118°.

The lustre of the principal fracture is splendent, of the cross fracture glistening or dull.

The principal fracture is foliated, and a three-fold cleavage is to be observed, of which two of the cleavages are parallel with the lateral planes, and the third nearly parallel with the terminal planes; the cross fracture uneven or conchoidal.

The fragments are sharp-edged.

It is translucent on the edges.

It scratches glass, but is scratched by quartz.

Specific gravity 3.451.

Constituent

Constituent Parts.

Alumina, -	53.25
Silica, -	32.28
Lime, -	0.88
Iron, -	0.85
Loss, by heating,	0.75
Further Loss,	11.36
	100.

Hisenger & *Berzelius.*

The loss in this analysis is ascribed to the escape of fluoric acid.

Geographic situation.

It is found at Finbo, near Fahlun, in Sweden.

Observations.

1. Haüy refers it to topaz; it appears to be interme diate between topaz and schorlite.

2. We still want a good description of this mineral. the above is copied from Steffens.

4. Euclasc.

Euclase, Journal des Mines, n. 28. p. 258.—Euclasius, Lin. Syst. Nat. ed. 13. Lipsiæ, 1793, t. iii. p. 442.—Euclase, *Daubenton,* Tabl. p. 6.—*Id. Hauy,* t. 2. p. 531.—*Id. Broch.* t. 2. p. 508.—*Id. Lud.* b. 1. s. 165.—*Id. Suck.* 1. th. s. 165. —*Id Lucas,* p. 45.—*Id. Brong.* t. 1. p. 413.—*Id. Brard.* p. 121.—*Id. Kid,* vol. i. p. 133.—*Id. Hauy,* Tabl. p. 32.— *Id. Steffens,* s. 47.

External

External Characters.

Its colour passes, on one side, from greenish-white through mountain-green, and celandine-green, into dark sky-blue ; on the other, side into apple-green, with a trace of blue.

It has been hitherto found only crystallised, and in the following forms.

1. Oblique four-sided prism, rather acutely acuminated by four planes, which are set on the lateral planes *.
2. The preceding figure, slightly truncated on the lateral edges.
3. The preceding figure, in which two of the acuminating planes meet under an obtuse angle, so that the prism appears with a very oblique bevelment.
4. The preceding figure, in which the bevelling planes are once broken.
5. Oblique four-sided prism, in which the lateral edges are bevelled, and the edges of the bevelment on the acute edges truncated. On the extremity of the prism there are three acuminations of four planes each, which are set on each other, and correspond to the lateral planes of the prism ; but these planes are modified by a suite of bevelments, which are placed on their obtuse lateral edges ; there is also a bevelment (with triangular planes) between

* According to Haüy, the primitive form of Euclase is a straight prism, with rectangular bases.

between these and the bevelment on the obtuse lateral edges of the prism ; and, lastly, a truncation on each of the superior edges of the truncations on the acute lateral edges of the prism. Fig. 34.

The lateral planes are more or less longitudinally streaked ; seldom smooth. The streaks give to the planes a rounded appearance, and the prisms then appear reed-like.

Externally the crystals are shining and splendent, and vitreous ; internally splendent.

The fracture is perfect and straight foliated, with a simple cleavage, parallel with the acute edges of the prism. By very careful examination, we can discover a second cleavage, which appears to cut the other at right angles, and to pass obliquely through the prism. The cross fracture is small conchoidal.

The fragments are tabular, and frequently almost cubical.

It alternates from transparent to translucent, and refracts double.

It is so hard as to scratch quartz.

It is very easily frangible.

Rather heavy.

Specific gravity 3.0625. *Hauy.*—2.907, *Lowry*, from a specimen in possession of Mr Rundell of London.

Chemical Characters.

Before the blow-pipe, it first loses its transparency, which indicates the presence of water of crystallization, and then melts into a white enamel.

Constituent Parts.

Silica,	-	35	to 36
Alumina,	-	18	19
Glucina,	-	14	15
Iron,	-	2	3
Loss,	-	31	27
		100	100

Vauquelin.

The loss in these analyses appears owing, partly to water of crystallization, and partly to an alkali.

Geognostic and Geographic Situation.

This rare and beautiful mineral was first found in Peru, from whence it was brought to Europe by the traveller Dombey. Very lately it has been brought from Brazil in isolated crystals, that appear to have been imbedded. The Brazilian euclase was discovered in the mine of Gerais near Casson.

Observations.

1. It is a very beautiful mineral; but, on account of its easy frangibility, cannot be used in jewellery.

2. It is named *euclase* by Haüy, on account of its very great frangibility.

3. The greater part of the description of this mineral, was drawn up by M. Verina from specimens in the possession of Mr Edmond Rundell, and Mr Henry Heuland of London.

5. Emerald.

5. Emerald.

This species is divided into two sub-species, viz. Emerald and Beryl.

First subspecies.

Emerald.

Schmaragd, *Werner.*

Gemma pellucidissima, Smaragdus, *Wall.* t. 1. p. 253.—Emeraude du Perou, *R. d. L.* t. 2. s. 245.—Schmaragd, *Wern. Cronst.* p. 102. *Ib. Wid.* p. 271.—Emerald, *Kirw.* vol. i. p. 247.—Smaragd, *Estner,* b. 2. p. 132.—Smeraldo, *Nap.* p. 122.—Emeraude, *Lam.* t. 2. p. 227. *Brochant,* t. 1. p. 217.—Emeraude verte, *Hauy,* t. 2. p. 516.—Smaragd, *Reuss,* b. 2. th. 1. s. 165. *Id. Lud.* b. 1. s. 69. *Id. Suck.* 1. th. s. 205. *Id. Bert.* s. 308. *Id. Mohs,* b. 1. s. 140. *Id. Hab.* s. 25.—Emeraude, *Lucas,* p. 44.—Beril emeraude, *Brong.* t. 1. p. 417. *Id. Brard.* p. 119.—Emerald, *Kid.* vol. i. p. 130.—Glatter smaragd, *Karsten,* Tabell.—Emeraude verte, *Hauy,* Tabl. p. 32.—Schmaragd, *Steffens,* b. 1. s. 41.

External Characters.

Its characteristic, and, we may almost say, its only colour, is emerald green, of all degrees of intensity, from deep to pale. The deep sometimes inclines a little to verdegris-green, and oftener to grass green: the pale varieties sometimes nearly pass into greenish-white.

It is said to occur massive, and in rolled pieces, but of such Werner has seen no specimens; he has only obser-

 ved

ved it crystallised in low, equiangular, six-sided prisms *. which present the following varieties:

1. Truncated on the lateral edges.
2. Truncated on the terminal edges.
3. On the terminal angles.
4. Terminal edges bevelled. When the truncations on the lateral edges increase, a twelve-sided prism is formed.

The lateral planes are smooth; the terminal planes rough.

The crystals are middle-sized and small, very rarely large; and occur imbedded, or in druses.

Internally the lustre is intermediate between shining and splendent, and is vitreous.

The fracture is small and imperfect conchoidal, yet it sometimes exhibits a foliated fracture, having a fourfold cleavage, of which the folia are parallel with the lateral and terminal planes, as is the case with beryl.

The fragments are angular, and more or less sharp-edged.

It alternates from transparent to translucent, and refracts double in a moderate degree.

It is hard, and scratches glass easily, but quartz with difficulty.

It is rather heavy.

Specific gravity 2.600, *Werner*.—2.775, *Brisson*.—2.7227 to 2.7755, *Hauy*.—2.692, of a cut specimen in possession of Mr Rundell, *Lowry*.

Chemical

* According to Haüy, the primitive form is a regular hexahedron.

Chemical Characters.

Before the blow-pipe it melts into a white, and rather vesicular glass.

Constituent Parts.

Silica,	64.5		68.50
Alumina,	16		15.75
Glucina,	13		12.56
Oxide of Chrome,	3.25		0.30
Lime,	1.6		0.25
Water,	2.0	Oxide of Iron,	1.0
		Loss,	1.70
	100.35		100.
Vauquelin, Jour. des Mines, N. 38, p. 98,			*Klaproth*, Beit. iii. 226.

Geognostic Situation.

It occurs in drusy cavities, along with iron pyrites, calcareous spar, and quartz, in veins that traverse clay-slate; also imbedded in mica-slate; and loose in the sand of rivers and other alluvial deposites.

Geographic Situation.

The most beautiful emeralds are at present brought from Peru. The most ancient mine is that of de Manta, which is now exhausted; the other emerald mine is situated in the valley of Tunca, in the jurisdiction of Santa Fe, between the mountains of New Granada and Popayan. It occurs imbedded in mica-slate in Salzburg.

The Romans are said to have procured it from Æthiopia and Upper Egypt.

Use.

The colour which characterises this fossil is extremely pleasing: the eye, after viewing the beautiful colours of the sapphire, oriental ruby, spinel and topaz, reposes with delight on the fresh and animating colour of the emerald, the charming emblem of the vegetable kingdom! It is rare, however, to fiad the colour pure and of good strength; hence such specimens are very highly valued, and are employed in the most expensive kinds of jewellery.

Observations.

1. Emerald and Beryl have a strong resemblance to each other: thus both are green, their crystallization differs but little, and fracture, hardness and weight are nearly the same. Notwithstanding these agreements, they are distinguished from each other by the following characters: Emerald occurs only green; but beryl, besides green, is also yellow and blue; the crystals of beryl are long, those of emerald are short; the lateral planes of beryl are streaked, those of emerald are almost always smooth; the terminal planes of beryl are smooth, those of emerald are rough; beryl is more distinctly foliated than emerald; beryl often presents distinct concretions, emerald never; beryl often shews a formation by acicular shoots, emerald never; beryl has transverse rents, emerald never; and the crystals of beryl are larger than those of emerald; and beryl is rather softer than emerald.

2. Many

2. Many of the emeralds described by the ancients appear to have been varieties of green fluor-spar; even in more modern times, fluor-spar has been preserved for emerald. Mr Coxe examined the famous emerald table in the Abbey of Reichenau near Constance, which he found to be a very fine green-coloured fluor-spar. The famous *sacro cattino di smeraldo orientale*, preserved at Genoa, and which could only be seen by an order from the senate, is a mass of cellular glass. Many fine Ethiopian emeralds, which were bequeathed to monasteries, appear to have been sold by the monks, and coloured glass substituted in their place.

3. This mineral was named *smaragdus* by the ancients. Pliny distinguished twelve species of the smaragdus; but under this title he includes, besides the true emerald, also green jasper, prase, malachite, fluor-spar, serpentine, and translucent varieties of gypsum. Theophrastus also mentions the true emerald, which he says occurs in small quantity, and very rarely: he enumerates along with it another mineral of a green colour, which, he says, is found in masses ten feet long, and which is probably a variety of serpentine. The emerald with which the hall of Assuerus was paved; the pillars of emerald in the temple of Hercules at Tyre, mentioned by Herodotus; and the large emeralds described by Pliny as having been cut into columns and statues, (thus, the statue of Serapis in Egypt, nine ells long, is said to be of emerald), cannot be referred to the true emerald. The confusion that prevails in the descriptions of this mineral in ancient authors, has led some mineralogists to believe, that the true emerald was not known in Europe until after the conquest of Mexico and Peru by the Spaniards. The following facts, however, are in opposition to this opinion;

(1.) The emerald was so highly prized by the Romans, that when the luxurious and rich Lucullus landed at Alexandria, he was presented by Ptolemy with an emerald, on which was engraven a portrait of the king of Egypt; and this was considered as the most valuable present that could be offered to him.

(2.) In the National Museum in Paris, there is a fine emerald, on which is engraved an eye, which is known to be a very common Egyptian hieroglyphic.

(3.) In the mitre of Pope Julius II. which was presented to Pius VII. by Buonaparte, there is a fine deep green coloured emerald. As he died in 1513, and Peru was not discovered and conquered by Pizarro before 1545, it is highly probable that this emerald was brought from Africa.

(4.) Werner has in his possession several antique emeralds; and Mr Hawkins informed the Abbé Estner, that he had seen a necklace of emeralds, which was found among the ruins of Portici near Naples.

4. The ancients attributed many virtues to the emerald; thus they maintained, that the sight of its animating and refreshing colour chased away melancholy; that it completely prevented the fatal effects of poison, and even cured the most obstinate diseases.

5. The Brazilian emerald is a variety of Tourmaline; and the Oriental emerald is green coloured Sapphire.

6. Emerald is one of the lightest and softest of the precious stones.

Sub-

Second Subspecies.

Beryl.

Edler Beril, *Werner.*

Smaragdus, Aqua marina, et Smaragdus Berillus, *Wall.* t. 1. p. 254.—Aigue marine de Siberie, *R. d. L.* t. 2. p. 252. *Id. Born.* t. 1. p. 71.—Edler Beril, *Wern. Cronst.* s. 100.—Beryl, *Kirw.* vol. i. p. 248.—Edler, Beril, *Wid.* s. 274.—*Id. Estner,* b. 2. s. 197.—Berillo, *Nap.* p. 125.—Aigue-marine, *Lam.* t. 2. p. 232.—Emeraude, *Hauy,* t. 2. p. 516 —Le Beril noble, *Broch.* t. 1. p. 220 —Gestreifter Smaragd, *Reuss,* b. 2. th. 1. s. 102. Edler Beril, *Lud.* b. 1. s. 70. *Id. Suck.* 1. th. s. 208. *Id. Bert.* s. 310. *Id. Mohs,* b. 1. s. 146.—Gestreifter Smaragd, *Hab.* s. 26.—Beril aigue-marine, *Brong.* t. 1. p. 415.—Emeraude vert-bleuatre, *Brard,* p. 12.—Beryl, *Kid,* vol. i. p. 128.—Emeraude vert-bleuatre & jaune-verdatre, *Hauy,* Tabl. p. 32.—Beril, *Steffens,* b. 1. s. 44.

External Characters.

Its principal colour is green, from which it passes on the one side into blue, and on the other into yellow. It is commonly mountain and celandine green: from these it passes through apple-green, asparagus-green, oil-green, into honey-yellow, which approaches to wine-yellow. From celandine-green it passes into smalt, sky, and, in rare instances, into azure-blue.

Almost all its colours are pale, seldom deep, and scarcely ever dark. Sometimes it has two colours at once, which alternate in layers, and sometimes it is iridescent.

It occurs massive, and crystallised in long equiangular six-sided prisms, which are either perfect, or truncated on the

the lateral edges, as fig. 35.; truncated on the terminal angles as in fig. 36.; or on the terminal edges and angles as in fig. 37.

The truncations on the terminal edges sometimes become so large as to form six-planed acuminations, in which the apices appear deeply truncated, as in fig. 37.

The lateral planes are deeply longitudinally streaked, but the terminal acuminating and truncating planes are smooth.

The lateral planes vary much in breadth; sometimes three planes are so large in comparison of the others, that the crystal appears almost trihedral: in other instances, four planes are so large, that the figure is almost tetrahedral. Sometimes the lateral planes are cylindrical convex, and then the crystals appear acicular or reed-like.

The crystals are sometimes jointed like basalt, having a concave surface at one extremity, and a convex surface at the other. They are seldom single; generally many occur together, and these cross each other in different directions, and are frequently superimposed and imbedded.

The crystals are small, large, and very large.

Transparent crystals occur a foot long, and four inches in diameter *.

Externally its lustre is shining and glistening; internally shining, which sometimes passes into glistening and splendent, and is vitreous.

The

* In Weiss's collection in Vienna there is a druse of very large beryl crystals; two of the crystals, which are of a mountain-green colour, and cross each other, are a foot and a half long, and one foot in diameter.

The principal fracture is foliated, with a fourfold cleavage: three of the folia or cleavages are parallel with the lateral planes, and the fourth with the terminal planes; and the cross fracture is imperfect conchoidal.

The fragments are angular, and more or less sharp-edged.

The massive varieties occur in straight and thin prismatic distinct concretions.

It is commonly transparent, but sometimes passes into translucent, and it refracts double. The translucent variety has cross rents.

It is hard: scratches quartz, but with difficulty, and is nearly equal in hardness to topaz, with which the mountain-green variety has been often confounded.

It is easily frangible.

Rather heavy.

Specific gravity 2.6500 to 2.7590, *Werner*.—2.682 to 2.683 to 2.722, *Brisson*.—2.664, *Lowry*.

Chemical Characters.

Before the blow-pipe it is difficultly fusible without addition, but with borax it melts easily.

Constituent Parts.

Silica,	69.50		68.0
Alumina,	14.00		15.0
Glucina,	14.00		14.0
		Lime,	2.0
Oxide of Iron,	1.00		1
	98.50		100
	Rose.	*Vauquelin*, Jour. des Mines, N. 43, p. 563.	

Geognostic

Geognostic Situation.

It occurs in veins, that traverse granite, along with rock-crystal, felspar, topaz, schorl, and iron-ochre; also imbedded in granite, and dispersed through alluvial soil.

Geographic Situation.

Europe.—It occurs in alluvial soil along with rock-crystal and topaz, in the upper parts of Aberdeenshire *. In Ireland, imbedded in granite, near Lough Bray, in the county of Wicklow, and near Cronebane in the same county †.

It occurs imbedded in granite on the south side of the Rathhausberg in Gastein in Salzburg, and on the highest summit of the Saualpe in Carinthia. In granite in the island of Elba, and in large crystals in veins of quartz that traverse granite near to Limoges; also at Marmagne, and a little to the west of Nantz in France.

Asia.—The finest beryls are found in veins that traverse the granite mountain Adon-Tschalon in Dauria; and it is from that quarter that nearly all the abundant supplies of Russian beryl are obtained. It also occurs, along with arsenical pyrites, in a kind of serpentine rock near Nertschinsk; in the mountain Tygirek (Mountain of Snow) in the Altain range; on the borders of the river

* Vid. Memoirs of the Wernerian Society, vol. i. p. 445,—452.

† Fitton, Stephens, and Weaver, Geological Transactions, vol. i. p. 275.

ver Lena; near the town of Ajatskaja in the Uralian range; and in the circle of Alepasski in Persia.

America.—In Brazil; and lately at Topham Maine *, and in the district of German Town, in the United States †.

Uses.

When pure, it is cut for ring-stones and necklaces, but is not so highly valued as emerald. Figures are sometimes cut on it. In the royal library at Paris there is a portrait of Julia, the daughter of Titus, engraved on a green-coloured beryl.

Observations.

1. The green-coloured varieties are sometimes named Aqua-marine, a term also applied to the green-coloured varieties of Topaz.

2. Beryl does not contain chrome; hence it has not the fine green colour of the emerald.

3. It was well-known to the ancients, who procured it from several places, where it is at present found. Pliny has given a good description of it ‡; yet in later times this description appears to have been forgotten; for we find it arranged with other precious stones, to which it had but little resemblance: thus the blue varieties were named Sapphire; the yellow, Topaz; and the green, Aqua-marine Topaz. Many years ago Werner obtained a complete suite of specimens of this mineral from Siberia,

* Greenough.

† Wister, in the American Mineralogical Journal, vol. i. p. 31.

‡ Plin. Hist. Nat. lib. xxxvii. c. 1.

Siberia, which enabled him to give it its proper place in the system.

4. It was Wallerius, and not Romé de Lisle, as mentioned by Haüy, who united emerald and beryl as one species.

6. Iolite.

Iolith, *Werner.*

Iolith, *Karsten,* Tabel.—Iolithe, *Hauy,* Tabl. p. 61, & 221.—Dichroite, *Cordier,* Journ. des Mines, t. 25. p. 129.—Iolith, *Steffens,* b. i. s. 369.

External Characters.

Its colour is muddy violet-blue. When viewed in the direction of the axis of its crystals, the colour is violet-blue; but when viewed perpendicular to the axis of the crystals, yellowish-brown.

It occurs massive, disseminated, and crystallised in small equiangular and equilateral six-sided prisms, which have rough surfaces.

Its internal lustre is shining, and intermediate between vitreous and resinous.

The fracture is sometimes small and imperfect conchoidal; but more frequently small-grained uneven. Very seldom imperfect foliated.

The fragments are sharp-angular.

It alternates from translucent to opaque.

It is hard; scratches glass easily, but quartz with difficulty.

Is

Is easily frangible.

Rather heavy.

Specific gravity 2.560, *Cordier.*

Chemical Characters.

It is not affected by acids.

Before the blow-pipe it melts with difficulty into a greenish-grey enamel. We obtain the same result with borax and carbonate of soda.

Geognostic and Geographic Situations.

It occurs imbedded in fragments of gneiss and compact felspar, contained in a variety of tuff, by some, as Cordier, said to be volcanic tuff.

It occurs most abundantly in the masses of compact felspar. The tuff forms a very thick horizontal bed, and contains, besides the masses of gneiss and felspar, also fragments and blocks of black and red scoria, of obsidian, and basalt or basaltic lava. It occurs at Cape de Gate near Valentia in Spain. It has been lately found imbedded in primitive trap at Arendal in Norway.

Observations.

This mineral was brought to France from Cape de Gate many years ago by the French mineral dealer Launoy. Owing to its colour and crystallization, it was for a time arranged with sapphire. Werner first established it as a distinct species, and arranges it in the system between the ruby and schorl families.

Karsten places it between azurite and andalusite ; but it is here placed immediately after emerald. on account of of resemblance to that mineral in several of its characters.

7. Schorl.

This species is divided into two subspecies, viz. Tourmaline or Precious Schorl, and Common Schorl.

First Subspecies.

Tourmaline.

Tourmalin.—*Werner.*

Zeolites electricus, Turmalin, *Wall.* t. i. p. 329.—Schorl transparent rhomboidal, *R. d. L.* t. ii. p. 344.—Brasilianischer Turmalin, *Wid.* p. 284.—Tourmaline, *Kirw.* t. i. p. 271.—Sorlo Brasiliano, *Nap.* p. 150.—Tourmaline, *Lam.* t. ii. p. 295.—Le Schorl electrique, t. i. p. 229.—Tourmalines vertes et blues, *Hauy,* t. iii. p. 31.—Edler Schorl, *Reuss,* b. ii. th. 1. s. 119. *Id. Lud.* b. i. s. 72. *Id. Suck.* 1. th. s. 221. *Id. Bert.* s. 191.—Tourmalin, *Mohs,* b. i. s. 163. *Id. Hab.* s. 33. *Id. Lucas,* p. 54.—Tourmaline crystallisée, &c. *Brong.* t. i. p. 405.—Tourmaline, *Brard,* p. 140. *Id. Kid,* vol. i. p. 233. *Id. Steffens,* b. i. s. 51.

External Characters.

Its principal colours are green and brown: from leek-green it passes into pistachio and olive-green, then into liver-brown, and yellowish and reddish-brown; further into hyacinth-red, crimson-red, violet-blue, azure-blue, dark Berlin blue, and, lastly, into indigo-blue *.

Its

* Honey-yellow is mentioned as one of its colours.

Its colours are almost all of them dark, often a little muddy, and when it is nearly opaque, on account of the darkness of the colour, it appears black.

It occurs very seldom massive; scarcely ever disseminated; oftener in rolled pieces; but most frequently crystallised.

Its fundamental crystallization is an equiangular three-sided prism, *flatly acuminated* on both extremities by three planes, which on the one extremity are set on the lateral edges, on the other on the lateral planes *.

The lateral edges are frequently bevelled, and thus a nine-sided prism is formed; when the edges of the bevelment are truncated, a twelve-sided prism is formed; and when the bevelling planes increase so much, that the original faces of the prism disappear, an equiangular six-sided prism is formed. The acuminations of these prisms exhibit very great variety of appearance; thus the angles edges and extremities, are frequently truncated, and the angles bevelled. Fig. 39. shews truncations on the angles of the acumination and bevelment of the lateral edges of the prism; PP, the original planes of the acumination; *o o*, the truncations on the angles. Fig. 40: in this figure the lateral edges of the prism are truncated. Fig. 41. the angles formed by the meeting of two acuminating planes and one lateral plane bevelled; *x x* are the bevelling planes. When the truncations *o o*, on the angles of the acuminations, increase very much, there is formed *a pretty acute acumination*, and the planes of the original

* According to Hauy, the primitive form of schorl is an obtuse rhomboid.

original acumination PP, appear as truncations on the edges formed by the meeting of the planes of this acumination, as in fig. 42. When the truncations on the edges formed by the meeting of the acuminating planes increase very much in size, *a very flat acumination* is formed, and the planes of the original acumination appear as truncations on the angles formed by the meeting of two of these planes, and one of the lateral planes of the prism. In fig. 43. PP are the original acuminating planes, truncated on the edges *n*. In fig 44. the truncations *n n n* form a very flat acumination, and the original planes PPP truncations on the angles of the acuminations, and the apex of the acumination is truncated, *k* being the truncating plane Sometimes the prism is nearly awanting, as in fig. 45. when a double three-sided pyramid is formed, in which the lateral planes of the one are set on the lateral edges of the other, and the remainder of the prism forms truncations on the edges of the common base.

The lateral planes are generally cylindrical convex, and deeply longitudinally streaked; the acuminating planes are mostly smooth and shining: sometimes the planes on one extremity are smooth, but on the other rough.

The crystals are seldom large, more commonly middle-sized and small, and sometimes scopiformly aggregated, as is the case with the red variety from Siberia.

The crystals are usually imbedded.

Internally its lustre is splendent and vitreous.

The fracture is perfect conchoidal; sometimes an indistinct foliated fracture is to be observed, and the folia are inclined to the axis of the crystal.

It rarely occurs in prismatic concretions; and these varieties generally verge on common schorl.

It

It alternates from nearly opaque to completely transparent.

Some varieties, when viewed in a direction oblique to the axis of the crystal, are transparent, but in the direction of the axis opaque: others again exhibit different colours, according to the direction in which they are held *.

It is hard; scratches glass, but is not so hard as quartz.

It is easily frangible.

Is intermediate, between rather heavy and heavy.

Specific gravity.—Green tourmaline, 3.086, *Werner.*—3.086, *Brisson.*—From 3.0863 to 3.3626, *Hauy.*——Blue tourmaline, 3.155, *Werner.*—3.130, *Brisson.*—Green tourmaline, 3.191, *Lowry.*

Chemical Characters.

Before the blow-pipe it melts into a greyish-white vesicular enamel; but the red-coloured Siberian tourmaline is infusible, which is conjectured to be owing to the considerable portion of manganese it contains.

Constituent

* Werner has in his possession a tourmaline which is sky-blue in the middle, but violet-blue on the sides.

Constituent Parts.

	Green Tourmaline, from Brazil.		Violet-coloured Tourmaline, from Siberia.		Red Tourmaline, from Rosena.
Silica,	40.		42.		43.5
Alumina,	39.		40.		42.25
Lime,	3.84	Soda,	10.	Soda,	9.
Oxide of iron,	12.5	Oxide of manganese, containing a little iron,	7.	Oxide of manganese,	1.5
Oxide of manganese,	2.			Lime,	0.1
				Water,	1.25
Loss,	2.66		1.		2.4
	100		100		100
	Vauquelin, Ann. de Chem. N. 88. p. 105.		*Vauquelin*, Ann. du Mus. T. iii. p. 243.		*Klaproth*, Journ. des Mines, N. 137, p. 383.

Black Tourmaline.

Silica,	35.
Alumina,	40.
Oxide of iron,	22.
Loss,	3.
	100

Klap. Karst. Tabel. p. 46.

Physical Characters.

By friction, it exhibits signs of vitreous electricity. by heating, vitreous electricity at one extremity, and resinous electricity at the other.

Geognostic

Geognostic Situation.

It scarcely occurs otherwise than imbedded in different kinds of mountain rock, and in them it is rather confined to single beds or strata, than disseminated through the whole mass of the mountain. The rocks in which it most commonly occurs are gneis, mica-slate, talc-slate, indurated talc; and the accompanying minerals are rock-crystal, common quartz, felspar and mica. It occurs also in granite, but has not hitherto been discovered in any rock of the transition or flœtz series. It does not occur in primitive trap, nor in general with any of the subspecies of hornblende; in short, it appears to have little or no geognostic affinity with hornblende, although they have been frequently confounded together.

In alluvial countries it occurs in rolled pieces.

Geographic Situation.

It was first discovered in the 16th century, in the island of Ceylon; afterwards in Brazil, and since that period in several other countries, as appears from the following enumeration.

Europe.—Langoe near Krageroe in Norway; near Uton at Anskop in Sweden; near Freyberg, and at Ehrenfriedersdorf, Dorfschemnitz, in the Saxon Erzgebirge, or Metalliferous Mountains; at Altsattel in Bohemia; in Silesia, Bavaria, Moravia, the Tyrol, Stiria, Swisserland, Austria, Italy, Spain, and France.

Asia.—The red tourmaline occurs in Siberia, Ava and Ceylon; and several of the other varieties in the same countries of Asia.

Africa.—Island of Madagascar.

North America.—Greenland; and at Goshen in Massachusets *.

South America —Brazil.

Use.

It is sometimes cut and polished, and worn as a jewel; but on account of the muddiness of its colours, it is not in general very much esteemed.

Observations.

1. *Distinctive Characters.*—*a* Between tourmaline and *common schorl.* Common schorl has but one colour, whereas tourmaline has a considerable suite of colours. The fracture of common schorl is small-grained, uneven; that of tourmaline conchoidal: common schorl is always opaque, tourmaline more or less translucent or semitransparent: common schorl generally occurs in distinct concretions, whereas tourmaline presents this character very rarely, and principally in those varieties which are passing into common schorl; and we may add, that schorl is very often massive and disseminated, but tourmaline very seldom.—*b.* Between tourmaline and *common* and *basaltic hornblende.* The colour suites and crystallizations of these minerals are very different from those of tourmaline; these minerals have a very distinct foliated fracture, whereas that of tourmaline is conchoidal, in some rare varieties indistinct foliated; and they are not so hard as tourmaline.

2. Different varieties of this mineral have received particular denominations. The following are some of these:

1. Green

* Weeks, in American Mineralogical Journal, vol. i. p. 123. It is the indigo-blue variety.

Names given to particular varieties of TOURMALINE.

1. Green Tourmaline, named	*Brazilian Emerald.*
2. Berlin-blue Tourmaline,	*Brazilian Sapphire.*
3. Indigo-blue Tourmaline,	*Indicolite.*
4. Honey-yellow Tourmaline,	*Peridot of Ceylon.*
5. Red Tourmaline,	*Rubellite, Siberite, Daourite, Tourmaline apyre, Red Schorl of Siberia.*

3. The red tourmaline is arranged as a distinct subspecies, under the title *Rubellite*, by Karsten and Steffens; and Karsten and Dandrada describe the indigo-blue variety as a separate species, under the title *Indicolite*.

4. In the Grevillian collection, there is a large and fine specimen of red tourmaline, which was presented to Colonel Simes by the King of Ava *; and in the beautiful collection belonging to Baron Racknitz at Dresden, I observed a specimen of this variety, nearly an inch in diameter, for which 400 rubles were paid. In Morgenbessers' cabinet at Vienna, there is a prism of red Siberian tourmaline which cost 800 rubles.

---

* This specimen consists of many crystals, and was valued at L. 500 by the commissioners who were appointed by Parliament to report on the value of the Greville collection, previously to its being purchased by Government for the British Museum.

Second Subspecies.

Common Schorl.

Gemeiner Schörl.—*Werner.*

Some of the varieties of Basaltes crystallizatus, *Wall.* t. 1. p. 333.—Schwartzer-stangen schorl, *Wid.* p. 279.—Schorl, *Kirw.* vol. i. p. 265.—Sorlo-nero, *Nap.* p. 146.—Tourmaline, *Lam.* t. ii. p. 295.—Le schorl noire, *Broch.* t. i. p. 226.—Tourmaline noire, *Hauy,* t. iii. p. 31.—Gemeiner schorl, *Reuss,* b. ii. th. i. s. 129. *Id. Lud.* b. i. s. 71. *Id. Suck.* 1. th. s. 217. *Id. Bert.* s. 193. *Id. Mohs,* b. i s. 177. *Id. Hab.* s. 33.—Tourmaline, *Lucas,* p. 54.—Tourmaline schorl, *Brong.* t. i. p. 407.—Tourmaline, *Brard,* p. 140.—Tourmaline, *Kid,* vol i. p 233.- Schorl opaque et noire, *Hauy,* Tabl. p. 39.—Gemeiner schorl, *Steffens,* b. i. s. 60.

External Characters.

Its colour is velvet-black, of various degrees of intensity.

It occurs often massive and disseminated, seldom in rolled pieces, and frequently crystallised, in three, six, and nine sided prisms, that present acuminations, truncations, and bevelments, of the same kinds as those that occur in tourmaline.

The crystals are mostly acicular; often broken, forming with the fragments a peculiar kind of fragmented stone or breccia; and are imbedded. The lateral planes are longitudinal streaked, and alternate from shining to glistening.

Internally

Internally its lustre is intermediate between shining and glistening, and is vitreous.

The fracture is intermediate between imperfect conchoidal, and small and coarse grained uneven, and inclines sometimes more to the one, sometimes more to the other.

The fragments are angular.

It occurs in distinct concretions of different kinds; rarely coarse and small granular; sometimes thin, or thick, and straight prismatic. Sometimes the prismatic concretions are so thin, that they verge on fibrous; and such varieties are sometimes parallel, but most frequently scopiform diverging fibrous. These prismatic concretions are sometimes again collected into others, which are thick and wedge-shaped.

It is opaque; but slightly translucent on the edges, when it passes into tourmaline.

It gives a grey streak.

It is hard; slightly inferior to quartz.

It is very easily frangible.

It is intermediate between rather heavy, and heavy.

Specific gravity, 3.092, *Brisson.*--3.150, *Gerhard.*—3.212, *Kirwan.*—3.0863 to 3.3636, *Hauy.*

Chemical Characters.

Before the blow-pipe it melts pretty easily, without addition, into a blackish slag. Melted with borax, it forms a greenish-coloured glass.

Constituent

Constituent Parts.

	Common Schorl from Eibenstock.	Common Schorl from the Spessart.
Silica,	36.75	36.50
Alumina,	34.50	31.0
Magnesia,	0.25	1.25
Oxide of iron,	21.0	23.50
Potash,	6.0	5.50
Trace of oxide of manganese.		
	98.50	97.75

Klaproth's Beiträge, B. 5. s. 148, 149.

Physical Characters.

By heating it exhibits positive electricity at one extremity, and negative at the other. Wiedenman remarks, that when it begins to cool, the positive end become negative, and the negative positive.

Geognostic Situation.

The geognostic situation of this subspecies agrees in part with that of tourmaline ; thus a part of it appears imbedded in crystals, or massive portions in granite, gneiss, mica-slate, and clay-slate ; but it occurs also disseminated, and forms an essential constituent part of certain rocks, as topaz-rock and schorl-rock.

It

It occurs in veins that traverse clay-slate and other rocks, along with quartz, felspar, garnet, mica, tinstone, &c. It has not been found either in flœtz or transition-rocks, or in any rock belonging to the trap formation.

Geographic Situation.

Europe.—Perthshire, Banffshire, Inverness-shire, Argylshire, in Scotland; in schorl rock and tinstone veins in Cornwall; Norway, Sweden, Saxon Metalliferous Mountains (Erzgebirge), Hartz, Bohemia, Franconia, Moravia, Silesia, Suabia, Bavaria, Swisserland, the Tyrol, Hungary, France, and Spain.

Asia.—Ceylon, borders of the lake Baikal, and different parts of the Uralian range.

America.—Greenland, Hudson's Bay, United States, Mexico.

Observations.

1. It was first found near the village of Schorlaw, in Saxony, whence its name.

2. It differs from *tourmaline* in colour, degree of lustre, fracture, transparency, and distinct concretions; also in its geognostic situation, for tourmaline occurs almost always imbedded and in single crystals; on the contrary, schorl is usually aggregated, and occurs in beds.

8. Epidote,

6. Epidote, or Pistacite.

Pistacit.—*Werner.*

Schorl vert du Dauphiny, *Romé de Lisle,* t. ii. p. 401.—Thallite, *Daubenton,* Tabl. p. 9.—Thallite, *La Metherie,* Theor. de la Terre, 2d edit. t. ii. p. 319.—Delphinite, *Saussure,* Voyage dans les Alpes, n. 1918.—Acanticone, *Dandrada,* Journ. Chem. von Schœrer, T. iv. p. 19.—Thallit, *Karsten,* Mineral. tabellen, p. 20.—Arendalite, *ib.* p. 34.—Thallite, *Reuss,* b. ii. th. i. s. 117..—Epidot, *Hauy,* T. iii. p. 102.—Thallite, Arendalite, *Lud.* b. ii. s. 136, 137.—Epidot, *Suck.* 1. th. s. 256.—Thallite, acanticone, *Bert.* s. 196. 173.—Epidot, *Mohs,* b. i. s. 57. *Id. Lucas,* p. 59. *Id. Brong.* t. i. p. 410. *Id. Brard,* p. 153.—Thallite, *Kid,* vol. i. p. 242.—Epidot, *Hauy,* Tabl. p. 43. *Id. Steffens,* b. i. s. 66.

External Characters.

Its principal colour is pistachio-green, which passes on the one side into blackish-green and greenish-black, on the other side into dark olive-green, oil-green, and siskin-green.

It occurs massive, and crystallised in the following forms:

1. Oblique four-sided prism, flatly bevelled on the extremities; the bevelling planes set on the acute lateral edges.

   This is the fundamental figure.
2. The preceding figure truncated on the acute edges, and flatly bevelled on the extremities; the bevelling

ling planes set on the truncating planes; or we may describe it as a broad unequiangular six-sided prism, flatly bevelled on the extremities; the bevelling planes set on the opposite smaller lateral planes, fig. 46.

3. The preceding six-sided prism acuminated on the extremities by four planes, which are set on the broader lateral planes, and the apex of the acumination, and the angles formed by the meeting of two acuminating planes, with a smaller lateral plane truncated, fig. 47.
4. The same prism acuminated by six planes, and the apex of the acumination deeply truncated, fig. 48.
5. The same prism truncated on the acute edges, and flatly acuminated on the extremities by six planes, which are set on the lateral planes, and the apices of the acuminations, and the edges formed by the meeting of two opposite acuminating planes with the lateral planes truncated, fig. 49.
6. Very oblique four-sided prism, truncated on the obtuse lateral edges, and doubly acuminated on the extremities by four planes, the apex of the acumination truncated, and simply bevelled by two planes set on the acute lateral edges, fig. 50.

Besides the crystallizations just described, others occur, which are to be viewed as varieties of the preceding, but are difficult of determination, on account of the multiplicity of truncations and bevelments they display, and the unequal increase of the different planes.

The

The crystals are sometimes reed-like, and are promiscuous, scopiform, and scalarwise aggregated; and are generally middle-sized.

The lateral planes of the crystals are deeply longitudinally streaked, and the terminal planes are sometimes spherical convex.

Externally, the lustre alternates from splendent to glistening, and is vitreous; internally, it is shining or glistening, and is resinous passing into vitreous.

The principal fracture is foliated, with at least a double cleavage; the cleavages are parallel with the lateral planes of the oblique four-sided prism: the cross fracture is small-grained uneven, and sometimes also concealed foliated. Some varieties have a scopiform, or stellular diverging radiated, or coarse fibrous fracture.

The fragments are angular.

It occurs in coarse and small granular, also in wedge-shaped distinct concretions.

It alternates from translucent to translucent on the edges, and to nearly transparent.

It is hard: scratches glass readily.

It is brittle.

Easily frangible.

It is rather heavy, approaching to heavy.

Specific gravity 3.4529, *Lucas.*—3.640, *Werner.*—Variety named Acanticone, from Norway, 3.407, *Lowry.*

Chemical Characters.

Before the blow-pipe it is converted into a brown-coloured scoria, which blackens by continuance of the heat.

Constituent

Constituent Parts.

	Pistacite from the Valais.	From Oisans.	From Arendal.
Silica,	37.0	37.0	37.0
Alumina,	26.6	27.0	21.0
Lime,	20.0	14.0	15.0
Oxide of iron,	13.0	17.0	24.0
Oxide of manganese,	0.6	1.5	1.5
Water,	1.8	3.5	1.5
Loss,	1.0	0	0
	100.0	100.0	100.0
		Descotils.	*Vauquelin.*

Laugier, Ann. du Mus. d'Hist. Nat. T. v. p. 149.

Geognostic Situation.

It occurs in beds and veins, and sometimes as an accidental constituent part of rocks. The beds in which it occurs are primitive, and contain augite, garnet, hornblende, quartz, calcareous-spar, and magnetic ironstone, as at Arendal in Norway; or, besides the epidote, they contain calcareous-spar, copper-pyrites, and variegated copper-ore, as in the Bannat and other places. The veins of which it forms a part are small, and of very old formation, usually traverse gneiss, and contain besides the pistacite, felspar, rock-crystal, axinite, chlorite, asbestus, prehnite, octahedrite, and several other minerals. The varieties that occur in veins, are distinguished from

from those that occur in beds, by their lighter colours, and the more needle-shaped aspect of the crystals. The rocks in which it occurs are sienite, porphyry, and undefined granitous rocks.

Geographic Situation.

Europe.—In Arran it occurs in flœtz sienite, and clay-slate: in the island of Icolmkill, in a rock composed of red felspar and quartz: in the island of Rona, also one of the Hebrides, in slender veins, traversing a rock composed of felspar and quartz, and felspar and hornblende: in similar rocks among the Malvern Hills in Worcestershire; in quartz, at Wallow Crag near Keswick in Cumberland; near Marazion in Cornwall; and in granitous rocks in the islands of Guernsey and Jersey *. Upon the Continent of Europe it occurs in magnificent crystals at Arendal; and in porphyry near Christiana in Norway; also in Sweden; imbedded in rolled masses of a granitous rock in Mecklenburg; Bavaria, France, Italy, and Swisserland.

Asia.—Imbedded in granular limestone in Siberia; and in India along with corundum.

Africa.—Found imbedded in common quartz on the banks of the Orange River, by Dr Sommerville.

America.—Upon the banks of Lake Champlaine, along with tremolite †; and in the mountains of South Carolina.

Observations.

---

* Horner, Geological Transactions, vol. i. p. 292.

† Greenough.

Observations.

1. *Distinctive Characters.—a.* Between epidote and *actynolite.* The colour of actynolite is in general lighter than that of pistacite; the massive varieties of actynolite are radiated, whereas those of epidote are compact or foliated: in actynolite, both the cleavages are distinctly seen; but, in epidote, frequently only one cleavage is to be seen; and the crystals of actynolite are generally imbedded, and their terminal edges and angles truncated, whereas the crystals of epidote are frequently superimposed, and their extremities are bevelled or acuminated: actynolite is softer than epidote: and actynolite, before the blow-pipe, melts into a greyish-white enamel; epidote into a black scoria.—*b.* Between epidote and *common hornblende.* The colour-suite of common hornblende is not the same with that of epidote; it has a pearly lustre, and is softer than epidote.—*c.* Between epidote and *asbestus.* Asbestus, when pounded, feels soft, whereas epidote feels rough; and asbestus fuses into an enamel, but epidote into a scoria.

2. The olive-green coloured variety, in four-sided prisms, has been described as a distinct species, under the name *Baikalite.*

3. Klaproth describes, under the title *Scorza,* a substance which undoubtedly belongs to this species. Its colour is intermediate between pistachio and siskin green: it occurs in fine, roundish, dull, and meagre grains, that scratch glass, and have the specific gravity of 3.135. It contains silica, 43.; alumina, 21.; lime, 14.; oxide of iron, 16.5.; oxide of manganese, 0.25. It is found in small nests, in a grey-coloured clayey stone, in a valley near the town of Muska, on the river Aranyos in Transilvania.

silvania. The Wallachian name for this substance, viz. Scorza, has been retained by Klaproth. It might be arranged as a subspecies of epidote, under the title *Arenaceous Epidote.* Karsten names it *Arenaceous Thallite.*

4. Hausman, according to Steffens, describes a mineral said to belong to this species, under the name *Earthy Epidote (erdiger epidote).* It has a pale siskin-green colour; occurs disseminated, and in membranes. Internally it is dull, and the fracture earthy; it is meagre to the feel, and soils. It occurs in granite, at Trolhatta in Sweden, and, I believe, in the island of Rona, and other parts of the Highlands of Scotland.

5. Hausman describes another mineral under the title *Capillary Epidote.* It is said to have a very dark pistachio-green colour, and to occur in very delicate capillary crystals, that have a lustre intermediate between silky and vitreous, and incrust small drusy cavities, at Hackedalen in Norway. This appears to be a variety of common epidote, and does not seem to have so strong a claim to be considered as a subspecies as the Scorza or the Earthy Epidote.

6. Karsten divides this species into three subspecies, viz. *Common*, *Splintery*, and *Arenaceous.* The arenaceous is the scorza already mentioned; the splintery includes all those varieties that have been described under the names Arendalite and Acanticone, and which seem to be but varieties of the common pistacite.

7. Epidote is very nearly allied to zoisite, and hence it is placed in the system beside it; it is also nearly connected with hornblende, actynolite, and augite.

8. It was named Green Schorl by Romé de Lisle; but La Metherie was the first who described it as a distinct species, and under the title Thallite. Saussure afterwards met with it in the Alps, and describes it under the name

name Delphinite; and other varieties found in Norway, were described by Dandrada, by the names Acanticone and Arendalite. Hauy and Werner, nearly about the same time, particularly examined this mineral. Hauy named it Epidote, and Werner Pistacite.

Hauy published a description of the species, which Werner has not done; therefore the name Epidote, given to it by Hauy, has been very generally adopted.

9. Zoisite.

This species is divided into two subspecies, viz. Common Zoisite and Friable Zoisite.

First Subspecies,

Common Zoisite.

Zoisit.—*Werner.*

Zoisit, *Karsten* in Klaproth, Beit. b. iv. s. 180.—Epidot, *Hauy,* Journ. des Mines, n. 113. p. 465. *Id. Hauy,* Tabl. p. 44.—Zoisit, *Steffens,* b. i. s. 74.

External Characters.

Its colours are yellowish-grey, and light bluish-grey, which approaches to smoke-grey.

It occurs massive, and in very oblique four-sided reed-like prisms.

The crystals are middle-sized, and deeply longitudinally streaked.

The principal fracture is perfect straight foliated, with a single cleavage, which is parallel with the axis of the crystal, and is splendent; the cross fracture is small-grained uneven, and glistening, and intermediate between pearly and resinous.

The fragments are angular.

It occurs in large and small longish granular distinct concretions; also in thin and straight lamellar concretions.

It is translucent.

Hard.

Very easily frangible.

Rather heavy.

Specific gravity 3.315.—*Klaproth.*

Constituent Parts.

Silica,	42.
Alumina,	29.
Lime,	21.
Oxide of iron,	3.
	98.

Klaproth, Beit. b. iv. s. 183.

Geognostic and Geographic Situations.

It was first observed in the Sau-Alp in Carinthia, where it occurs imbedded in a bed of quartz along with kyanite, garnet, and augite; or it takes the place of felspar, in a granular rock, composed of quartz and mica. It also occurs imbedded in a coarse granular granite from Thiersheim near Wunsiedel, in Bareuth in Franconia; and in Bavaria, Salzburg, the Tyrol, Carniola, and Swisserland. I have it from Glenelg in Inverness-shire, and from Shetland, I believe the island of Unst.

Observations.

1. It is named Zoisite, in honour of a zealous cultivator of mineralogy, Baron von Zoïs of Laybach. It was described and named at the same time by Werner and Karsten.

2. Hauy considers it but as a variety of epidote; from which, however, it differs in colour, lustre, distinct concretions, and chemical composition.

3. It bears the same relation to tremolite, that epidote does to actynolite; indeed it was at first arranged as a variety of tremolite, until Werner and Karsten ascertained its distinguishing characters.

4. It would probably be an improvement, to arrange zoisite as a subspecies of epidote.

Second Subspecies.

Friable Zoisite.

Murber Zoisit,—*Karsten.*

Murber Zoisit, *Karsten,* Magazin. de Berlin, Geselch, 2. Jahrg. 3. quart. 1808, s. 187.—*Id. Steffens,* b. i. s. 76.—*Id. Klaproth,* Beit. b. v. s. 41.

External Characters.

Its colour is reddish-white, which is spotted with pale peach-blossom red.

It is massive.

 It

It is very feebly glimmering.

The fracture is intermediate between earthy and splintery.

The fragments are not very sharp-edged.

It occurs in very fine granular concretions, which are very loosely aggregated together.

It is translucent on the edges.

It is semihard.

Brittle.

Rather heavy.

Specific gravity, 3.300, *Klaproth.*

Constituent Parts.

Silica,	44.
Alumina,	32.
Lime,	20.
Oxide of iron,	2.50
	98.50

Klaproth, Beit. b. v. s. 43.

Geognostic and Geographic Situation.

It occurs imbedded in green talk at Radelgraben in Carinthia.

10. Axin-

10. Axinite, or Thumerstone.

Thumerstein.—*Werner.*

Schorl transparent lenticulare, *R. de L.* t. ii. p. 353.—Glass-schorl, or Glastein, *Wid.* p. 294.—Thumerstone, *Kirwan,* vol. i. p. 273.—Glastein, *Klap.* b. ii. s. 118.—Tumite, *Nap.* p. 158.—Janolite, *Lam.* t. ii. p. 316.—La pierre de Thum. *Broch.* t. i. p. 236.—Axinite, *Hauy,* t. iii. p. 22.—Axinit, *Reuss.* b. ii. th. 1. s. 200. *Id. Suck.* 1. th. s. 230.—Thumerstein, *Bert.* s. 184. *Id. Mohs,* b. i. s. 180 *Id. Lud.* b. i. s. 73. *Id. Hab.* s. 22.—Axinite, *Lucas,* p. 53. *Id. Brong.* T. i. p. 389. *Id. Brard.* p. 138. *Id. Kid.* vol. i. p. 240. *Id. Steffens,* b. i. s. 77. *Id. Hauy,* Tabl. p. 37.

External Characters.

Its most common colour is clove-brown, of various degrees of intensity; from which it passes on the one side into plum-blue, on the other into pearl-grey, ash-grey, and greyish-black *.

It is seldom found massive, often disseminated, but most frequently crystallised,

1. In very flat rhombs, in which the two opposite acute lateral edges are generally truncated. Fig. 51.
2. Oblique four-sided table, in which two opposite terminal planes are set on obliquely; the other two bevelled;

* It has sometimes a green colour, owing to intermixed chlorite, and crystals of this colour are alleged to be the most regular. Brong. t. i. p. 390.

bevelled; and two opposite acute angles truncated. The truncating planes are smooth, but the others are streaked. Fig. 52.

The crystals sometimes intersect one another, forming a kind of cellular aggregation.

Externally, its lustre is generally splendent; internally, it alternates from glistening to shining, and is vitreous.

The fracture is fine-grained uneven; in the translucent varieties it sometimes approaches to splintery; in the transparent varieties, to small and imperfect conchoidal.

The fragments are angular, and sharp-edged.

The massive varieties occur in curved lamellar distinct concretions, whose surface is shining and streaked.

It alternates from perfectly transparent to feebly translucent.

Rather hard; scratches glass, and is harder than felspar, but not so hard as quartz.

It is very easily frangible.

Rather heavy, approaching to heavy.

Specific gravity, from 3.213 to 3.2956, *Hauy.*—3.295, *Kirwan.*—3.250, *Gerhard.*

Chemical Characters.

Before the blow-pipe it intumesces, and melts into a greyish-black coloured glass.

Constituent

Constituent Parts.

Silica,	52.70	44.0	50.50
Alumina,	25.79	18.0	16.
Lime,	9.39	19.0	17.
Oxide of iron,	8.63	14.0	9.50
—— of manganese,	1.0	4.0	5.25
			Potash, 0.25
	97.51	99.0	98.50
	Klaproth, t. 2. p. 126.	*Vauquelin*, Journ. de Mines. n. 23.	*Klaproth*, t. v. p. 28.

Geognostic Situation.

The greater proportion of massive thumerstone occurs in beds; as is the case in the Saxon Metalliferous Mountains; and probably also at Kongsberg in Norway. At Thum and Ehrenfriedersdorff it occurs with massive calcareous spar, common chlorite, magnetic pyrites, iron pyrites, arsenical pyrites, copper pyrites, blende, and probably also with actynolite and hornblende. Mohs is of opinion, that this aggregation of minerals forms a bed or beds in the primitive-trap formation; and he thinks it probable, that one portion of the Kongsberg thumerstone occurs in a similar situation; the other, which occurs along with native silver, galena, slaty glance-coal, &c. in veins of newer formation. In the Felberhal in Salzburg, it occurs in mica-slate; and in the Hartz, along with quartz and asbestus: at Arendal in Norway, along with calcareous spar, common actynolite, common iron pyrites,

pyrites, felspar, epidote, and common titanite. The thumerstone from Dauphiny, Savoy, and other places, occurs in small and very old veins, that traverse gneiss, in which it is generally the newest mineral. In these veins it is associated with crystallised felspar, rock-crystal, asbestus, epidote, octahedrite, mica, and chlorite.

Geographic Situation.

Europe.—It occurs in Carrarach-mine, two miles north of St Just's Church in Cornwall *. Upon the continent of Europe, at Arendal and Kongsberg in Norway. At Thum near Ehrenfriedersdorf, Schneeberg, and Siebenschlien in Upper Saxony; and at Treseburg in the Hartz in Lower Saxony; also in the Black Forest (Schwartzwald) in Swabia; in the valley of Lauterbrun in the Canton of Bern, the valley of Ferrera in Graubunden, and the valley of Chamouny in Swisserland; at Ayarsun in Guipuscoa in Spain; and in Dauphiny and Elsass.

Africa.—In Mount Atlas.

Observations.

1. *Distinctive Characters.*—Between axinite and *common felspar*. Felspar has a different suite of colours from that of axinite: felspar has a foliated fracture, axinite a compact fracture; felspar does not occur in curved lamellar concretions, as is the case with axinite; felspar is softer than axinite; felspar has a specific gravity of 2.4 to 2.7, axinite 3.21 to 3.29; and felspar melts before the blow-pipe into a white-coloured enamel; axinite to a blackish-coloured glass.

2. The

* Greenough.

2. The first crystal of axinite was described by Rome de Lisle, but he arranged it with schorl: it was Werner who first established it as a distinct species.

3. Werner named it Thumerstone, from Thum in Saxony, where it was first found: Hauy's name Axinite, is derived from the shape of the crystals, which somewhat resembles that of an axe.

4. It differs considerably from the other species of the schorl family, and would seem to form a member of a distinct family.

V. GARNET FAMILY.

This family contains the following species: Leucite; Vesuvian, Grossular; Melanite, Allochroite, Garnet, Grenatite, Pyrope; Cinnamon-stone.

1. Leucite.

Leuzit.—*Werner.*

Grenat d'un blanc cristallin, et grenat dicolore, *R. de L.* p. 330. —Grenat d'un blanc mat a 24 facettes, *Born.* t. i. p. 436. —Leuzit, *Wid.* s. 292.—Vesuvian or White Garnet, *Kirw.* vol. i. p. 285.—Leuzit, *Estner,* b. ii. s. 188. *Id. Emm.* b. i. s. 348. *Id. Lam.* t. ii. p. 259. *Id. Broch.* t. i. p. 188 — Amphigene, *Hauy,* t. ii. p. 559.—Leucit, *Reuss,* b. ii. th. 1. s. 396. *Id. Lud.* b. i. s. 63. *Id. Suck.* 1. th. s. 202. *Id. Bert.* s. 175. *Id. Mohs,* b. i. s. 74. *Id. Hab.* s. 21.—Amphigene, *Lucas,* s. 47. *Id. Brong.* t. i. p. 364. *Id. Brard.* p. 126. *Id. Kid,* vol. i. p. 254. *Id. Hauy,* Tabl. p. 33.—Leucit, *Steffens,* b. i. s. 80.

External Characters.

Its colours are yellowish and greyish white; these, although rarely, pass into light ash-grey, or yellowish-grey; and it very seldom occurs reddish-white.

It occurs in round and angular grains; also crystallised, in acute double eight-sided pyramids, in which the lateral planes of the one are set on the lateral planes of the other, and the summits are deeply and flatly acuminated by four planes, which are set on the alternate edges.

The crystals are all around crystallised; commonly small, and seldom middle-sized.

The surface of the grains is rough, and dull, or feebly glimmering; that of the crystals is smooth, seldom slightly streaked, and glistening. Internally the lustre is shining, approaching to glistening, and is vitreous, inclining to resinous.

The fracture is imperfect and flat conchoidal, and sometimes foliated.

The fragments are indeterminate angular, pretty sharp edged.

It is translucent, semitransparent, and some varieties approach to transparent.

It is hard in a low degree; scratches glass with difficulty.

It is brittle.

Easily frangible.

Rather heavy, approaching to light.

Specific gravity 2.468, *Brisson*.—2.464, *Kirwan*.—2.455 to 2.490, *Klaproth*.—2.461, *Karsten*.

Chemical Characters.

Before the blow-pipe it is infusible without addition: with borax it forms a brownish transparent glass. According to Lampadius, when exposed to a stream of oxygen gas, it melts easily into a white transparent glass *.

---

* Lampadius, Samml. prakt. chem. Abhandl. b. ii. s. 62.

Constituent Parts.

Mean of different analyses.

Silica,	54	56
Alumina,	24	20
Potash,	21	20
Lime,		2
Loss,	1	2
	100	100
	Klaproth.	*Vauquelin.*

Geognostic Situations.

This mineral occurs in lava, and in rocks belonging to the newest trap formation. Also in ejected unaltered stones at Vesuvius, where it is accompanied with black mica, hornblende, garnet, vesuvian, and calcareous spar. Lelievre mentions leucite as occurring in granite in the Pyrenees; Dolomieu, in an auriferous vein in Mexico; and Schumacher and others, in syenite in Norway. These three latter facts are of a dubious nature; it would appear from the description published by Schumacher, that the leucite of Norway at least is cubicite.

Geographic Situation.

Europe.—In Italy, where it occurs most frequently, it is found at Borgheto, Monte Albano near Rome, Bolsena, Capraruola, Civita Castellana, Ronciglione, Viterbo, Vesuvius, Monte Somma, Pompeii, and the island of Ischia.

Ischia. In mica-slate in the Pyrenees; and at Tekeroe, in Transilvania.

America.—In an auriferous vein in Peru.

Observations.

1. *Distinctive Characters.*—*a.* Between leucite and *garnet.* The colours of garnet are red or brown; that of leucite white or grey: garnet has a considerable variety of crystallizations; leucite but one form, technically denominated the *leucite figure:* garnet scratches quartz, leucite scarcely glass. The specific gravity of garnet extends from 3.5 to 4.2; but the specific gravity of leucite does not exceed 2.46. And, lastly, garnet is fusible before the blow-pipe, leucite infusible.—*b.* Between leucite and *cubicite.* Cubicite has in general an uneven or foliated fracture; that of leucite is conchoidal, but rarely foliated: cubicite is harder than leucite; cubicite occurs covering the walls of drusy cavities, leucite imbedded; and, lastly, cubicite is fusible before the blow-pipe, leucite is infusible.

2. It was named by Bergman White Garnet. Werner named it Leucite, from its white colour; and it was he who first established it as a distinct species.

3. It sometimes weathers to a white earth, in the manner of felspar; a change which is probably owing to the abstraction of its alkali.

4. Von Buch is of opinion that this mineral is an ignigenous production; whereas Werner, Mohs, &c. view it as an aquatic production.

5. Karsten subdivided it into three subspecies, viz. Conchoidal, Uneven, and Earthy; but these appear to be but varieties produced by the action of volcanic fire.

2. Vesuvian.

2. Vesuvian.

Vesuvian, *Werner.*

Hyacinth du Vesuve, *R. de L.* t. ii. p. 291.—Vulcanischer Schorl, *Wid.* s. 290.—Vesuvian, *Estner,* b. ii. s. 177. *Id. Emm.* b. i. s. 342.—Hyacinthe, *Lam.* t. ii. p. 323.—La Vesuvienne, *Broch.* t. i. p. 184.—Idocrase, *Hauy,* t. ii. p. 574. —Vesuvian, *Reuss,* b. ii. th. 1. s. 91. *Id. Lud.* b. i. s. 63. *Id. Suck.* 1. th. s. 197. *Id. Bert.* s. 156. *Id. Mohs,* b. i. s. 68. *Id. Hab.* s. 27.—Idocrase, *Lucas,* p. 48. *Id. Brong.* t. i. p. 391. *Id. Brard,* p. 128.—Vesuvian, *Kid,* vol. i. p. 252. —Idocrase, *Hauy,* Tabl. p. 34.—Vesuvian, *Steffens,* b. i. s. 358.

External Characters.

Its principal colour is liver-brown, which passes into reddish-brown and blackish-brown, and into oil-green and blackish-green.

It occurs seldom massive, most frequently crystallised. Its crystallizations are as follow:

1. Rectangular four-sided prism, flatly acuminated by four planes, which are set on the lateral planes; the lateral edges and summits of the acumination truncated. Fig. 53.
2. Preceding figure, in which the lateral edges are bevelled, and the edges of the bevelments truncated. Fig. 54.
3. Preceding figure, in which the acuminating edges are truncated. Fig. 55.

4. When the four-sided prism, acuminated by four planes, becomes so low, that the acuminating planes touch other, a flat octahedron is formed, in which the summits and the angles on the common base are truncated *.

The crystals are generally middle-sized; sometimes all around crystallised, at other times superimposed.

The lateral planes of the prism are longitudinally streaked; all the other planes are smooth.

Externally the crystals are splendent; internally glistening, approaching to shining, and the lustre is vitreo-resinous.

The fracture is small-grained uneven.

The massive varieties appear to occur in coarse granular distinct concretions.

It alternates from translucent to translucent on the edges; and refracts double.

It is hard; scratches glass.

Brittle; and rather easily frangible.

Rather heavy.

Specific gravity, 3.409, *Werner*.—3.0882 to 3.409, *Hauy*.—3.365 to 3.420, *Klaproth*.—3.4412, *Karsten*.

Chemical Characters.

Before the blow-pipe it melts without addition into a yellowish and faintly translucent glass.

Constituent

* Professor Beauvois observed in Piedmont vesuvian crystals having cylindrical and very deeply longitudinally streaked lateral planes: but these according to M. Verina, appear to be groupes of acicular crystals.

Constituent Parts.

	Vesuvian of Vesuvius.	Of Siberia.
Silica,	35.5	42.
Lime,	33.0	34.
Alumina,	22.25	16.25
Oxide of Iron,	7.5	5.5
Oxide of manganese,	0.25	
Loss,	1.5	2.25
	100	100

Klaproth, Beit. b. ii. s. 32, & 38.

Geognostic and Geographic Situations.

Europe.—It was first found in the vicinity of Vesuvius, where it still occurs in considerable abundance, in unaltered ejected rocks, composed of granular limestone, mica, hornblende, melanite, garnet, quartz, epidote, felspar, chlorite, and specular iron-ore.

These rocks are supposed to be part of the primitive mass in which that celebrated volcanic mountain is situated; and are probably disposed in beds.

It occurs in small irregular veins traversing gneiss, in the vicinity of Monte Moro, eastward of Monte Rosa; and in a rock named Testa Ciarva, in the plain of Musa in Piedmont, in veins traversing serpentine. In large crystals upon Mount St Gothard; and near Pitigliano in the district of Sienna. In gneiss, along with hornblende, precious garnet, and magnetic ironstone, at San Lorenzo in Spain. In Ireland, (according to my

 friend

friend and pupil Dr Fitton *), it occurs at Kilranelagh in primitive country, in a rock composed of garnet, quartz, and felspar; also at Donegal, in a rock composed of quartz, granular limestone, and a fibrous substance supposed to be tremolite.

Asia.—It occurs in Kamschatka, at the mouth of the rivulet Achtargada, in a pale greenish-grey coloured clay-stone, which contains crystallised magnetic iron-stone; also in serpentine, and in a rock composed of chlorite and calcareous spar.

Uses.

At Naples, it is cut into ring stones, and is sold under various names: the green-coloured varieties are denominated Volcanic Chrysolite; and the brown Volcanic Hyacinth.

Observations.

1. *Distinctive Characters.*—Between vesuvian and *garnet.* The planes of garnet are glistening, or glistening inclining to shining; those of vesuvian splendent: garnet is heavier than vesuvian: the most frequent forms of garnet, are the garnet dodecahedron and leucite figure, neither of which occur in vesuvian: lastly, garnet is not so fusible as vesuvian, and yields rather a black scoria than a translucent glass.

2. It has been described under a variety of names, as Volcanic Schorl, Chrysolite, Hyacinth, and Topaz. Werner

* Transactions of the Geological Society, vol. i. p. 274.

Werner first established it as a distinct species, and gave it its present name.

3. The Peridote Idocrase of Bonvoisin, which is found in the Alps of Musa in Piedmont, along with garnet in serpentine, appears to be but a variety of vesuvian.

3. Grossular.

Grossular, *Werner*.

Grossular, *Steffens*, b. i. s. 93.——Olivengrunen Granats aus Siberien, *Klaproth*, b. iv. s. 319.

External Characters.

Its colour is asparagus-green, approaching to mountain-green.

It has been hitherto found only crystallised, and in the following figures :

1. Garnet dodecahedron, in which all the edges are deeply truncated.
2. Leucite crystallization ; or the acute double eight-sided pyramid, flatly acuminated on both extremities by four planes ; the acuminating planes set on the alternate edges of the double eight-sided pyramid.

The planes of the crystals are smooth ; and this is characteristic for the species.

Externally it is shining ; internally shining, and the lustre vitreous, inclining to resinous.

The fracture is small conchoidal.

It is strongly translucent.

It is hard.

Rather easily frangible.

Rather heavy.

Specific gravity 3.351, *Werner.*—3.372, *Klaproth.*

Constituent Parts.

Silica,	44.
Lime,	33.50
Alumina,	8.50
Oxide of Iron,	12.
Loss,	2.
	100

Klaproth, Beit. b. iv. s. 323.

Geognostic and Geographic Situations.

It occurs imbedded in small crystals, along with vesuvian, in a pale greenish-grey claystone, near the river Wilui in Siberia.

4. Melanite.

Melanit, *Werner.*

Melanit, *Broch.* t. 1. p. 191. *Id. Reuss,* b. ii. th. 1. s. 136. *Id. Suck.* 1. th. s. 194. *Id. Bert.* s. 162. *Id. Mohs,* b. i. s. 76. *Id. Lud.* b. 1. s. 64.—Grenat Melanite, *Brong.* t. 1. p. 397.—Schlackiger Granat, *Karsten,* Tabel.—Grenat noire, *Hauy,* Tabl. p. 33.—Melanit, *Steffens,* b. i. s. 92.

External

External Characters.

Its colour is velvet-black, which sometimes inclines to greyish-black.

It occurs crystallised; probably also in grains.

Its crystalline figure is the garnet dodecahedron, or six-sided prism flatly acuminated by three planes, which are placed on the alternate lateral edges; and the edges are more or less deeply truncated.

It is all around crystallised.

The crystals are middle-sized and small.

Externally it is always smooth and shining, sometimes approaching to splendent. Internally it is shining, inclining to glistening; and is resinous.

The fracture is imperfect flat conchoidal: sometimes imperfect foliated, with traces of a threefold cleavage parallel with the acuminating planes of the six sided prism.

The fragments are angular and sharp-edged; sometimes rhomboidal.

It is opaque.

Hard; scarcely scratches quartz.

Rather easily frangible.

Rather heavy.

Specific gravity 3.800, *Werner.*—3.691, *Karsten.*

Constituent Parts.

Silica,	35.5
Alumina,	6.
Lime,	32.5
Oxide of Iron,	25.25
Oxide of Manganese,	0.4
Loss,	0.35
	100

Klaproth's Beiträge, b. v. p. 168.

Geognostic and Geographic Situations.

It is found in a rock at Frescati near Rome; the rock contains besides melanite, felspar, vesuvian, and basaltic hornblende. At Monte Somma near Naples it occurs in granular limestone; and in the Pyrenees in a compact limestone. It is said to occur in grains in the basalt of Bohemia; and it is found in crystals, imbedded in basalt, in Poland.

Observations.

1. *Distinctive Characters.—a.* Between melanite and *precious garnet.* Red is the only colour of precious garnet, and it exhibits several varieties of it; whereas velvet-black is the only colour of melanite: In precious garnet, the suite of crystals extends from the garnet or rhomboidal dodecahedron to the double eight-sided pyramid acuminated by four planes, or the leucite form; whereas the melanite has but one figure, that is the garnet dodecahedron

cahedron truncated on its edges. The internal lustre of precious garnet is vitreous, that of melanite resinous; precious garnet alternates from transparent to translucent, melanite is opaque: precious garnet scratches quartz more readily than melanite; and he specific gravity of precious garnet is 4.2, that of melanite only 3.6 or 3.8.—*b*. Between melanite and *common garnet*. The colours of common garnet are green and brown, colours that do not occur in melanite: common garnet occurs most commonly massive, melanite never: the suite of crystallizations of common garnet is the same as in precious garnet, therefore very different from melanite: the fracture of common garnet is uneven, that of melanite conchoidal: common garnet occurs in granular concretions, which is never the case with melanite; and common garnet is more or less translucent, melanite always opaque.

2. It used formerly to be considered as a variety of schorl or of garnet: Werner first established it as a distinct species, and gave it its name and place in the system.

5. Allochroite.

Allochroit, *Werner*.

Allochroit, *Dandrada*, Journ. de Phys. t. 51 p. 235. *Id. Broch.* t. ii. p. 552. *Id. Reuss*, b. ii. th. 2. s. 478. *Id.* Lud. b. ii. s. 159. *Id. Suck.* 1. th. s. 716. *Id. Bert.* s. 163. *Id. Lucas*, p. 330. *Id. Brong.* t. i. p. 401. *Id. Brard*, p. 399.—Splittriger Granat, *Karst.* Tabel.—Allochroite, *Haüy*, Tabl. p. 57. *Id. Steffens*, b. i. s. 98.

External Characters.

Its colours are greenish-grey and yellowish-grey; and the greenish-grey passes into a colour intermediate between asparagus and oil green.

It occurs massive.

Externally it is glimmering; internally glistening and resinous.

The fracture is sometimes small-grained uneven; sometimes even, or even passing to flat conchoidal.

The fragments are rather blunt-edged.

It is translucent on the edges.

It gives sparks with steel, but does not scratch quartz.

It is rather easily frangible.

It is rather heavy.

Specific gravity 3.5754, *Dandrada*.—3.58, *Brard*.—3.50, *Brongniart*.

Chemical Characters.

According to Vauquelin, it is fusible without addition, before the blow-pipe, into a black, smooth and opaque enamel.

Constituent

Constituent Parts.

Silica,	35.	37.
Lime,	30.5	30.
Alumina,	8.	5.
Oxide of Iron,	17.	18.5
Carbonat of Lime,	6.	
Oxide of Manganese,	3.5	6.25
Loss,		3.25
	100	100

Vauquelin, in Lucas, Tabl. p. 30. *Rose*, in Karsten, Tabel. 33.

Geognostic and Geographic Situations.

It has hitherto been found only in Virums iron-mine at Drammen in Norway, where it is associated with calcareous spar, reddish-brown garnet, and magnetic ironstone.

Observations.

1. This mineral was first particularly described and named by Professor Schumacher of Copenhagen and M. Dandrada.

2. It is distinguished from *common garnet*, the only mineral with which it could be confounded, by its lighter colours, inferior lustre and transparency, general appearance of the fracture, and inferior hardness and weight.

3. Werner

3. Werner considers it as a distinct species; but Karsten arranges it as a subspecies of garnet, under the title *splintery garnet*.

6. Garnet.

THIS species is divided into two subspecies, viz. Precious Garnet and Common Garnet.

First Subspecies.

Precious Garnet.

Edler Granat, *Werner.*

Granatus, *Wall.* t. 1. p. 262.—Grenat, *Romé de Lisle,* t. ii. p. 316. *Id. Born.* t. i. p. 147.—Oriental Garnet, *Kirw.* vol. i. p. 258.—Edler Granat, *Emm.* b. i. s. 358.—Almandin, *Karst.* Tabl.—Grenat, *Hauy,* t. ii. p. 540.—Le Grenat noble, *Broch.* t. i. p. 193.—Almandin, *Reuss,* b. i. th. i. s. 69.—Edler Granat, *Lud.* b. i. s. 64.—Almandin Granat, *Suck.* 1. th. s. 173.—Edler Granat, *Bert.* s. 271. *Id. Mohs,* b. i. s. 79.—Grenat, *Lucas,* p. 46.—Grenat noble, *Brong.* t. i. p. 395.—Grenat, *Brard,* p. 123.—Garnet, *Kid,* vol. i. p. 147.—Grenat, *Hauy,* Tabl. p. 33.—Edler Granat, *Steffens,* b. i. s. 84.

External Characters.

All the colours of this subspecies are red; the principal colour is columbine-red, which passes into cherry-red, and into a colour intermediate between cherry-red and blood-red, and it appears even to run into brownish-red.

All

All these colours contain an intermixture of blue, but very seldom of black.

It occurs very seldom massive, more often disseminated, and in original roundish grains and small pieces. It occurs most commonly crystallised, and has two principal crystallizations, viz.:

1. The garnet dodecahedron. Fig. 56.
2. The double eight-sided pyramid, or leucite figure. Fig. 57.

a. Garnet dodecahedron. This figure is either perfect, or more or less truncated on all its edges. Fig. 58. When these truncations increase, and cause the original faces of the dodecahedron to disappear, an

b. Acute double eight-sided pyramid (the leucite figure) is formed, in which the lateral planes of the one are set on the lateral planes of the other, and the summits are deeply and flatly acuminated by four planes, which are set on the alternate lateral edges. Fig. 59.

c. The same figure as the preceding, in which the eight acute angles, formed by the meeting of the acuminating and lateral planes, and the alternate angles on the common basis, and all the edges, are truncated. Fig. 60.

The crystals are usually small, also middle-sized, seldom large, and are all around crystallised. The surface of the crystals is sometimes smooth, particularly the truncating planes: the acuminating planes are sometimes diagonally streaked. The surface of the grains is rough or granulated.

Externally

Externally the lustre of the crystals and grains is glistening; internally it is shining, bordering on splendent; and is vitreous, inclining slightly to resinous.

The fracture is perfect conchoidal, which often passes into imperfect conchoidal. Sometimes a concealed foliated fracture is to be observed *.

Its fragments are angular, and more or less sharp-edged.

It sometimes occurs in lamellar distinct concretions †.

It alternates from completely transparent to translucent, according to the kind of fracture ‡, and refracts single.

It is hard; scratches quartz.

It is brittle.

Rather difficultly frangible.

Heavy §.

Specific gravity, 4,230, *Werner*.—4.085, *Klaproth*.—4.352, *Karsten*.—4.188, *Brisson*.—4.1888, *Hauy*.

Constituent

---

* The conchoidal variety has greatest, and the coarse-grained uneven the least lustre.

† The distinct concretions occur most frequently in the garnet of Greenland.

‡ The transparent varieties are often impure in the middle.

§ After Zircon it is the heaviest of the precious stones.

Constituent Parts.

Silica,	35.75		36.
Alumina,	27.25		22.
Oxide of Iron,	36.0	Lime,	3.
Manganese,	0.25	Oxide of iron,	41.
Loss,	0,75		
	100		102
	Klaproth, b. ii. s. 26.		*Vauquelin.*

Chemical Characters.

Before the blow-pipe it melts pretty easily into a black scoria or enamel.

Geognostic Situation.

It occurs almost always in primitive rocks; most frequently in mica-slate; seldomer in gneiss, hornblende-slate, chlorite-slate, and granular hornblende rock, and but sparingly in granite, and talc-slate. It occurs imbedded in serpentine; in this situation resembling the pyrope, and certain varieties of common garnet. It has not been observed in beds or veins *.

Sometimes rolled grains occur in sandstone.

Geographic

* According to Ramond, black, red and white garnets occur imbedded in a compact limestone in the Pic d'Eredlitz in the Pyrenees.

Geographic Situation.

Europe.—In Scotland, it occurs in Perthshire, Aberdeenshire, Ross-shire, the outer range of the Hebrides, as Harris and Lewis; and in several of the Shetland islands, as Mainland and Unst.

Upon the continent of Europe, it occurs in Norway, Lapland, Sweden, Saxony, Bohemia, Silesia, Swisserland, Stiria, the Tyrol, Salzburg, Hungary, and France.

Asia.—It is found in many parts of Siberia, also in Armenia, Pegu, and Ceylon.

Africa.—Ethiopia and Madagascar.

America.—United States, Mexico and Brazils.

Use.

It is cut as a precious stone. The larger kinds are used as ring-stones, and, after cutting and polishing, are set either *à jour*, or are provided with a violet-blue foil. Jewellers, in order to heighten the colour and transparency of certain garnets, form them either into doublets, by attaching to the lower part of the stone a thin plate of silver *, or hollow them underneath. The smaller kinds are used for necklaces and bracelets. Many fine pieces of engraving have been executed on this mineral. In the National Museum in Paris, there are several beautiful engraved garnets, and among others, a very fine head of Louis XIII. One of the finest engraved garnets is

* This practice appears to be very ancient, for it is mentioned by Pliny in his Hist. Nat. lib. 37.

is that executed by the celebrated artist Cali, in the possession of Lord Duncannon, which represents the Dog Sirius.

Crystals sometimes occur the size of the fist or even larger: these are cut into small vases, which are very highly valued, particularly if they are free of flaws, and possess a good colour, and considerable degree of transparency.

The coarser kinds are used as emery, for polishing other minerals; for this purpose, they are previously repeatedly heated and quenched in water, reduced to powder in an iron-mortar, and, lastly, diffused through water, poured into other vessels, and allowed to settle, in order to obtain an uniform powder. In this state it is known to artists by the name of *red emery*.

This gem is successfully imitated by the following composition, which, when well and judiciously cut and polished, equals the garnet in lustre and transparency:

Purest white glass, 2 ounces.
Glass of antimony, 1 ounce.
Powder of Cassius, 1 grain.
Manganese, 1 grain.

Observations.

1. *Distinctive Characters.*—Between precious garnet and *common garnet.* Brown and green are the most common colours of common garnet, but red is the only colour of precious garnet: the lustre of common garnet is resino-vitreous and glistening, but that of precious garnet is vitreous, slightly inclining to resinous; and is shining inclining to splendent: the fracture of common garnet is fine-grained uneven, that of precious garnet conchoidal;

conchoidal: common garnet is only translucent, whereas precious garnet is semitransparent and transparent: common garnet occurs in granular concretions, precious garnet never: common garnet is usually small, and very small, seldom middle-sized, whereas precious garnet is sometimes large, and often middle-sized: common garnet has a specific gravity of 3.7, that of precious garnet is 4.2. —*b*. Between dodecahedral garnet and *dodecahedral zircon*. If the dodecahedral garnet be viewed as a six-sided prism, the dodecahedral zircon will appear as a four-sided prism; and in the garnet, the adjacent planes meet under angles of 120°, but in the zircon under angles of 124° 12′, and 117° 54′.

2. It would appear from Pliny, that the Carthaginian Ruby of the ancients, and also their Carbuncle, which were said to shine in the dark like a glowing coal, were varieties of precious garnet.

3. Karsten considers precious garnet as a distinct species, and places it in his system between zircon and garnet, under the name *Almandine;* he considering it as identical with the *alabandicus* (Pliny, Hist. Nat. lib. xxxvii. sect. 25.) of the ancients.

4. The precious garnet is sometimes named Syrian Garnet, not from Syria, but from Syrian, a town in Pegu, now destroyed, where it was met with in great beauty.

Second

Second Subspecies.

Common Garnet.

Gemeiner Granat, *Werner.*

Le Grenat commun, *Broch.* t. i. p. 198.—Granat, *Reuss,* b. ii. th. i. s. 79.—Gemeiner Granat, *Suck.* 1. th. s. 181. *Id. Bert.* s. 160. *Id. Lud.* b. i. s. 65. *Id. Mohs,* b. i. s. 85.—Grenat commun, *Brong.* t. i. p. 396.—Grenat brun, rougeatre, verdatre, &c. *Hauy,* Tabl. p. 33.—Gemeiner Granat, *Steffens,* b. i. s. 87.

External Characters.

Brown and green are its most common colours. Of brown it occurs liver-brown, yellowish-brown, and reddish brown; and of green, blackish-green. From liver-brown it passes into olive, pistachio, blackish and leek-green, and from this even into mountain-green: from yellowish-brown it passes into isabella yellow: from reddish-brown into a middle colour between hyacinth and blood-red: from blackish green into greenish black. In many specimens different colours occur together.

It occurs most commonly massive, but never in grains or angular pieces: Sometimes crystallised, and possesses all the figures of the precious garnet *.

The crystals are always simply aggregated, and form druses. They are commonly small and very small, seldom middle-sized.

The surface of the crystals, and particularly that of the dodecahedron, is diagonally streaked; and the lustre is shining and glistening.

* According to Haüy, the primitive form of garnet is the rhomboidal dodecahedron; but that of leucite is either a cube or rhomboidal dodecahedron.

Internally the lustre is glistening, seldom shining, and is intermediate between resinous and vitreous.

The fracture is fine grained uneven, sometimes slightly inclining to imperfect conchoidal.

The fragments are angular, and not particularly sharp-edged.

It occurs in small, and fine angulo-granular distinct concretions, which sometimes pass into coarse granular.

It is more or less translucent; the black nearly opaque.

It is a little softer than precious garnet.

It is easily frangible.

Heavy in a middling degree.

Specific gravity, 3,757 to 3.754, *Werner*.—3.668, *Karsten*.

Chemical Characters.

It melts more easily before the blow-pipe than precious garnet.

Constituent Parts.

	Wiegleb.	Merz.
Silica,	26 46	40
Alumina,	22.70	20
Lime,	17.91	8
Iron,	16.25	20
Loss,	16.68	12
	100	100

Voigt's Mineralog. und Bergmann. Abhandlungen, Th. i. s. 15 & 22.

Geognostic

Geognostic Situation.

It occurs massive or crystallised, in drusy cavities, in beds, in mica-slate, clay-slate, chlorite-slate, and primitive-trap, where it is accompanied by different ores, as magnetic ironstone, red ironstone, magnetic pyrites, common iron pyrites, arsenical pyrites, copper pyrites, vitreous copper ore, blende and galena, and by various earthy minerals, as actynolite, hornblende, epidote, augite, coccolite, tremolite and schaalstone. It sometimes also forms the whole mass of beds. It also occurs imbedded in serpentine. Humboldt mentions it as occurring in veins in Mexico

Geographic Situation.

Europe.—According to Dr Fitton and Mr Stephens, it occurs at Kilranelagh and Donegal in Ireland *. Upon the Continent, it occurs at Arendal and Drammen in Norway, where it is accompanied with granular limestone, common quartz, felspar, augite, hornblende, mica, axinite, and apatite; less frequently with epidote, coccolite, arcticite, and fluor-spar. At Kemi, in Russian Lapland, in chlorite-slate. In Sweden it occurs in beds in mica-slate at Langhannshytta and Sunnerskog; it is also found at Dannemora, Fahlun, and Garpenberg, in Sweden. It occurs in several places in the Saxon Erzgebirge, as at Berggiesshübel, along with brown blende, calcareous spar, and copper pyrites;

 at

* Transactions of the Geological Society, vol. i.

at Breitenbrun, with quartz, common actynolite, and magnetic ironstone; and at Geier, along with quartz, hornblende, common iron pyrites, and magnetic ironstone, also mixed with quartz, brown blende and copper pyrites; at Hohenstein, Kupferberg, Presnitz, &c. in Bohemia; Moravia; in the Boberthal in Silesia, in a bed along with calcareous spar, actynolite, quartz, malachite, and copper and iron pyrites; at Dobschau in Hungary, in dodecahedral crystals, imbedded in a serpentine.

Asia.—In New Holland.

Use.

On account of its easy fusibility and richness in iron, it is frequently employed as a flux in smelting rich iron ores, and as an addition to poor ores. In some countries it is named Green Iron-ore. It is seldom cut or polished for ornamental purposes.

Observations.

1. It is distinguished from the *precious garnet* by colour, degree of transparency, lustre, kind of fracture, distinct concretions, druses, aggregation of crystals, specific gravity, occurring in beds, and very rarely imbedded.

2. The mineral named *Aplome* by Hauy, appears to be crystallised common garnet. Weiss remarks, that there is no reason why this substance should be considered a distinct species; for although the streaks, parallel with the shortest diagonal of the rhomboidal planes, observed in the aplome, point out a cubical primitive form, it only follows, that garnet, like leucite, may have a double primitive form.

7. Grenatite.

7. Grenatite.

Granatit, *Werner*.

Grenatite, *Saussure*, sect. 1900. *Id. Lam.* t. ii. p. 290. *Id. Broch.* t. ii. p. 406.—Staurotide, *Hauy*, t. iii. p. 93.—Staurolith, *Reuss*, b. i. th. 1. s. 196.—Grenatit, *Lud.* b. i. s. 66. *Id. Suck.* 1. th. s. 227. *Id. Bert.* s. 289. *Id. Mohs*, b. i. s. 94.—Staurolith, *Hab.* s. 34.—Staurotide, *Lucas*, p. 58. *Id. Brong.* t. i. p. 402. *Id. Brard*, p. 151.—Staurolite, *Kid*, vol. i. p. 251. *Id. Hauy*, Tabl. p. 54.—Staurolith, *Steffens*, b. i. s. 191.

External Characters.

Its colour is dark reddish-brown.

It occurs only crystallised, and the following are the figures which it presents:

1. Very oblique four-sided prism, truncated on the acuter lateral edges; or it may be named an *unequiangular six sided prism* *. Fig. 61.
2. The preceding figure, acutely bevelled on the extremities; the bevelling planes set on the obtuse lateral edges, and the edge of the bevelment truncated. Fig. 62.
3. Twin-crystal, formed by two perfect six-sided prisms crossing each other at right angles. Fig. 63.
4. Twin-crystal, formed by two perfect six-sided prisms crossing each other obliquely. Fig. 64.

The crystals are small and middle-sized.

The surface of the crystals is sometimes smooth, sometimes uneven, and the lustre is glistening and vitreous.

 Internally

* According to Haüy, the primitive form of grenatite is a rectangular prism whose bases are rhombs, with angles of 120½° and 50½°.

Internally its lustre is glistening, and vitreo-resinous.

The principal fracture is imperfect foliated and the folia parallel with the axis of the crystals; the cross fracture is intermediate between small-grained uneven and imperfect conchoidal.

The fragments are angular, and not very sharp-edged.

It is very often opaque, sometimes translucent, and very rarely semitransparent.

It is hard; scratches quartz feebly.

It is brittle, and easily frangible.

Rather heavy.

Specific gravity 3.286, *Hauy*.

Chemical Characters.

Infusible before the blow-pipe.

Constituent Parts.

	From Morbihan.	St Gothard.		St Gothard.
Alumina,	44.	52.25		41.
Silica,	33.	27.		37.5
Lime,	3.84			
Oxide of Iron,	13.	18.5		18.25
Oxide of Manganese,	1.	0.25	Magnesia,	0.5
Loss,	5.16	2.		2.75
	100	100		100
	Vauq. Jour. du Min. n. 53.	*Klap.* Bullet. des Scien de la Soc. Phil. t. i. p. 171.		*Klap.* ibid.

Geographic

Geognostic Situation.

The geognostic relations of this mineral are nearly the same with those of precious garnet, with this difference, that precious garnet occurs in a greater variety of rocks. It has been hitherto found only imbedded in mica-slate, and sometimes in gneiss, and very generally accompanied with kyanite and precious garnet.

Geographic Situation.

Europe.—It occurs in a micaceous rock at the Glenmalur lead-mines in the county of Wicklow in Ireland *. Upon the Continent of Europe, it occurs in the Tyrol, in Swisserland, as at St Gothard, imbedded in mica-slate, with kyanite and precious garnet; on the north side of the Glacier of Gries in the Vallais, in mica-slate; and in the Piora Alp, also in mica-slate; Transilvania; at St Jago di Compostella in Gallicia in Spain; in Brittany, and other places in France.

America.—In Cayenne. Also in North America, according to Mr Greenough.

Observations.

1. *Distinctive Characters.*—Between grenatite and *precious garnet.* The red colours of precious garnet are more or less intermixed with blue, those of grenatite

* Dr Fitton, in Geological Transactions, vol. i. p. 275.

with brown: precious garnet occurs in grains; grenatite only crystallised: precious garnet exhibits a suite of crystallizations, extending from the rhomboidal dodecahedron to the leucite form; grenatite occurs only in the form of a six-sided prism, or twin crystal: precious garnet is harder and heavier than grenatite: and, lastly, precious garnet is fusible before the blow-pipe; grenatite is infusible.

2. It is more nearly allied to precious garnet and pyrope than to common garnet.

8. Pyrope.

Pyrop, *Werner.*

Pyrop, *Broch.* t. ii. p. 498. *Id. Lud.* b. i. s. 67. *Id. Hab.* s. 28. *Id. Mohs,* b. i. s. 97. *Id. Lucas,* p. 265.—Grenat Pyrope, *Brong.* t. i. p. 396.—Pyrop, *Karst.* Tabl.—Grenat, rouge de feu, granuliforme, *Hauy,* Tabl. p. 33.—Pyrop, *Steffens,* b. i. s. 94.

External Characters.

Its colour is dark blood-red, which, when held between the eye and the light, falls strongly into yellow *.

It occurs in small and middle sized roundish and angular grains.

It

* Pyrope and garnet, when cut and polished, are easily distinguished from spinelle and sapphire, by the dark tinge which their colours possess. Hauy, t. ii. p. 544.

Its lustre is splendent and vitreous.

The fracture is perfect conchoidal.

The fragments are angular and sharp-edged.

It is completely transparent, and refracts double *.

It is hard; scratches quartz more readily than precious garnet.

Rather heavy, approaching to heavy.

Specific gravity 3.941, *Werner*.—3.718, *Klaproth*.

Constituent Parts.

Silica,	40.0
Alumina,	28.50
Magnesia,	10.00
Lime,	3.50
Oxide of iron,	16.50
—— of manganese,	0.25
Loss,	1.25
	100

Klaproth, t. ii. p. 21.

The magnesia which it contains, distinguishes it in a chemical view from precious garnet. Its fine red colour is conjectured to be owing to chromic acid.

Geognostic

* Haberle.

Geognostic Situation.

It occurs imbedded in trap-tuff, wacke, and primitive serpentine.

Geographic Situation.

At Chrasstian, Podsedlitz, Scheppenthal, and Meronitz in Bohemia, it occurs imbedded in trap-tuff and wacke; also in alluvial soil, formed from these rocks by decomposition, where it is associated with sapphire, zircon, melanite, olivine, and iron-sand. At Zœblitz in Saxony it is imbedded in primitive serpentine.

Use.

It is employed in almost every kind of jewellery, and is generally set with a gold foil. The small and very small grains are pounded, and used in place of emery in cutting softer stones.

Observations.

1. *Distinctive Characters.*—Between pyrope and *precious garnet.* Precious garnet possesses a considerable colour suite; pyrope but one colour, and that is blood-red: precious garnet occurs crystallised, which is never the case with pyrope: the internal lustre of precious garnet is shining; that of pyrope is splendent: precious garnet exhibits several varieties of fracture; pyrope is only conchoidal: precious garnet refracts single; pyrope double: precious garnet is softer than pyrope, and has a higher specific gravity: and, lastly, pyrope contains

tains 10 *per cent.* of magnesia, an earth that does not occur in precious garnet; and it is more difficultly fusible.

2. It used to be considered as a variety of precious garnet, and was generally known by the name of Bohemian Garnet, from its occurring in that country in great beauty and perfection.

3. It is the Carbo-pyropus of the ancients.

4. Hauy conjectures it to be precious garnet with an accidental portion of magnesia, derived from the matrix in which it was formed: but pyrope occurs in wacke, a rock which is said to contain little or no magnesia.

9. Cinnamon-stone.

Kanelstein, *Werner.*

Hiacint, *Mohs.*

Hiacint, *Mohs,* Journ. des Mines, n. 130. p. 139.—Kanelstein, *Klap.* in Karst. Tabel. also in Beit. b. v. p. 138. *Id. Hauy,* Tabl. p. 62. *Id. Steffens,* b. i. s. 97.

External Characters.

The principal colour of this mineral is hiacinth-red, which on the one side inclines to blood red, and on the other passes through dark honey-yellow into a particular kind of orange-yellow The colours are pure.

It has been hitherto found only in small original blunt-angular and roundish pieces.

Its surface is uneven, and the inequalities are filled with a greyish coloured earth.

Externally

Externally it is glistening; internally shining, approaching to splendent; and the lustre is vitreous.

The fracture in every direction is rather imperfect, and flat conchoidal.

The fragments are angular, and very sharp-edged.

It does not occur in distinct concretions.

It is transparent and semitransparent; but it is generally so impure and full of cracks, that faultless specimens rarely occur.

Hard; scratches quartz, but with difficulty.

Brittle.

Rather difficultly frangible.

When cut, it feels somewhat greasy, and rather heavy.

Its specific gravity is 3.602 to 3.640, *Mohs* and *Hauy.*

Chemical Characters.

Before the blow-pipe it fuses into a blackish-brown enamel.

Constituent Parts.

Silica,	38.8
Alumina,	21.2
Lime,	31.25
Oxide of Iron,	6.5
Loss,	2.25
	100

Klaproth, Beit. b. v. s. 138.

Geognostic

Geognostic and Geographic Situations.

It is found in the sand of rivers in the island of Ceylon.

Use.

It is cut as a precious stone, and, when free of flaws, is of considerable value.

Observations.

1. *Distinctive Characters.—a.* Between cinnamon-stone and *pyrope.* The only colour of pyrope is blood-red; but cinnamon-stone has several colours, and none of them are distinct blood-red: pyrope scratches quartz more readily than cinnamon-stone: cinnamon-stone occurs in granular concretions, which is not the case with pyrope: pyrope has not the greasy feel which is observed in cut and polished cinnamon-stone; and we do not observe in pyrope the numerous rents and flaws that occur in cinnamon-stone.—*b.* Between cinnamon-stone and *vesuvian.* Vesuvian has a different colour-suite from hyacinth: vesuvian occurs crystallised; cinnamon stone never: the internal lustre of vesuvian is vitreo-resinous, and the fracture small-grained uneven; whereas that of cinnamon-stone is vitreous, and the fracture conchoidal.

2. Quisst distinguishes two sorts of hyacinth; one whose specific gravity is but 3.6, and is fusible; another which is infusible, and has a specific gravity of 4.3. The first is evidently the cinnamon-stone; the other the hyacinthine zircon.

3. The

3. The analysis of hyacinth given by Bergman, which is as follows, I refer to the cinnamon-stone:- Alumina, 40.; Silica, 25.; Carbonat of Lime, 20.; Iron, 13.

VI. QUARTZ FAMILY.

This family contains the following species: Quartz, Iron Flint, Hornstone, Flinty-slate, Flint, Calcedony, Heliotrope; Siliceous Sinter, Hyalite, Opal, Menelite; and Jasper.

1 Quartz.

This species is divided into six subspecies, viz. Amethyst, Rock-crystal, Rose or Milk Quartz, Common Quartz, Prase, and Cats-eye.

First Subspecies.

Amethyst.

This subspecies is subdivided into two kinds, viz. Common Amethyst, and Thick Fibrous Amethyst.

First

First kind.

Common Amethyst.

Gemeiner Amethyst, *Werner*.

Gemeiner Amethyst, *Reuss*, b. ii. th. 1. s. 205. *Id. Lud.* b. i. s. 74. *Id. Suck.* 1. th. s. 280. *Id. Bert.* s. 225. *Id. Mohs*, b. i. s. 193 —Amethyst-Quartz, *Hab.* s. 4. *Id. Karst* Tabel. —Quartz-hyalin Amethyste, *Brong.* T. i. p. 279.—Quartz-hyalin Violet, *Hauy*, Tabl. p. 25.—Amethyst, *Steffens*, b. i. s. 110.

External Characters.

Its principal colour is violet-blue, of all degrees of intensity. It passes on the one side from dark violet-blue, through plum-blue into clove-brown, and a particular kind of brownish-black; on the other side, from pale violet-blue through pearl grey, ash-grey, greyish white, greenish white, olive-green, into pistachio-green, which latter is uncommonly rare.

In the massive varieties, several colours occur together, and these are disposed in stripes, or fortification wise.

Besides massive, it occurs in rolled pieces, in angular pieces, and very frequently crystallised.

The following are its crystallizations:

1. Equiangular rather acute six-sided pyramid, which is sometimes double, sometimes single; when double, it is either perfect, or more or less deeply truncated on the common basis, and the lateral planes of the one are set on those of the other.

2. Six-

2. Six-sided prism, rather acutely acuminated by six planes, which are set on the lateral planes; and the edges between the acuminating and lateral planes deeply truncated, as in fig. 65. Sometimes the acuminating planes disappear, and then the truncations on the edges appear as an acute six-sided pyramid; and in other varieties, the alternate planes of this acute six-sided pyramid are very small, and then the pyramid appears three-sided.

The crystals occur in drusy cavities, and are commonly superimposed, or side by side. They are middle-sized and small.

The planes of the pyramids are smooth; those of the common base streaked.

Externally its lustre is splendent; internally it passes from splendent, through shining, to glistening, and is vitreous.

The fracture is intermediate between perfect and imperfect conchoidal; it is sometimes uneven, and even coarse splintery.

The fragments are angular and sharp-edged.

The massive varieties are commonly in distinct concretions, which are straight and thick prismatic, obliquely transversely streaked, and when free at the extremities, shoot into the pyramidal form.

These concretions are generally intersected by others which are lamellar, and fortification-wise bent; and the colour delineation arranges itself in the direction of these lamellar concretions.

Sometimes the prismatic concretions, when they are very short, (which is very seldom the case), approach in shape to coarse granular concretions.

It alternates from translucent to transparent.

It is hard.

Brittle,

Brittle.
Pretty easily frangible.
Rather heavy.
Specific gravity 2.632, deep blue, *Lowry*.

Chemical Characters.

Lampadius exposed it for four hours to the strongest heat of a wind furnace, when it suffered no other change but the loss of its colour, and about one and a quarter *per cent*. of its weight * According to Eherman, when exposed to a stream of oxygen gas, it loses its colour, and melts into a transparent bead.

Constituent Parts.

Silica,	97.50
Alumina,	0.25
Oxide of Iron,	0.50
Trace of Manganese.	
	98.25

Rose, Karsten's Tabell, s. 23.

Geognostic Situation.

It occurs sometimes in agate-balls in amygdaloid, greenstone and porphyry, and in veins in primitive and flœtz rocks. In the agate balls, it is associated with layers

* Lampadius, Samml. pract. chem. Abhandl. b. i. s. 225.

layers of calcedony, carnelian, flint, &c. and is usually the uppermost or newest layer of the series. When it occurs in veins, it is either along with ores of particular kinds, with agate, or with fibrous amethyst. It occurs also in rolled pieces in alluvial country.

Geographic Situation.

Europe.—In veins and drusy cavities in the flœtz greenstone of Fifeshire, particularly in those varieties that occur in the vicinity of Burntisland; also near Montrose, Hill of Kinnoul, and many other parts of Scotland. Near Cork in Ireland.

It occurs in the flœtz-trap rocks of Iceland, and the Faroe Islands.

Upon the Continent of Europe, it occurs at Dannemora in Sweden; in the Clausthal and other parts of the Hartz; at Annaberg, Kunnersdorf, &c. in Upper Saxony; Bohemia; Silesia; Bavaria; Stiria; Salzburg; Carniola; Swisserland; Hungary; Transilvania; Spain; and France.

Asia.—At Catharinenburg, Nertschinsk, Mursinska, and other places in Siberia; Cambay in India; Persia; and the Island of Ceylon.

America.—Real del Monte in Mexico; and in the United States.

Use.

The most highly prized varieties of amethyst were those imported by the Danish East India Company from Cambay in India. Formerly the Saxon and Bohemian amethysts were highly esteemed in Turkey, and were exported

ported by the way of Venice to Constantinople. At present, very beautiful varieties are found at Catharinenburg in Siberia; near the town of Vique in Murcia in Spain; and sometimes in the Val-Louise in the High Alps.

When the colour is good, and uniformly diffused, and the transparency considerable, it is cut into necklaces, bracelets, ear-rings, ring-stones, and seals. When less pure, it is fashioned into snuff-boxes. As its hardness is not considerable, it is cut on a copper-wheel with emery, and is polished on a tin-plate with tripoli. It should be set in gold rather than in silver, as the yellow colour forms with it a much more agreeable contrast than the white.

When the colour is not uniformly diffused, jewellers expose it for a short time in a mixture of sand and iron-filings, to a moderate heat, by which it is rendered more uniform. When we wish to give it a red colour, it is inclosed in a piece of charcoal, which is ignited, and allowed to consume gradually, and the amethyst will be found in the ashes of the charcoal of the desired colour.

Observations.

1. *Distinctive Characters.—a.* Between amethyst and *rock-crystal.* The colour-suite of amethyst does not agree with that of rock-crystal: amethyst occurs principally in the pyramidal form; whereas rock-crystal is generally prismatic: the fracture of amethyst is imperfect conchoidal or splintery; that of rock-crystal is perfect conchoidal or foliated: the lustre of amethyst is lower than that of rock-crystal: the crystals of amethyst do not attain the same magnitude as rock-crystals: amethyst

is not so perfectly transparent as rock-crystal: and, lastly, amethyst occurs in prismatic concretions, a form very seldom assumed by rock-crystal; and in lamellar concretions, a form never observed in rock-crystal.—*b.* Between amethyst and *rose* or *milk quartz.* The colour-suites of the two minerals are very different: rose quartz occurs only massive, whereas amethyst occurs both massive and disseminated; and more frequently crystallised than massive: the lustre of rose quartz is shining and vitreo-resinous; but that of amethyst extends from splendent to glistening, and is vitreous: rose quartz is translucent, approaching to semitransparent: amethyst alternates from semitransparent to transparent.

2. The ancients very frequently engraved upon this mineral. In the collection in the Royal Library at Paris, there are many very fine engraved amethysts; one of the most beautiful represents an Achille Cytharide. One of the largest engraved amethysts preserved in this collection, is that on which is represented the bust of Trajan, and which was formerly in the Royal Collection in Berlin.

3. The green variety of amethyst is the Chrysolite of some authors; and the Oriental Amethyst is a variety of Sapphire.

Second kind.

Thick Fibrous Amethyst.

Dickfasriger Amethyst, *Werner.*

Dickfasriger Amethyst, *Reuss,* b. ii. th 1. s. 210. *Id. Mohs,* b. i. s. 148.—Fasriger Quartz, *Steffens,* b. i. s. 125.

External

External Characters.

It has generally a pretty dark violet-blue colour, which when pale and light borders on pearl grey, and from this latter passes into milk and yellowish white.

It occurs only massive and in rolled pieces.

Internally its lustre is glistening and vitreous.

It has a double fracture; the principal fracture is thick fibrous, and is straight, and scopiformly diverging fibrous; the cross fracture is imperfect conchoidal, and sometimes splintery.

The fragments are angular or wedge-shaped, and sharp-edged.

It occurs in coarse angulo-granular concretions; and sometimes inclines to wedge-shaped concretions.

It is generally translucent; some varieties incline to semitransparent.

It is hard.

It agrees in the remaining characters with the preceding kind.

Geognostic Situation.

It is found in agate veins, and is generally accompanied with common amethyst. When both kinds occur together in the same vein, the fibrous is always the oldest, or adheres to the wall of the rent.

Geographic Situation.

Nearly the same as the former.

Second Subspecies.

Rock or Mountain Crystal.

Bergcrystal, *Werner.*

Quartzum pellucidum cristallizatum, *Wall.* p. 226.—Cristal de roche, *R. de L.* t. ii. p. 70.—Bergcrystal, *Wern. Cronst.* p. 111. *Id. Wid.* p. 296.—Mountain Crystal, *Kirw.* p. 241. —Berg-cristal, *Emm.* b. i. p. 217. *Id. Estner,* b. ii. s. 318. —Quarzo, *Nap.* p. 170.—Quartz, *Lam.* t. ii. p. 119.—Le Cristal de Roche, *Broch.* t. ii. p. 243.—Quartz, *Hauy,* t. ii. p. 406.—Berg Krystal, *Reuss,* b. ii. s. 212. *Id. Lud.* b. i. s. 75, 76.—Edler Quarz, *Suck.* 1. th. s. 284. *Id. Bert.* s. 253.—Berg Crystal, *Mohs,* b. i. s. 200–220. *Id. Hab.* s. 4.—Quartz hyalin, *Lucas,* p. 32. *Id. Brong.* t. i. p. 273. *Id. Brard,* p. 90.—Transparent Quartz, *Kid,* vol. i. p. 195.—Quartz hyalin, *Hauy,* Tabl. p. 24.—Berg Crystal, *Steffens,* b. i. s. 105.

External Characters.

Its principal colour is white; it occurs often also brown. From snow-white it passes into greyish-white, yellowish-white, and reddish-white; from greyish-white it passes into pearl grey; from yellowish-white it passes through pale ochre-yellow, wine-yellow, yellowish-brown, clove-brown, which falls into red, into brownish-black; from yellowish-brown it passes into orange-yellow and hyacinth red.

Of these colours, white and brown are the most common. It is characteristic of this fossil, that the yellow and brown colours are not unoften disposed in striped delineations.

It

It occurs very seldom massive: often in rolled pieces, and very often in crystals. Its crystallizations are the following:

1. Equiangular six-sided prism, rather acutely acuminated on both ends by six planes, which are set on the lateral planes *. Fig. 66.
2. When the prism becomes shorter, a double six-sided pyramid is formed, in which the lateral planes of the one are set on the lateral planes of the other, and the remains of the prism form a truncation on the common basis; or the truncation is entirely wanting. Fig. 67.
3. N° 1. in which the alternate angles formed by the meeting of the acuminating and lateral planes are truncated. Fig. 68.
4. N° 1. in which all the angles formed by the meeting of the acuminating and lateral planes are truncated. Fig. 69.
5. The prism is sometimes so broad, that it resembles a rectangular four-sided table, in which the terminal planes are bevelled.
6. N° 2. in which three alternate planes in each pyramid become larger than the others, and thus a figure approaching to the cube is formed.
7. N° 1. in which the lateral planes converge towards one extremity.
8. An acute simple six-sided pyramid.
9. N° 1. in which the alternate and unconformable lateral planes converge towards both extremities of the prism.

* The primitive figure of rock-crystal, according to Hauy, is a slightly obtuse rhomboid of 94° 4′, and 85° 56′.

10. An acute double three-sided pyramid; originates from the preceding figure.

The proportional size both of the lateral and acuminating planes vary so much, as to render it at first sight somewhat difficult to determine the form of the crystal *.

The crystals are from uncommonly large to small, but are most frequently middle-sized and large. The prisms are in general larger than the pyramids.

The lateral planes of the crystals are transversely streaked, but the acuminating planes are smooth †.

Externally, the crystals are generally splendent, the rolled pieces are only glistening, passing into glimmering; internally they are splendent and vitreous.

The fracture is almost always perfect conchoidal, or concealed foliated; some rare varieties shew a perfect foliated fracture, and the folia are parallel with the planes of the six-sided pyramid, or with the acuminating planes of the six-sided prism.

The fragments are angular, and very sharp-edged.

It very rarely occurs in granular and prismatic distinct concretions ‡.

It is always transparent §, and it refracts double when viewed

---

* We sometimes observe one crystal penetrating another longitudinally, forming a kind of twin-crystal.

† The streaks point out the cleavage.

‡ The rare variety in prismatic concretions is one of the links which connects rock-crystal with amethyst.

§ In transparent crystals we generally find the basis a point of adherence nearly opaque.

viewed through a pyramidal and lateral plane at the same time.

It is hard; scratches glass.

Rather easily frangible.

Rather heavy.

Specific gravity 2.650, rock-crystal from Madagascar, *Brisson.*—2.605, clove-brown crystal, *Karsten.*—2.888, snow-white transparent, from Marmerosch, *Karsten.*—2.5813, *Hauy.*

Chemical Characters.

Completely infusible by the blow-pipe, and, according to Lavoisier, it remains unaltered even when exposed to a stream of oxygen gas. Coloured rock crystal, if carefully exposed to a gentle heat, loses its colour, but retains its transparency.

Constituent Parts.

Silica,	$99\frac{3}{8}$
Trace of ferruginous alumina.	
	100

Bucholz, Gehlen's Journ. 1808, p. 150.

It appears from this analysis, that rock-crystal is an anhydrous silica.

Physical

Physical Characters.

When two rock-crystals are rubbed together, they are phosphorescent in the dark, and exhale a peculiar empyreumatic odour. Its inflammability has not been proved, although, from its oryctognostic affinity to diamond, it is not improbable that it is an inflammable body. The experiments of Davy are in favour of this opinion.

Geognostic Situation.

1. Although rock-crystal occurs more frequently, and in more numerous geognostic relations than amethyst, yet it is not the most common subspecies of quartz. It appears most frequently, and in the largest and most transparent crystals, in primitive rocks, where it occurs in beds, veins, and large drusy cavities.

These veins often contain large drusy cavities, and these, as well as the other parts of the vein, contain other minerals besides the rock-crystal: thus in Swisserland, France and Scotland, they contain also adularia, common felspar, epidote, chlorite, and calcareous spar; and in Siberia and other countries, topaz, beryl, and mica. In these veins we very seldom meet with ores, and when they do occur, it is but in small quantities; and almost the only species hitherto observed, are specular iron ore and octahedrite.

In other situations, however, it is associated with considerable variety and abundance of ores, and not only in veins, but also in beds. Thus in Hungary, Transilvania, Saxony, and other countries, it occurs in veins along with

with galena or lead glance, blende, copper-pyrites and iron-pyrites; and in beds in the tin formation of Zinnwald.

These are its principal geognostic relations in primitive mountains; and there it occurs more frequently and abundantly in granite and gneiss than in the newer rocks of the same class.

It has not been observed in transition rocks.

It occurs rarely in flœtz rocks, and only in limestone, marl and greenstone.

It may also be noticed, that rock-crystal forms one of the constituent parts of the variety of granite named *graphic granite.*

2. Various substances are found inclosed in rock-crystal; Thus it occasionally contains cavities, which are either wholly or partially filled with air, water, or petroleum; when the cavity is partially filled with water or petroleum, the air-bubble, or space unoccupied by the water, is visible in moving the crystal in different directions. The following are other minerals observed inclosed in rock-crystal; epidote, schorl, chlorite, asbestus, actynolite, hornblende? fluor-spar, heavy-spar, native silver, specular iron-ore, grey antimony ore, arsenical pyrites, and rutile. Sometimes the crystals of grey antimony ore and rutile are decomposed and carried away by some agent unknown to us, and then the rock-crystal appears traversed by a number of hollow prismatic canals.

Geographic

Geographic Situation.

Europe.—Crystals of great size and beauty are found in different parts of Scotland; the rock-crystals of the Island of Arran, which occur in drusy cavities in granite, are well known; but the largest and most valuable are found in the neighbourhood of Cairngorm, in the upper part of Aberdeenshire, where they occur in alluvial soil, along with beryl and topaz. I am not acquainted with many of its English localities; Cornwall may be mentioned as one. Others are given by authors, but these in general refer to common quartz, not to rock crystal. On the Continent of Europe it is very widely, and often abundantly distributed. Thus it is found at Kongsberg in Norway, along with native silver; and in the same country, in beds of quartz, in primitive greenstone; likewise in drusy cavities in granite, in the Hartz; in Upper Saxony, in a tinstone formation; Bohemia; in Silesia, in granite; in clay-slate in Bavaria; Tyrol; Carinthia; Carniola; Italy; Hungary; Transilvania; Swisserland, particularly in Mont Blanc; Spain; and France, particularly in Dauphiny, where very magnificent groupes of crystals are found

Asia.—Island of Ceylon; Catharinenburg; Adon-Tschelon, along with beryl; at Nertschinsk.

Africa.—Large and beautiful crystals, which are sometimes traversed by crystals of rutile, are found in Madagascar

America—Beautiful small crystals are found at Cape Diamond near Quebec; in West Greenland, and many other parts in North America.

Large and beautiful crystals are found in the Brazils, in the Caraccas, and other districts of South America.

Use.

Use.

Rock-crystal is cut and polished as an inferior kind of gem or ornamental stone. It receives the required shape by sawing, splitting, and grinding. The sawing is effected by means of an extended copper-wire fixed to a bow; the wire is coated with a mixture of oil and emery, and is drawn backwards and forwards until the operation is finished. A this process is a very tedious one, particularly when the mass is large, a more expeditious, although less certain, method is sometimes followed: The crystal is heated red-hot, and then a wet cord is drawn across in the direction we intend to split it; by the rapid cooling thus effected in the direction of the cord, the stone easily splits, and generally in the desired direction, by a single blow of a hammer. The grinding is done by means of emery; and the polishing with tin-ashes and tripoli.

Different kinds of work in rock-crystal must be perforated, and the perforation is executed by means of the diamond-splitter and a drill machine. It is cut into ring-stones, seal-stones, necklaces and ear-pendants; and when the masses are large, into snuff-boxes, vases, and ornaments for chandeliers. The ancients valued vases of this stone very highly, particularly when of considerable size. Such were the two cups which the tyrant Nero broke into pieces in a fit of despair, when he was informed of the revolt which caused his destruction. One of these was estimated at 15,000 livres. At Briançon there was formerly a manufacture, where the rock-crystal of Dauphiny was worked into ornaments for chandeliers.

Different

Different colours may be communicated to the white varieties of rock-crystal: Thus, if they are heated, and plunged into a solution of indigo, they acquire a blue colour; if into a decoction of cochineal, a red colour; or if into a solution of copper, a blue tint. A clove-brown colour may be given to white coloured crystals, by exposing them to the vapour of burning wood. Artists sometimes communicate beautiful colours to rock-crystals, by forming them into *doublets*. Two modes are followed. In the one, we take a semibrilliant of rock-crystal, and hollow it underneath, and fill the hollow with a liquor of the colour we wish the stone to exhibit, and then inclose it by a plate of glass. If this kind of doublet be dexterously executed, we do not readily discover that the stone is hollow underneath, and only coloured in the middle, but the whole mass appears of an uniform tint. The second kind of doublet is formed, by cementing a coloured plate of glass on the base of a roset or brilliant cut rock-crystal, by means of which the whole stone acquires the colour of the plate.

Rock-crystal is sometimes imitated by artificial pastes; but these can be distinguished from the true stone by their inferior hardness, and their containing roundish air-vesicles irregularly distributed throughout the mass.

Of these pastes, the most celebrated is that known under the name of *Strass paste*.

Observations.

1. Transparent snow-white rock-crystal resembles glass in its general appearance; but, independent of other characters, the vesicles and rents that occur in both afford an easy mode of distinguishing them; the air-bubbles

of

or vesicles in glass being irregularly diffused, and nearly of a globular shape; but in rock-crystal they are disposed in the same plane or parallel planes; and generally in the form of clouded specks.

2. We sometimes observe on the surface, or in the interior of rock-crystals an iridescence. It owes this property, when superficial, to a slight coating of metallic oxide; but when internal, it is caused by the refraction of light in consequence of fissures. This appearance may be artificially produced by heating rock crystal nearly red-hot, and then plunging it into hot water.

Brongniart conjectures that the iridescent varieties of rock-crystal may be the *Iris* of Pliny.

3. The ancients were of opinion that rock-crystal was congealed water, and that this was the cause of its often flying in pieces when exposed to a high temperature.

4. The largest masses of rock-crystal hitherto found are those from the Island of Madagascar. Faujas St Fonde mentions masses weighing more than one hundred and fifty pounds weight, imported into France from Madagascar *.

Single crystals of considerable size and great purity have been found in the Highlands of Scotland. An Edinburgh lapidary had in his possession a pure crystal weighing upwards of nineteen pounds; and another artist one crystal, but less pure, weighing seventy-eight pounds.

5. The wine and orange yellow coloured varieties of rock-crystal are sometimes sold as Topaz; the clove-brown

* Essai de Geologie, T. II. part. ii. p. 92.

† The yellow varieties are sometimes named Spanish Topaz.

brown is sold under the name *Smoke Topaz;* and in Scotland these varieties are named *Cairngorm-stones,* and are in considerable repute as articles of ornamental dress, being cut into ring stones and seal-stones, and sometimes into elegant snuff-boxes.

5. The varieties of rock crystal that contain crystals of actynolite, are named *Thetis hair stones;* those with crystals of rutile, *Venus hair-stones;* and when cut and polished, are generally worn as ring-stones.

Third Subspecies.

Rose or Milk Quartz.

Milch Quartz, *Werner.*

Rosen rother Quarz, *Wid.* p. 301.—Rosy red Quartz, *Kirw.* vol. i. p. 245.—Milch Quarz, *Emm.* b. i. s. 136.—Quartz laiteux, *Lam.* t. ii. p. 123.—Quartz laiteux ou Quartz Rose, *Broch.* t. i. p. 246.—Quartz-hyalin-rose, *Hauy,* t. ii. p. 418.—Milch Quarz, *Reuss,* 2. b. i. th. s. 221. *Id. Lud.* b. i. s. 76. *Id. Suck.* 1. th. s. 283. *Id. Bert.* s. 255. *Id. Mohs,* b. i. s. 220. *Id. Hab.* s. 4.—Quartz rose, *Lucas,* p. 32.—Quartz-hyalin-rose, *Brong* t. i. p. 278.—Quartz rose, *Brard,* p. 93.—Transparent rose-red Quartz, *Kid,* vol. i. p. 199.—Milch Quarz, *Steffens,* b. i. s. 112.

External Characters.

Its most common colour is rose red, and next in frequency is milk-white. The rose red colour sometimes inclines to flesh-red, and passes into crimson red, reddish-white.

white, pearl-grey, and, lastly, into milk-white, which reflects a yellowish light.

It occurs only massive.

Internally its lustre is shining, sometimes passing to splendent, and is vitreous, slightly inclining to resinous.

The fracture is more or less perfect conchoidal.

The fragments are angular and sharp-edged.

Some varieties shew a tendency to straight and thick lamellar distinct concretions.

It is more or less translucent, even approaching to semitransparent.

It is hard.

Easily frangible.

The other characters are the same as those of rock-crystal.

Constituent Parts.

It is supposed to be silica coloured with manganese.

Geognostic Situation.

It occurs in beds in granite and gneiss; and in veins that traverse flinty-slate.

Geographic Situation.

Europe.—It was first discovered in Bavaria, where it occurs in beds, in granite, at Rabenstein *, Zwiesel, and Hörlberg.

* Brard, Traite des Pierres Precieuse, t. i. p. 89., says, that it occurs in veins along with manganese at Rabenstein.

Hörlberg. It is found in rolled masses at Hüttschlag in Salzburg; in the Harzburg Forest in veins that traverse flinty-slate; in the Hochwald near Stolpen, in the Electorate of Saxony; at Chateau-neuf in Auvergne, and Miosin in France; Arendal in Norway; and in the Island of Coll, one of the Hebrides.

Asia.—At Sljudanka, in the vicinity of the Lake Baikel: in large beds at Kalywan, and in the snowy mountains of Tigerateky in the Altain range.

America.—In Greenland.

Uses.

1. It is employed in jewellery, and the larger masses are cut into vases *. It takes a good polish, and when the colour is good, the ornaments made of it are beautiful. When cut and polished, and of a good colour, it is sold for spinel; yet its deficiency in hardness, transparency, and fire is so great, that the deception is easily detected.

Observations.

1. This subspecies is distinguished from Rock-crystal, the only one of the subspecies of the quartz species with which it could be confounded, by its colour, massive external shape, fracture, lustre, and lamellar concretions.

2. The

* M. Dedrée has in his possession a beautiful vase of rose quartz.

2. The milk or bluish-white variety of this mineral, is by some jewellers named *water sapphire*, *false sapphire*, or *occidental sapphire.* The ancients are said to have known it under the title *leuco-sapphire.*

3. The finest bluish white varieties, and which in commerce are always cut *en cabachon*, are said to be from Macedonia.

4. It loses its colour by keeping, particularly in a warm place.

Fourth Subspecies,

Common Quartz.

Gemeiner Quarz, *Werner.*

Quartzum rude, *Wall.* t. i. p. 220. *Id. Rome de Lisle* —Gemeiner Quarz, *Wid.* p. 300.—Quartz, *Kirwan,* vol. i. p. 242. *Estner,* b. ii. s. 265. *Id Emm.* b. i. s. 125.– Quarzo, *Nap.* p. 170. *Lam.* t. ii. p. 119.—Quartz-hyalin amorphe, *Hauy,* t. ii. p. 423.—Le Quartz commune, *Broch.* t. i. p. 248.—Gemeiner Quarz, *Reuss,* b. ii. s. 44. *Id. Lud.* b. i. s. 76. *Id. Mohs,* b. i. s. 222–245. *Id. Bert.* s. 250. *Id. Suck.* 1. th. s. 290. *Id. Hab.* s 5.—Quartz hyalin opaque, *Lucas,* p. 32. —Quartz hyalin amorphe, *Brong.* t. i. p. 274.—Quartz hyaline opaque, *Brard,* p. 94.—Quartz amorphe, *Hauy,* Tabel. p. 25.—Gemeiner Quarz, *Steffens,* b. i. s. 119.

External Characters.

Its most common colours are white and grey. Of white, the following varieties have been observed: snow-white, greyish-white, yellowish-white, greenish-white, and reddish-white: from greenish-white it passes into a colour intermediate between verdigris-green and celandine-green. The varieties of grey are ash-grey, smoke-grey, yellowish-grey, bluish grey, and pearl-grey. From yellowish-grey it passes into wax and honey yellow; from pearl grey into flesh-red, which sometimes approaches to blood and brick red, and further into hyacinth-red, reddish-brown, and pale chesnut-brown.

Of these colours, white and grey are the most frequent; next in frequency is red, the others are rare. Smoke-grey in some varieties is so dark, that it passes into greyish-black.

It occurs massive, disseminated, in blunt-edged pieces, in roundish grains, in plates, and rolled pieces: also stalactitic, reniform, globular, specular, corroded, vesicular, ramose, amorphous, cellular, and with impressions; of the cellular it presents the following varieties, hexagonal, polygonal, and circulo-cellular; and of this latter form, the parallel, double and spongiform varieties.

The impressed forms are tabular, cubical, pyramidal, and conical.

It occurs in true and supposititious crystals.

The following are the true crystals:

1. Six-sided prism acuminated on both extremities by six planes. It is either crystallised on both extremities, and then it is imbedded, or crystallised only at one extremity, and then it adheres.
2. Simple six-sided pyramid. The crystals of this figure are either single, resting on each other, or are aggregated in form of a bud.
3. Double six sided pyramid, which is sometimes aggregated in rows.

The

The crystals occur of every size, from very small to very large, but they never attain the magnitude of rock-crystal. The prisms are generally larger than the pyramids.

The surface of the crystals is the same as in rock-crystal.

The following are the supposititious crystals:

1. Double three-sided pyramid, hollow, and the surface drusy. Originates from calcareous spar.
2. Single three-sided pyramid, hollow, and surface drusy. Originates from calcareous spar.
3. Rhomboid, hollow, and surface drusy. Originates from calcareous spar.
4. Double six-sided pyramid, hollow, and surface drusy. Originates from calcareous-spar.
5. Regular octahedron, sometimes hollow, and surface drusy. Originates from fluor spar.
6. Cube. Originates from fluor-spar.
7. Rectangular four-sided tables, hollow, and surface drusy. Originates from heavy-spar.
8. Oblique four-sided table, surface drusy. Originates from heavy-spar.
9. Eight-sided table, sometimes hollow, sometimes partly filled with straight lamellar heavy-spar, and the surface drusy. Originates from heavy-spar.
10. Lens, hollow, and surface drusy. Originates from gypsum.

Externally the lustre of the true crystals varies from splendent to glistening; that of the rolled pieces is glimmering, passing into dull.

Internally it is shining, which sometimes borders on glistening, and sometimes approaches to glimmering, and is vitreous.

The fracture passes from perfect conchoidal into coarse and fine splintery. Some rare varieties shew a parallel fibrous, and others a concealed foliated fracture.

The fragments are angular, and rather sharp-edged.

It is generally compact; sometimes, however, it occurs in distinct concretions, and these are prismatic granular, and very rarely lamellar. The granular concretions are generally fine, seldom small, and very rarely large. The fine granular variety has a slaty fracture, and is flexible when in thin tables. The prismatic concretions are thin and thick, and sometimes parallel, sometimes diverging; and the lamellar concretions are thick and straight.

It is generally translucent, seldom semitransparent *, and the darker varieties are only translucent on the edges.

The other characters the same as those of rock-crystal.

Chemical Characters.

It is infusible without addition before the blowpipe, but when exposed to a stream of oxygen gas, according to Ehrmann, it melts into a milk-white porcellanous ball.

Geognostic Situation.

This is one of the most abundant minerals in nature, and appears in many different geognostic situations. It occurs

* It is only in the crystallised varieties that semitransparency occurs.

occurs in primitive, transition, flœtz, and alluvial rocks, and either as a constituent part of these rocks, or associated with them in the form of beds, lying masses, mountain masses, and veins. Thus it forms a principal constituent part of granite, the oldest, and one of the most frequent and abundant of rocks; it is also one of the component parts of gneiss, mica-slate, and topaz-rock: occurs imbedded in grains and crystals in porphyry, and accidentally intermixed with clay-slate and limestone, Beds of quartz occur in granite, gneiss, mica-slate, and clay-slate, but more abundantly in mica-slate and clay-slate than in granite or gneiss. Lying masses are connected with mica-slate and clay-slate; and the mountain masses, which form whole hills, and constitute what is called the Quartz Formation, occur in an overlying position in regard to clay-slate, mica-slate and other rocks. Most of the veins that traverse primitive rocks, with the exception of those that afford fluor-spar, heavy-spar, and some other minerals, contain common quartz. These veins are frequently entirely composed of the quartz, and are of great width and extent; indeed they are so large, that Dolomieu and others maintain, although erroneously, that quartz occurs more abundantly in veins than in any other kind of repository.

It also occurs in metalliferous beds, along with ores of different kinds, as galena or lead-glance, tinstone, and various pyritical minerals.

Such are the general geognostic relations of common quartz in Primitive mountains.

In Transition mountains it occurs less abundantly than in rocks of the primitive or flœtz series. In the form of veins it traverses grey-wacke, one of the most abundant of the transition rocks: it also occurs in many of the numerous metalliferous veins that traverse grey-wacke and grey-wacke slate; and rolled masses of it are contained in grey wacke.

In Flœtz mountains it occurs in vast abundance: thus several of the great sandstone formations are principally composed of fragments of this substance, which appear to have been derived from older previously existing rocks, either of the transition or primitive classes. But this is not its only form in flœtz mountains: it occurs also forming original beds, imbedded in various rocks, as gypsum, limestone, &c. and as a constituent part of the numerous veins that traverse the different rocks of the flœtz series, from the oldest to the newest rock of the class.

In Alluvial country, it occurs in vast abundance, in the form of rolled pieces, gravel, and sand.

Geographic Situation.

Europe.—The highest mountains in the island of Jura, one of the Hebrides, are principally composed of common quartz; the upper part of the mountain of Scuraben in Caithness, and several mountains in Sutherland, are composed of the same mineral. At Portsoy in Banffshire, there is a hill entirely composed of common quartz; and in the islands of Islay, Coll, Tiree, Icolmkill, Arran, Bute, Seil, and Skye, it occurs frequently either in beds or veins, and generally in primitive country. It also occurs in the Zetland islands, and either in beds or veins in primitive country, or as a constituent part of sandstone. In the Orkney islands it occurs principally in flœtz country, as a constituent part of sandstone.

On

On the mainland of Scotland, it occurs in the form of beds and veins in all the Highland districts where primitive rocks prevail, as in Sutherland, Inverness-shire, Ross-shire, Banffshire, Perthshire, Argyleshire, &c. It also occurs in the transition rocks both to the north and the south of the Frith of Forth; in the flœtz rocks that abound in various parts of Scotland, particularly the middle and southern quarters, where it occurs in great abundance as a constituent part of sandstone, in beds in a coal formation, and in veins that traverse many different kinds of flœtz rocks, as sandstone, greenstone, &c.

All around the coasts of Scotland, and on the shores of several of the Hebrides, Orkney and Zetland islands, it occurs in the form of sand; and in some places has been blown from the coast into the interior.

It is also a frequent mineral in the primitive and transition rocks of England; but it does not occur so abundantly as in Scotland, owing to the small extent of primitive rocks in that part of the island. It also forms a principal constituent part of the different sandstones that occur in the lower and flatter parts of England, and occurs in veins that traverse not only sandstone but also limestone, trap, and other flœtz rocks *.

It is also abundant in the primitive and flœtz rocks of Ireland.

In the alluvial districts in the different parts of Britain and Ireland, it abounds in the form of gravel and sand.

On

* The *Bristol Diamonds*, are crystals of common quartz inclining to rock-crystal.

On the Continent of Europe it is very abundantly and widely distributed; indeed we cannot name a country, from the coast of Norway to the Black Sea, and from the Arctic Ocean to the Mediterranean Sea, that does not contain much common quartz.

Asia.—A beautiful indigo-blue variety, along with the common varieties, occurs in the island of Ceylon. In Siberia it occurs in vast abundance, either as a constituent part of mountain rocks, or in beds alternating with them, or in veins traversing them. In the peninsula of India it occurs as a constituent part of granite, gneiss, mica-slate, also accidentally mixed with clay-slate, and as a constituent part of various sandstones.

Africa.—At the Cape of Good Hope it occurs in veins, and as a constituent part of granite; and the great ranges of mountains to the north of that promontory are formed of sandstone, of which this mineral forms a principal ingredient. The vast sandy desarts that occupy so great a portion of the surface of Africa, contain much common quartz.

America.—The great tracts of primitive and transition country in the northern and southern parts of North America abound in common quartz. The limestone districts afford smaller portions of it, but in some places on the coasts of the ocean it abounds in the form of sand.

In South America it is an abundant mineral, appearing in its usual geognostic situations. in the great tract of primitive, transition, flœtz, and alluvial rocks that traverse that vast continent. The flexible variety was first found in Brazil, and has been lately observed in North America *.

Use.

* Greenough.

Use.

It is employed in the manufactory of glass; also in the preparation of smalt, and as an ingredient in porcelain and different kinds of earthen-ware. The vesicular and corroded variety forms a most excellent millstone, known in commerce under the name of *Buhr-stone.* This buhr-stone has hitherto been found only in France; but it is so much esteemed in this country, that the Society of Arts of London have for many years past offered a considerable annual reward for its discovery in Great Britain.

In the form of sand, it is used, with quicklime, in the composition of mortar. In agriculture, for the improvement of particular kinds of soil; sometimes as a paving-stone, and in the coarser kinds of masonry.

Some varieties of common quartz exhibit numerous points or spots that glitter like gold. This appearance is sometimes owing to the intermixture of scales of mica; in other instances it is caused by reflection from numerous small rents or fissures in the stone. These varieties have received the name Aventurine, from the following circumstance: A French workman having by accident *(per aventure)* dropped filings of brass or copper into a vitreous mixture in a state of fusion, gave the name *Aventurine* to the glittering mixture thus formed, and of which artists make vases and other ornamental articles. Mineralogists have applied the same name to those varieties of common quartz that exhibit a nearly similar appearance. These are cut into various ornamental articles, and are sometimes sold at a very high price. The natural aventurine is found in Arragon in Spain, at Face-bay

bay in Transilvania, in the vicinity of Quimper in Brittany.

Mr Greenough found it near Fort-William, in the Highlands of Scotland; and I observed it in Mainland, one of the Zetland islands.

Observations.

1. It is distinguished from Rock-crystal by its colour-suite; its various particular and supposititious forms; the greater regularity of its crystallizations; its lower degree of lustre and transparency; splintery or imperfect conchoidal fracture; and distinct concretions.

2. Leonhard describes as a new species, under the name *Siderite*, what Werner considers but as a variety of common quartz.

3. There has been lately discovered, near Nantz in France, a variety of common quartz, (imbedded in granite), of a grey colour, conchoidal, passing to splintery fracture, and semitransparent, which when broken exhales a disagreeable smell, somewhat resembling that of sulphurated or carbonated hydrogen. It is by Steffens arranged as a distinct subspecies of common quartz, under the title *Stink-quarz*.

4. The flexible variety of common quartz is arranged as a separate subspecies, under the title *Gelenk-quarz*, by Steffens.

5. The red crystallised variety of common quartz, found near Compostella in Spain, used to be described under the title of *Compostella Hyacinth*.

Fifth

Fifth Subspecies.

Prase.

Prasem, *Werner.*

Prasem, *Wern. Cronst.* s. 116.—Lauchgruner Quarz, *Wid.* p. 301.—Prasium, *Kirw.* vol. i. p. 249.—*Id. Estner,* b. ii. s. 207. *Id. Emm.* b. i. s. 103.—Quarzo verde di porro, *Nap.* p. 171.—La Prase, *Broch.* t. ii. p. 252.—Quarz-hyalin verd obscur, *Hauy,* t. iii. p. 419.—Prasem, *Reuss,* b. ii. s. 235. *Id. Lud.* b. i. s. 76–77. *Id. Suck.* 1. th. s. 299–300. *Id. Bert.* s. 171. *Id. Mohs,* b. i. s. 163 *Id. Hab.* s. 5.—Quartz-hyalin vert obscur, *Lucas,* p. 32 —Quartz Prase, *Brong.* t. i. p. 280.—Quartz-hyalin vert obscur, *Brard,* p. 93 Prase, *Kid,* vol. i. p. 203. *Id. Hauy,* Tabl. p. 25.—Prase, *Steffens,* b. i. s. 113.

External Characters.

Its colour is leek-green, of various degrees of intensity

It occurs generally massive, seldom crystallised.

Its crystallizations are the following:

1. Six-sided prism, acuminated by six planes, like quartz.
2. Six-sided pyramid, but this is rare.

The crystals are small, and middle-sized, and have always a drusy surface.

Its lustre is shining, approaching to glistening, and is resino-vitreous.

The

The fracture is imperfect conchoidal, passing into coarse splintery.

The fragments are angular, and more or less sharp-edged.

The massive varieties occur in distinct concretions, which are cuneiform, prismatic, and sometimes coarse granular.

The surface of the concretions is rough and transversely streaked.

It is translucent.

Hard.

Rather difficultly frangible.

Rather heavy.

Geognostic Situation.

It occurs in mineral beds, which are composed of magnetic ironstone, magnetic pyrites, iron-pyrites, copper-pyrites, galena or lead-glance, blende, quartz, calcareous spar and common actynolite. These beds are probably connected with primitive trap.

It also occurs in small quantity in clay-slate. It does not occur as a constituent part of any rock, nor has it been hitherto found in veins.

Geographic Situation.

Europe.—It is found in small quantity in the island of Bute, in the Frith of Clyde: also in Borrodale, and elsewhere, in the neighbourhood of the English lakes *. On the

* Greenough.

the Continent, it occurs in metalliferous beds at Breitenbrun near Schwartzenberg in Saxony; at Mummelgrund in Bohemia; at Bojanowitz in Moravia: at Kupferberg in Silesia; island of Elba in the Mediterranean; and near the Lake Onega in Finland.

Asia.—Siberia.

Use.

It is sometimes cut and polished as an ornamental stone, but is not highly esteemed. When set, it should have a gold foil.

Observations.

It is an intimate mixture of quartz and actynolite.

Sixth Subspecies.

Cat's-eye.

Katzenauge, *Werner.*

Pseudopalus opacus radios—Oculus cati, *Wall.* t. i. p. 296.—Oeil de chat, *R. de L.* t. ii. p. 145. —Variety of Mondstein, or Adularia, *Wid.* p. 344.—Cat's-eye, *Kirw.* vol. i. p. 301.—Katzenauge, *Emm.* b. i. s. 188.—Occhio di gatto, *Nap.* p. 225. —Oeil de chat, *Lam.* t. ii. p. 152. *Id. Broch.* t. i. p. 292.—Quartz agathe chatoyant, *Hauy,* t. ii. p. 427.—Katzenauge, *Reuss,* b. ii. th. 1. s 47. *Id. Lud.* b. i. s. 86.–87. *Id. Suck.* 1. th. s. 319–321. *Id. Bert.* s. 263. *Id. Mohs,* b. i. s. 185–187.—Quartz agathe chatoyant, *Lucas,* p. 33.— Quartz hyalin chatoyant, *Brong.* t. i. p. 277.—Quartz agathe chatoyant, *Brard,* p. 96.—Cat's-eye, *Kid,* vol. i. p. 229.—Schiller Quarz, *Karst.* Tabel.—Quartz agathe chatoyant, *Hauy,* Tabl. p. 27.—Katzenauge, *Steffens,* b. i. s. 122.

External

External Characters.

Its principal colour is grey, of which it presents the following varieties: yellowish, greenish, and ash grey; from yellowish grey it passes into yellowish brown, and into a kind of isabella-yellow; and further, into a yellowish, reddish, and hair brown, and into a colour intermediate between hyacinth and brick red. From greenish-grey it passes into mountain-green and olive-green; and from ash-grey into greyish-black.

It exhibits a beautiful common opalescence, particularly when cut in a spherical form *.

It is found in blunt-edged pieces, in rolled pieces, and also massive.

Internally it is shining, and the lustre is vitreo-resinous.

The fracture is small and rather imperfect conchoidal, sometimes approaching to uneven.

The fragments are angular, and more or less sharp-edged.

It is generally translucent, sometimes also semitransparent, and translucent on the edges.

It is hard.

Easily frangible.

Rather heavy.

Specific gravity from 2.625 to 2.600, *Klaproth.*—2.647, *Lowry.*

Chemical

* It is usually brought into Europe cut in a spherical form.

Chemical Characters.

By exposure to the heat of a porcelain furnace it loses its hardness, lustre, and transparency, and partly its colour, but is not melted. Before the blow-pipe, according to Saussure, it melts with great difficulty.

Constituent Parts.

Silica,	95.00	94.50
Alumina,	1.75	2.00
Lime,	1.50	1.50
Oxide of Iron,	0.25	0.25
Loss,	1.50	1.75
	100	100

Klaproth, Beit. t i. p. 90.

Geognostic Situation.

In the Hartz it occurs in cotemporaneous veins along with quartz, amianthus, asbestus, axinite, and calcareous spar, in primitive trap. Its geognostic situation in other countries is still unknown.

Geographic Situation.

Europe.—In the Hartz in Hanover.

Asia.—Island of Ceylon, coast of Malabar, island of Sumatra, Persia, and Arabia.

Africa.—It is said to occur in Egypt.

Use.

It is generally cut into ring-stones; and the most advantageous form for displaying its peculiar lustre is the oval, with a convex surface. The red and olive-green varieties are the most highly prized.

Observations.

1. It has been by some mineralogists referred to Opal, by others to Felspar: it is, however, sufficiently distinguished from opal by its hardness and weight; and its fracture distinguishes it from felspar.

2. Werner is of opinion, that the beautiful opalescence which it exhibits, and from which it derives its name, is owing to an intermixture of some other mineral. This has been proved to be the case by the observations of Karsten and Cordier. Ribbontrop found in the Hartz a pale greenish-grey coloured mineral, which he sent to Karsten. He examined it, and found it to be a mixture of quartz and amianthus, and which had the characters of the Indian cat's-eye. The same observation has been made by Cordier.

3. It has received its name from the property it possesses, of reflecting, when cut in particular directions, a peculiar whitish or yellowish lustre, resembling that of the eye of the cat in the dark.

2. Iron-Flint.

Eisenkiesel, *Werner.*

Le Cailloux ferrugineux, *Broch.* t. i. p. 238.—Quartz rubigineux, *Hauy,* Tabl. p. 25.—Eisenkiesel, *Reuss,* b. ii. th. 1. s. 300. *Id. Lud.* b. i. s. 73. *Id. Suck.* 1. th. s. 347. *Id. Bert.* s. 270. *Id. Mohs,* b. i. s. 187.—Quartz rubigineux, *Brong.* t. 1. p. 281.—Eisenkiesel, *Hab.* s. 6. *Id. Steffens,* b. i. s. 126. *Id. Hoff.* b. ii. s. 60. *Id. Oken.* b. i. s. 270.

External Characters.

The principal colours are brown and red. The brown colours

colours are yellowish-brown, which sometimes approaches to ochre-yellow; farther, a colour intermediate between chesnut and liver brown. The only red colour is one intermediate between brownish-red and blood-red.

It occurs most commonly massive, but also crystallised in small equiangular six-sided prisms, which are acuminated on both extremities by three or six planes, which are set on the lateral planes.

Externally its lustre is shining, approaching to glistening; internally it is glistening, and is vitreo-resinous.

The fracture is imperfect and small conchoidal, which, in some varieties, approaches to uneven.

The fragments are angular, and rather sharp-edged.

It occurs almost always in small angulo-granular distinct concretions, which approach sometimes to the fine, and more rarely to the coarse granular.

It is opaque.

Gives sparks with steel.

Rather difficultly frangible

Rather heavy, approaching to heavy.

Specific gravity, 2.627 2.691 2.814 2.838, *Haberle.*
2.576 2.618 2.746, *Hoff.*

Chemical Characters.

Is infusible without addition before the blow-pipe.

Constituent Parts.

Yellow Iron-Flint.	
Silica, -	93.5
Oxide of iron, -	5.0
Volatile matter,	1.0
Bucholz,	99.5

Yellowish-Brown Iron-Flint.	
Silica, -	92.00
Oxide of iron,	5.75
Oxide of manganese,	1.00
Volatile matter,	1.00
Bucholz,	99.75

Red Iron-Flint.

Silica, - -	$76\frac{5}{6}$
Alumina, - -	$\frac{1}{4}$
Red oxide of iron, -	$21\frac{4}{6}$
Volatile matter, -	1
Bucholz,	$99\frac{3}{4}$

Geognostic Situation.

It occurs in veins of ironstone; the red variety in red ironstone, the brown, in brown ironstone; and also in trap rocks.

Geographic Situation.

Europe.—In rocks near Bristol; in trap rocks that lie over white limestone, island of Rathlin, off the coast of Ireland *; and in trap rocks at Dunbar in Scotland. At Orpes; Hohenstein and Sedlitz in Bohemia; in the Fichtelgebirge in Franconia; in brown ironstone veins at Ilfeldt and Fischbach in the Hartz; in ironstone veins at Altenberg, Eibenstock in Upper Saxony, and at Oberstein in France.

Asia.—According to M. Von Moll, it occurs in Siberia.

Observations.

1. It appears to be a chemical compound of quartz and iron-ochre.

2. It renders the iron-ore along with which it occurs, very difficult of fusion.

3. Hornstone.

* Greenough.

3. Hornstone.

Hornstein, *Werner.*

THIS species is divided into three subspecies; Splintery Hornstone, Conchoidal Hornstone, and Woodstone.

First Subspecies.

Splintery Hornstone.

Splittriger Hornstein, *Werner.*

Splittriger Hornstein, *Wern.* Pabst. t. i. p. 247.—Petrosilex squamosus, *Wall.* t. i. p. 280.—Splittriger Hornstein, *Emm.* b. i. s. 251. *Id. Estner.*—Le Hornstein ecailleux, *Broch.* t. i. p. 255.—Splittriger Hornstein, *Reuss,* b. ii. th. 1. s. 325. *Id. Lud.* b. i. s. 77. *Id. Suck.* 1. th. s. 356–360. *Id. Bert.* s. 234. *Id. Mohs,* b. i. s. 248. *Id. Hab.* s. 14.—Silex corné, *Brong.* t. i. p. 319.—Quartz agathe grossier, *Hauy,* Tabl. p. 27.—Splittriger Hornstein, *Steffens,* b. i. s. 167. *Id. Hoff.* b. ii. s. 65. *Id. Lenz.* b. i. s. 366. *Id. Oken.* b. i. s. 299.

External Characters.

Its principal colours are grey, red, and green, of which the following varieties occur: Of grey; bluish, greenish, yellowish, smoke and pearl grey: From pearl-grey it passes into flesh-red, brick-red, brownish-red, and reddish-brown: it seldom inclines to ochre-yellow: from greenish-grey it passes into mountain and pale olive green; and from pale smoke-grey into greyish-white, and yellowish-white.

It occurs generally massive, sometimes also in large balls, and seldom with pyramidal impressions from calcareous spar.

It occurs in lenticular, and six-sided prismatic suppositititious crystals.

The internal lustre is always dull.

The fracture is splintery, generally small, and fine, seldom coarse splintery; the latter sometimes approaches to large conchoidal.

The fragments are angular, and rather sharp-edged.

The globular varieties occur in thick concentric lamellar distinct concretions.

It is more or less translucent on the edges; but some varieties that incline to quartz, and which are coarse splintery, are translucent.

It is hard, but not in so high a degree as quartz or flint.

Difficultly frangible; and

Rather heavy, in a low degree.

Specific gravity 2.536, 2.602, 2.626, 2.635, *Hoff.*

Chemical Characters.

Infusible without addition before the blow-pipe.

The fusible varieties mentioned by some mineralogists, are compact felspar.

Geognostic Situation.

It occurs in veins in primitive country, along with ores of silver, lead, zinc, copper, and iron; also in the shape of balls in pitchstone and limestone, and forming the basis of hornstone porphyry

Geographic

Geographic Situation.

Europe.—In Scotland it occurs sometimes in veins, but most frequently in the form of porphyry, as in the island of Arran, in Perthshire, Argyleshire, Fifeshire, Mid Lothian, and the Zetland islands. It is a frequent mineral on the Continent of Europe, occurring more or less abundantly in the various countries that extend from Scandinavia to the shores of the Black Sea. In Sweden it occurs in veins, also forming the basis of porphyry, as at Dannemora and Garpenberg; at Drammen and other parts in Norway; in the Hartz; Lusatia; in the Saxon Metalliferous Mountains (Erzgebirge), where it occurs in veins associated sometimes with ores of silver, galena, and zinc, and sometimes with grey copper-ore, and frequently in veins of red ironstone; in the same country it occurs in balls in pitchstone, and forming the basis of hornstone porphyry: Bohemia; Silesia; Franconia; Bavaria; Moravia; Salzburg; Austria; Hungary; Transilvania; and France.

Asia.—In the silver-mine of Zmeof in the Altain range, and in many places in the Uralian Mountains.

America.—Mexico.

Observations.

1. *Distinctive Characters.*—*a.* Between splintery hornstone and *compact felspar*. Compact felspar has a foliated fracture conjoined with the splintery, whereas splintery hornstone is simply splintery: compact felspar has a glimmering, inclining to glistening, lustre, whereas splintery hornstone is dull: compact felspar is not so hard as splintery hornstone: and, lastly, compact felspar melts

 without

without addition before the blowpipe, whereas splintery hornstone is infusible.—*b.* Between splintery hornstone and *conchoidal hornstone.* In splintery hornstone, the colours are duller than in conchoidal hornstone, and are always simple; internally, splintery hornstone is dull, whereas conchoidal hornstone is glimmering or glistening; the fracture of splintery hornstone is splintery, but that of conchoidal hornstone, conchoidal; and splintery hornstone is softer, and more difficultly frangible, than conchoidal hornstone.

2. It passes into compact felspar and claystone; also into quartz, and common jasper; and into calcedony, flint, and flinty-slate.

3. It appears to contain more silica than compact felspar, but less than quartz; hence it is harder than compact felspar, but softer than quartz.

4. Some of the varieties of the petrosilex of Dolomieu, Lelievre, Brongniart, and Hauy, appear to be splintery hornstone; others seem to be compact felspar.

5. The Paliopetre and Neopetre of Saussure, appear to include both the splintery hornstone and common flinty-slate of Werner.

6. We are still without an accurate analysis of this mineral.

7. The name Hornstone was given to this substance by reason of its resemblance to horn, in colour, fracture, and transparency.

Second Subspecies.

Conchoidal Hornstone.

Muschlicher Hornstein, *Werner.*

Muschlicher Hornstein *Wern.* Pabst. b. i. s. 250.—Petrosilex æquabilis, *Wall.* t. i. p. 281.—Le Hornstein conchoide, *Broch.* t. i. p. 250.—Muschlicher Hornstein, *Reuss,* b. ii. s. 328. *Id. Lud.* b. i. s. 78. *Id. Suck.* 1. th. s. 360. *Id. Bert.* s. 236. *Id. Mohs,* b. i. s. 255. *Id. Hab.* s. 14. *Id. Steffens,* b. i. s. 169. *Id. Hoff.* b. ii. s. 69. *Id. Lenz,* b. i. s. 368. *Id. Oken.* b. i. s. 360

External Characters.

Its principal colour is grey, of which it exhibits the following varieties, yellowish-grey, greenish-grey, and pearl-grey; from yellowish-grey it passes into isabella yellow, yellowish-white, and greyish-white; from pearl-grey into flesh-red and cherry-red; and from greenish-grey into mountain-green. The colours are almost always light, and sometimes they occur in spotted, clouded, and striped delineations.

It occurs most frequently massive, often also in globular forms, and very seldom in the following supposititious crystals:

1. Flat double three-sided pyramid.
2. Acute double six-sided pyramid.
3. Six-sided prism, acuminated with three planes.
4. Perfect six-sided prism.

These figures originate from calcareous spar.

Internally it is glimmering, sometimes approaching to glistening, and the lustre is vitreous.

The

The fracture is more or less perfect and flat conchoidal.

The fragments are angular, and pretty sharp-edged.

It never occurs in distinct concretions.

It is translucent, but in a lower degree than splintery hornstone.

It is hard; it is harder than splintery hornstone, but no so hard as quartz.

It is rather difficultly frangible.

Rather heavy.

Specific gravity 2.572, 2.580, 2.601, *Hoff.*

Geognostic Situation.

It occurs in metalliferous veins and agate veins; also, in imbedded portions, in pitchstone porphyry, and in striped jasper. The metalliferous veins contain, besides the hornstone, sometimes ores of silver, of lead, or of cobalt, but never of red ironstone, the ore which frequently accompanies splintery hornstone. In agate veins it is associated with calcedony, &c. Haberle says, that it sometimes forms the basis of a porphyry which constitutes whole mountains *; and further, that it passes into pitchstone, and occurs in beds and kidneys in claystone.

Geographic Situation.

It is found along with claystone in the Pentland Hills near Edinburgh; also in Saxony and Bohemia.

Observations.

It is nearly allied to striped felspar.

Third

* Hoffman says that it never forms whole beds or mountain masses.

Third Subspecies.

Woodstone.

Holzstein, *Werner.*

Holzstein, *Wid.* p. 329.—Woodstone, *Kirw.* vol. i. p. 315.—Le bois petrifie ou le Holzstein, *Broch.* t. i. p. 259—Holzstein, *Reuss,* b. ii. th. 1. s. 322. *Id. Lud.* b. 1. s. 78. *Id. Bert.* s. 236. *Id. Mohs,* b. i. s. 256. *Id. Hab.* s. 14.—Quartz-agathe xyloide, *Hauy, Tabl.* p. 28. *Id. Steffens,* b. i. s. 171. *Id. Hoff.* b. ii. s. 72. *Id. Lenz,* b. i. s. 370. *Id. Oken.* b. i. s. 300.

External Characters.

Its most common colour is ash-grey, from which it passes into a greyish-black, and into greyish-white; further into yellowish-grey, sometimes into smoke-grey and pearl-grey, flesh-red, blood-red, and brownish-red The yellowish-grey passes into wood-brown and hair-brown, and ochre-yellow. It occurs rarely greenish-grey, and mountain-green.

In general, several colours occur together, and these are arranged in irregular clouded and striped delineations.

It occurs in rolled pieces, and in the shape of trunks, branches, and roots.

Its external surface is uneven and rough.

Internally it is sometimes dull, sometimes glimmering and glistening, according as it is more or less of the nature of the two preceding subspecies.

The cross fracture is imperfect conchoidal; the longitudinal fracture is splintery and fibrous.

The

The fragments are angular, and rather sharp-edged sometimes splintery.

It is generally translucent on the edges; sometimes feebly translucent.

It is hard in a low degree.

It is rather difficultly frangible.

Rather heavy.

Specific gravity, 2.561, 2.624, 2 636, *Hoff.*

Geognostic Situation.

It is found imbedded in sandy loam in alluvial soil; and it is said also in a kind of sandstone-conglomerate and claystone.

Geographic Situation.

Europe.—It occurs at Loch Neagh in Ireland: at Chemnitz and Hilbersdorf in Upper Saxony In the year 1752, the whole under part of the trunk of a tree with branches and roots, in the state of woodstone, was found near Chemnitz. In the Electoral cabinet at Dresden there is a specimen of woodstone from Chemnitz: it is a portion of the trunk of a tree, and measures five feet in length, and as many in diameter. In Bohemia; Franconia; Silesia; Swabia; Bavaria; Austria; Hungary; Transilvania; and France.

Asia.—Kamtschatka; where whole trees and branches are found in the state of woodstone: also near Irkuzk, and Catharinenburg.

Use.

It receives a good polish, and hence is sometimes cut as an ornamental stone.

Observations.

Observations.

1. As woodstone exhibits characters different from all other minerals, it is very properly arranged in the system as a distinct substance.

2. We must be careful not to confound together all the varieties of petrified wood that occur in nature; for wood is sometimes petrified with hornstone, forming Woodstone; sometimes with opal, forming Wood-opal; at other times with quartz or calcareous earth.

4. Flinty-Slate.

This species is divided into two subspecies, viz. Common Flinty-Slate, and Lydian-Stone.

First Subspecies.

Common Flinty-Slate.

Gemeiner Kieselschiefer, *Werner.*

Id. Wid. s. 380.—Siliceous Schistus, *Kirw.* vol. i. p. 306.—Kieselschiefer, *Estner,* b. ii. s. 343. *Id. Emm.* b. i. s. 178.—Schisto silicea, *Nap.* p. 244.—Schiste silicieux commun, *Broch.* t. i. p. 283.—Gemeiner Kieselschiefer, *Reuss,* b. 2. s. 332. *Id. Lud.* b. i. s. 84. *Id. Mohs,* b. i. s. 259. *Id. Bert.* s. 168. *Id. Suck.* 1. th. s. 361.—Jaspe schisteux, *Brong.* t. i. p. 327.—Gemeiner Kieselschiefer, *Steffens,* b. i. s. 175. *Id. Hoff.* b. ii. s. 75. *Id. Lenz,* b. i. s. 373. *Id. Oken,* b. i. s. 297.

External Characters.

Its principal colour is grey, and most frequently ash-grey. It passes on the one side into smoke-grey and greyish-

greyish-black, on the other into pearl-grey, from which it passes into flesh-red, and into a colour intermediate between brownish-red and cherry-red.

The colours sometimes occur in flamed and striped, also in spotted and clouded delineations.

It is often traversed by quartz veins.

It occurs massive, in mountain masses, and frequently in rolled pieces, and in blunt-edged pieces.

Internally it is faintly glimmering, almost dull.

The fracture in the great is slaty, and in the small splintery.

The fragments are angular, and more or less sharp-edged, sometimes tabular.

It sometimes occurs in lamellar concretions.

It is more or less translucent, and passes into translucent on the edges.

It is hard.

Uncommonly difficultly frangible.

Rather heavy.

Specific gravity, 2.613, 2.628, 2.644, *Hoff.* 2.641, *Kirwan.*

Chemical Characters.

It is to be regretted that this interesting substance has not been hitherto chemically examined.

Geognostic Situation.

M. Verina observed it in an unconformable and overlying position over gneiss and mica-slate, in Sáxony. It also occurs in beds in transition and flœtz country.

Geographic

Geographic Situation.

It occurs in differeut parts of the great tract of transition mountains which extends from St Abb's Head to New Galloway; also in the Pentland and Moorfoot Hills near Edinburgh.

It is also found in Norway, Saxony, Bohemia, Silesia, France, and other countries.

Observations.

1. It is distinguished from *Splintery Hornstone*, with which it has been confounded, by its colours being in general darker, its glimmering internal lustre, its slaty fracture, its lamellar concretions, and its geognostic relations. Colour, lustre, translucency, and more difficult frangibility, distinguish it from *Lydian-stone*.

2. In early writings it is named Horn-Slate, (Hornschiefer); under which denomination mineralogists included a variety of slaty rocks, as Porphyry-slate, and Greenstone. Werner first accurately described it, and gave it its present name and place in the system.

Second

Second Subspecies.

Lydian-Stone

Lidischerstein, *Werner*.

Lapis Lydius, *Wall.* t. 1. p. 353.—L. Stein, *Wid.* p. 360.—Basanite, *Kirw.* vol. i. p. 307.—Lidischerstein, *Estner*, b. ii. s. 346. *Id. Emm.* b. i. s. 181.—Schisto silicea, *Nap.* p. 244. —Lydienne, *Lam.* t. ii. p. 384.—La pierre de Lydie, *Broch.* t. i. p. 286.—Lydischerstein, *Reuss*, b. ii. s. 337. *Id. Lud.* b. i. s. 85. *Id. Mohs*, b. i. s. 262. *Id. Bert.* s. 168. *Id. Suck.* 1. th. s. 363.—Jaspisartiger Kieselschiefer, *Hab.* s. 13.—Jaspe schisteux, *Brong.* t. i. p. 328.—Lydischerstein, *Steffens*, b. i. s. 176. *Id. Hoff.* b. ii. s. 79.—Lydit, *Lenz*, b. i. s. 374. *Id. Oken.* b. i. s. 297.

External Characters.

Its colour is greyish-black, which passes into velvet-black.

It occurs massive, and also in trapezoidal-shaped rolled pieces, which have smooth and glistening surfaces.

It is, like the preceding subspecies, traversed by quartz veins.

Internally it is glimmering.

The fracture is generally even, and approaches sometimes to flat conchoidal.

The fragments are angular, more or less sharp-edged, and sometimes approach to the cubical shape.

It is opaque.

It is hard, but not in so high a degree as flint.

Rather difficultly frangible.

Rather heavy.

Specific gravity, 2.596, *Kirwan.* 2.629, *Karstein.* 2.585, *Hoff.*

Geognostic

Geognostic Situation.

It occurs very frequently along with common flinty-slate in beds in primitive clay-slate; but it has not been found in any of the older primitive rocks. It occurs in masses of various sizes, imbedded in grey-wacke, and in beds that alternate with strata of that rock. A rock very nearly allied to it occurs in beds in the oldest coal formation, viz. that associated with the old red sandstone, and in some newer coal formations.

Geographic Situation.

It is found near Prague and Carlsbad in Bohemia; at Hainchen near Freyberg in Saxony; in the Hartz; and in the Moorfoot and Pentland Hills, near Edinburgh.

Use.

This mineral is sometimes used as a touchstone, for ascertaining the purity of gold and silver. When we wish to determine the relative purity of different kinds of gold and silver alloys, we draw the alloy across the surface of the stone, and compare the colour of its trace with that of the pure metals, or of known compounds of these metals, and we thus obtain by simple ocular inspection a pretty correct knowledge of the purity of the alloy.

A good touchstone should be harder than the metals or metallic compounds to be examined: if softer, the powder of the stone mixes with the trace of the metal, and obscures it. It must also possess a certain degree of roughness on its surface, in order that the metal may leave a sufficiently distinct trace or streak; it must not, how-

 ever,

ever, be too rough, otherwise the particles of the metal will be hid amongst inequalities, and no distinct or continuous trace will be formed.

Lastly, a good touchstone must have a black colour, as this tint shews the colour of the streak better than any other.

Those varieties of Lydian-Stone which are neither too hard nor too soft, and which have a kind of velvety feel, and are not traversed by quartz veins, are those which are preferred for touchstones.

They are cut into tables by means of pumice; then ground with sandstone, and, lastly, rubbed with charcoal-powder or ivory-black.

Observations.

1. Compact varieties of Clay-slate and of Basalt are sometimes used as touchstones.

2. According to Humboldt, it contains a small portion of carbon.

3. It was first noticed in Lydia, whence its name. It is described by Pliny and Theophrastes under its present name.

5. Flint.

5. Flint.

Feurstein, *Werner*.

Silex igniarius, *Wall.* t. 0. p. 275.—Feurstein, *Wid.* p. 308.—Flint, *Kirwan,* vol. i. p. 301.—Feurstein, *Estner,* b. ii. s. 360. *Id. Emm.* b. i. s. 143.—Pietra focacia, *Nap.* p. 180.—Silex ou Pierre à fusil, *Lam.* t. i. p. 137. *Id. Broch.* t. i. p. 268. —Quartz-agathe pyromaque, *Hauy,* t. ii. p. 427.—Feurstein, *Reuss,* b. ii. s. 295. *Id. Lud.* b. i. s. 79. *Id. Suck.* 1. th. s. 343. *Id. Bert.* s. 260. *Id. Mohs,* b. i. s. 264. *Id. Hab.* s. 9.—Silex pyromaque, *Brong.* t. i. p. 313.—Quartz-agathe pyromaque, *Lucas,* p. 83. *Id. Brard,* p. 96.—Black Flint, *Kid,* vol. i. p. 211.—Feurstein, *Steffens,* b. i. s. 163. *Id. Hoff.* b. ii. s. 83. *Id. Lenz,* b. i. s. 377. *Id. Oken.* b. i. s. 265.

External Characters.

Its most common colour is grey, of which the following varieties occur: ash-grey, yellowish-grey, and smoke-grey. From smoke-grey it passes on the one side through ash-grey, into greyish-black; on the other, into yellowish-grey, and into a colour intermediate between ochre and wax yellow; further, into yellowish-brown, reddish-brown, and into a colour intermediate between blood-red and brownish-red.

It sometimes presents zoned, striped and flamed coloured delineations.

Besides massive, in regular plates, in angular grains and pieces; it occurs also in globular and elliptical rolled pieces, in the form of sand, and tuberose and perforated.

It sometimes, although rarely, occurs in supposititious crystals. These are:

1. Acute double six-sided pyramid.
2. Flat single or double three-sided pyramid.
3. Six-sided prism, very flatly acuminated by three planes, which are set on the alternate lateral planes.
4. Table.

These crystals are internally hollow: the pyramidal and prismatic forms originate from calcareous spar; the tabular from heavy-spar.

It occurs in extraneous external shapes, viz. in the form of echinites, corallites, madreporites, fungites, belemnites, mytilites, &c.: of these, the echinites are the most frequent, and the mytilites the rarest.

The external surface of the angular pieces is smooth and glistening, that of the other forms is sometimes rough, sometimes uneven.

Internally its lustre is glimmering.

The fracture is perfect and large conchoidal.

The fragments are angular, and sometimes incline to very sharp edged.

It occurs generally compact, sometimes in lamellar distinct concretions, which are either straight or concentrically curved.

It is translucent; the blackish varieties are seldom more than translucent on the edges.

It is hard; rather harder than quartz.

It is easily frangible; and

Rather heavy.

Specific gravity, 2.594, *Blumenbach.* 2.581, *Geller.* 2.581, *Brisson.* 2.594, 2.592, *Hoffman.* 2.580, 2.630, *Kirwan.*

Chemical

Chemical Characters.

Before the blow-pipe, it is infusible without addition.

Constituent Parts.

Silica,	98.0	98.
Lime,	0.50	0.50
Alumina,	0.25	0.25
Oxide of Iron,	0.25	0.25
Loss,	1.00	1.
	100	100
	Klaproth, Beit. b. i. s. 46.	*Vauquelin*, Journ. de Mines, n. xxxiii. p. 702.

Physical Characters.

When two pieces of flint are rubbed together in the dark, they phosphorise very much, and emit a peculiar smell.

Geognostic Situation.

1. It occurs in primitive, transition, flœtz, and alluvial rocks. In primitive and transition rocks, it occurs in metalliferous and agate veins. The metalliferous veins contain in some instances ores of silver or of cobalt, in others ores of iron; and the agate veins are composed of flint, calcedony, jasper, amethyst and quartz. In flœtz countries, where this mineral occurs in greatest abundance, it is found in sandstone, limestone, chalk, and amygdaloid. The sandstone is that variety known under the name of *Puddingstone:* the limestone is one of the newer flœtz rocks, in which the flint appears in beds, imbedded masses, and veins: in chalk it occurs

in

in great abundance in beds, also imbedded in angular and tuberose-shaped masses, and in various extraneous shapes: and in amygdaloid it forms one of the constituent parts of agate, or occurs in veins. In alluvial country, it appears in the form of rolled masses, or gravel, and sometimes as coarse sand.

2. The imbedded masses of flint are frequently hollow, and the walls of the cavities are lined with crystals of quartz, and sometimes with crystals of sulphur.

3. The beds of flint contained in chalk have been deposited in the same manner as other beds. The imbedded tuberose, and other shaped masses, according to Werner, have been formed by infiltration: he conjectures, that during the deposition of chalk, air was evolved, which, in endeavouring to escape, formed irregular cavities, that were afterwards filled up, by infiltration, with flint *.

4. It is often covered with a whitish crust, which is usually produced by weathering; but in other instances, appears to be an original formation.

Geographic Situation.

Europe.—In Scotland it occurs imbedded in flœtz limestone in the island of Mull, and near Kirkcaldy in Fifeshire; and in veins and agates in primitive, transition, and flœtz rocks in various parts of the country. In England, it abounds in the form of gravel, also contained in chalk, in flœtz limestone, and in veins that traverse both transition and flœtz rocks.

In Ireland it occurs in considerable quantity in flœtz limestone.

On

* This ingenious idea of Werner may not be altogether correct, as facts afterwards to be stated, render it probable that tuberose flint is of cotemporaneous formation with the chalk.

On the Continent of Europe it is not unfrequent: thus it occurs imbedded in chalk in the islands of Rugen and Zeeland; in flœtz limestone in Swabia, Bavaria, Saxony, Prussia, Franconia, Austria, Galicia, France, Spain, and Swisserland.

Asia—In the Uralian mountains it occurs in beds in flœtz limestone; also in veins that traverse both primitive and flœtz rocks. It has been found on the shores of the lake Baikal; and on the banks of the river Tura, also in Siberia; and in different parts of China.

America.—It occurs in North America, either imbedded in rocks, or in rolled pieces.

Uses.

The principal use of this mineral is for gun-flints, for which purpose it is excellently fitted, on account of its hardness, the abundance of sparks it affords with steel, and the sharp fragments it gives in breaking *. The most celebrated manufactures of gun-flints are those in England, Muesnes near Berry in France, in Gallicia, and of Avio in the Tyrol. The operation of making them is so simple and easy, that a good workman will make 1500 flints in a day. The whole art consists in striking the stone repeatedly with a kind of mallet, and breaking off at each stroke a fragment, sharp at one end, and thicker at the other. These fragments are afterwards shaped at pleasure, by laying the line at which it is wished they should break, upon a sharp iron instrument, and then giving it repeatedly smart blows with a mallet. During the whole operation, the workman holds the stone in his left hand,

 or

* Flint was first used as for muskets in the year 1670.

or merely supports it on his knee. All the varieties of flint are not equally well fitted for gun-flints: the best are the yellowish-grey; the dark smoke and ash grey varieties are also used, but they are neither so easily split, nor do they afford such thin fragments as the other, and, owing to their greater hardness, they wear the lock sooner. In Prussia, an attempt was made to substitute a kind of porcelain for flint, and such flints were for some time used by the Prussian soldiers. In ancient times, flint was fashioned into cutting instruments; and it is conjectured that the stone knives used by the Hebrews for circumcision were of this mineral; and hence probably also originated the word Silex, which is derived from *scindere*. It also forms a principal ingredient in that species of pottery named *Flint-ware*; it is used as a mill-stone, particularly in smalt works; sometimes it is employed as a building-stone; and by chemists for mortars.

Observations.

1. Flint is distinguished from *Common Calcedony*, by its colour suite, its glimmering lustre, its perfect conchoidal fracture, its inferior translucency, and its inferior hardness.

2. It passes into Hornstone, Carnelian, Calcedony, and even into a kind of Flinty-slate.

3. It occurs frequently in extraneous external shapes, a character which distinguishes it from *Hornstone*.

4. Hacquet has endeavoured to shew that it originates from chalk, and is daily forming.

5. Flint, when dug out of its repository, is very generally enveloped in a thin white opaque crust: if this crust be removed, and the flint exposed to the influence of the weather, it will, in the course of time, become opaque, and of a whitish colour.

Calcedony.

Calcedony.

Kalzedon, *Werner*.

THIS species is divided into four subspecies, viz. Common Calcedony, Chrysoprase, Plasma, and Carnelian.

First Subspecies.

Common Calcedony.

Gemeiner Kalzedon, *Werner*.

Achates chalcedonius, *Wall.* t. 1. p. 287.—Calcedoine, *R. de L* t. 2. p. 145.—Gemeiner Chalzedon, *Wid.* p. 317.—Common Chalcedony, *Kirw.* vol. i. p. 298.—Chalcedon, *Estner,* b. ii. s. 368. *Id. Emm.* b. i. s. 151.—Calcedonia, *Nap.* p. 183.—La Calcedoine, *Lam.* t. ii. p. 142. *Id. Broch.* p. 268.—Quartz-agathe calcedoine, *Hauy,* t. ii. p. 425.—Gemeiner Chalcedon, *Reuss,* b. ii. p. 271. *Id. Lud. Suck. Bert. Mohs,* b. i. s. 273. *Id. Karsten,* Tabell. s. 24. *Id. Leonhard,* Tabell. s. 10.—Quartz-agathe calcedoine, *Lucas,* p. 33.—Silex calcedoine, *Brong.* t. i. p. 298.—Quartz-agathe calcedoine, *Brard.* p. 96.—Calcedony, *Kid,* vol. i. p. 217.—Calcedoine, *Hauy,* Tabl. p. 26.—Gemeiner Kalcedon, *Steffens,* b. i. s. 153. *Id. Hoff.* b. ii. s. 108 —*Id Lenz,* b. i. s. 385.—Chalcedon, *Oken.* b. i. s. 266.

External Characters.

Its most common colour is grey, of which the following varieties occur: smoke-grey, bluish-grey, pearl-grey, greenish-

greenish-grey, and yellowish-grey. The bluish-grey passes into milk-white *, and smalt-blue; the pearl-grey, into pale violet-blue and plum-blue; the greenish-grey, into a colour which is intermediate between grass and apple green; the yellowish-grey passes into honey-yellow †, wax-yellow, and ochre-yellow; from this into yellowish-brown, blackish-brown and brownish-black ‡.

The two last-mentioned colours are very dark, and when held between the eye and the light appear blood-red.

The colours occur in clouded, striped, dendritic and moss-like delineations.

The bluish-grey varieties, in concentric lamellar concretions, when cut across into thin tables, and held between the eye and the light, exhibit an iridescent appearance, and hence have been named *rainbow calcedony*. When cut parallel to the concretions, they exhibit a clouded delineation.

It occurs massive, in blunt-edged pieces, rolled pieces, with smooth surfaces, plates, crusts, balls, (which sometimes contain water, forming what are called *enhydrites*, and more rarely, as at Irkutsk, mineral oil), reniform, botryoidal, corralloidal, stalactitical, cellular, with impressions, (generally from cubes of fluor-spar), and crystallised.

The

* Leucachates of Pliny.

† Cerachates of Pliny.

‡ The green and blue varieties are the rarest. M. De Dree mentions an *azure-blue* variety, under the name of *sapphirine*, and which is much prized on account of its beauty and rarity. It is found in Transylvania, and at Nertschinski in Siberia.

The following are its crystallizations:

Supposititious Crystals.

1. Single three-sided pyramid.
2. Double three-sided pyramid.
3. Six-sided pyramid.
4. Six-sided pyramid, sometimes acuminated with three planes.
5. Rhomb.

True Crystals

1. Cube, in which the sides are either drusy or granulated, with a glimmering lustre; or smooth, with a splendent lustre. They occur aggregated in druses.

It occurs also in extraneous external shapes, in the form of ammonites, turbinites, echinites, madreporites, and of petrified wood.

Internally it is dull; the splintery varieties exhibit a faint degree of lustre.

The fracture is even; it sometimes passes into splintery, sometimes into imperfect and flat conchoidal. The reniform varieties exhibit a very delicate fibrous fracture.

The fragments are angular, and pretty sharp edged.

It occurs in lamellar distinct concretions, varying in thickness, and which are sometimes reniformly curved, sometimes globular and concentrically curved; and very rarely occurs in prismatic concretions, which are thin, and composed of coarse, long, angulo-granular concretions.

It is generally semitransparent; but the black and white varieties are only translucent.

It

It is hard; rather harder than flint.

It is brittle.

It is easily frangible.

Rather heavy.

Specific gravity, 2.600 to 2.655, *Kirwan*. 2.615, *Blumenbach*. 2.618, 2.643, *Karsten*. 2.583, 2.586, *Hoff*. 2.664, *Brisson*.

Chemical Characters.

Infusible before the blowpipe without addition.

Constituent Parts.

Silica,	99	
Loss,	I	
	100	*Tromsdorf**.

Geognostic Situation.

It occurs in primitive, transition, flœtz, and alluvial rocks, in balls, kidneys, angular pieces, short and thick beds, and veins. The balls and kidneys occur most frequently in flœtz amygdaloid, and contain, besides the calcedony, also flint, &c. and in their interior exhibit beautiful reniform, botryoidal, stalactitical, and other particular external shapes. The angular pieces are most frequent in flœtz amygdaloid; occur also in transition amygdaloid,

* In the early analyses of Bergman, Gerhard and Lampadius, alumina, in the proportion of from 12 to 16 *per cent* is stated as occurring in calcedony; and Guyton and Bindheim mention lime as one of its constituent parts.

amygdaloid, and in primitive porphyry. The beds occur in primitive porphyry, but more abundantly, in flœtz-trap rocks, as in amygdaloid, basalt, greenstone, &c. The veins are of two kinds: agate veins, which occur in primitive, transition and flœtz rocks; and metalliferous veins, that occur principally in primitive and transition rocks. The agate veins contain, besides the calcedony also flint, hornstone, opal, and amethyst. The metalliferous veins are of three formations: In the first the calcedony is associated with silver and lead ores, and brown-spar as in the Saxon Erzgebirge; in Lower Hungary, and Transylvania: In the second, with brown and black ironstone, sparry ironstone, hornstone, and other fossils, as in the Saxon Erzgebirge; in Voightland; at Huttenberg, in Carinthia, &c.: And in the third, along with ores of copper, as in the Trevascus mine, in Cornwall. The calcedony in these veins exhibits many different external forms, as stalactitical, botryoidal, coralloidal, reniform &c. In alluvial country, it occurs only in rolled pieces.

Geographic Situation.

Europe.—This mineral occurs more or less frequently in the flœtz-trap rocks of Scotland. Thus, it forms contemporaneous veins in greenstone rocks in Fifeshire; and occurs in balls, kidneys, and veins, either singly or along with other minerals forming agate, in the trap rocks of the Pentland Hills, near Edinburgh, those in West Lothian, Mid Lothian, East Lothian, Dumfriesshire, Lanarkshire, Dunbartonshire, Stirlingshire, Perthshire, Angushire, &c. The trap rocks of Mull, Rume, Canna Eigg and Skye, afford calcedony; and small portions of it occur in the trap rocks of Zetland. The most singular varieties of English calcedony are those found in

Trevascus

Trevascus mine, in Cornwall. It abounds in the amygda loidal rocks of the island of Iceland and the Faroe islands It is a rare mineral in the Scandinavian peninsula; and occurs more or less frequently in the Hartz; in the electorate of Saxony; Silesia; Bohemia; Franconia; Suabia; Gallicia; France; Switzerland; Italy; Austria; Hungary; and Transilvania.

Asia.—It occurs in rolled pieces in the island of Ceylon; in the trap rocks of Dauria; in Calmuck Tartary; in the Altain and Uralian mountains; and on the shores of the sea of Ochotsk.

Africa.—It occurs in rolled pieces on the banks of the river Nile. The ancients are said to have procured their finest calcedonies from the mountains in the neighbourhood of Thebes.

America—It occurs in Greenland; in different parts of the United States, in porphyry and amygdaloid; in Mexico; and at Panama, in New Granada, in South America.

Uses.

As it is hard, susceptible of a fine polish, and exhibits beautiful colours, and considerable transparency, it is employed as an article of jewellery. It is cut into ring and seal stones, necklaces, ear-pendents, small vases, cups, and snuff-boxes. The finer varieties, particularly those named Onyx, were much prized by the ancients, and were by them cut into cameas.

The Camea is a kind of engraving in relief, in which the figure is of a different colour from the ground. When the colours are good, and distinctly separated from each other, and the layers equal and parallel, the onyx is much

much prized. Many fine engraved cameas of this kind are preserved in collections. The National Museum in Paris, and the magnificent collection of M. De Dree, also in Paris, are rich in cameas.

The concentrically striped onyxes, which are very rare, were much prized by the ancients and they cut upon them very beautiful figures in demi-relief. The *vasa murrhina* are said to be of this kind. One of the most beautiful works cut in this variety of calcedony, is the celebrated *Mantuan Vase*, which was seized by the Germans at the storming of Mantua, and ever since has been preserved in the Ducal collection in Brunswick. Several beautiful plates of onyx are preserved in the Electoral cabinet in Dresden: there is one, $\frac{1}{4}$th foot broad and $\frac{1}{4}$th foot long, valued at 44,000 dollars.

The dendritic calcedonies, or mocha-stones, are much prized as ornamental stones. The arborisations are black, red, brown, or green. The black are the most common, and most distinct: the red, on the contrary, are rarer, and are less distinct, and are named *corallines*, from the resemblance of the dendritic delineations to coral. These arborisations appear in some cases to be owing to iron, in others to manganese, and some are conjectured to be true vegetables. This latter opinion is maintained by Dutens *, Von Moll, Daubenton, and lately by Lenz, Blumenbach, and Dr Macculloch. Dutens says, that if the plants contained in calcedony be extracted, and the fragments thrown on burning charcoal, a bituminous smell is exhaled; and Von Moll maintains, that calcedony sometimes contains brown and green moss.

Lenz

* Vid. Traite des Pierres Precieuses, p. 70, 71.

Lenz affirms, that the calcedony found in the amygdaloid of Deuxponts contains musci of different kinds, such as lichen rangiferinus, confervæ, byssi, and brya. And Blumenbach says, in a letter to Baron Von Moll, that though he had hitherto disbelieved the occurrence of vegetable bodies in the dendritic variety of calcedony named *mocha-stone*, he must now admit that it does sometimes contain plants, apparently of the nature of conferva. He observed these in specimens from Iceland and Catherinenburg. The same celebrated naturalist found, in the interior of an agate, the fructification of an unknown plant, somewhat resembling the *sparganium erectum*. Dr Macculloch informs me, that, after examining several hundred specimens of mocha-stone, he is of opinion that they contain cryptogamous plants.

Observations.

1. Its dull and even fracture distinguish it from *Flint*; and the same characters distinguish it from *Carnelian*.

2. It passes into Opal and Flint; probably also into Siliceous Sinter and Hyalite.

3. It was first accurately described by Werner.

4. The name of this subspecies is by some said to be derived from a town in Asia Minor, named *Chalcedon*, where it was collected by the ancients.

5. The dendritic variety is supposed to have been originally brought from Arabia, by the way of Mocha; and hence the name *Mocha-stone* given to it by jewellers.

6. The following are the names given to the different varieties of calcedony by antiquaries and collectors.

a. Onyx, where white or grey stripes alternate with brown.

b. Chalcedonyx,

b. Chalcedonyx, in which we observe an alternation of brown, black, white and grey stripes.

c. Cáma-huia, or *Gemma-huia* ; a name for onyx, and appears to be derived from the words gemma and onychia, or onychina.

d. Memphites ; a variety of calcedony with green stripes.

e. Mocha-stone ; dendritic calcedony.

f. St Stephen's stone ; calcedony, with small red points of carnelian or jasper. This variety was formerly in great esteem, as it was believed that the red spots were the blood of St Stephen.

Second Subspecies.

Chrysoprase.

Krisopras, *Werner.*

Achates-prasius, *Wall.* t. i. p. 292.—Chrysoprase, *Rome de L.* t. ii. p. 167.—Krisopras, *Wern.* Cronst. p. 99. *Id. Wid.* p. 356.—Chrysoprasium, *Kirw.* vol. i. p. 284.—Crysopras, *Estner*, b. ii. s. 349. *Id. Emm.* b. i. 174.—Crisoprasio, *Nap.* p. 195.—*Lam.* t. ii. p. 177.—La Chrysoprase, *Broch.* t. i. p. 280.—Quartz-agathe prase, *Hauy*, t. ii. p. 426.—Crysoprase, *Reuss*, b. ii. s. 270. *Id. Lud. Suck. Bert. Mohs*, b. i. s. 304. *Id. Hab. Id. Karsten*, Tabel. *Id. Leonhard.* Tabel. —Silex Chrysoprase, *Brong* t. i. p. 298.—Quartz-agathe prase, *Lucas*, p. 33. *Id. Brard*, p. 96.—Chrysoprase, *Kid*, vol. i. p. 204.—Quartz-agathe prase, *Haüy*, Tabl. p. 26.—Chrysopras, *Hoff.* b. ii. s. 98. *Id. Steffens*, b. i. s. 157. *Id. Lenz*, b. i. s. 395. *Id. Oken.* b. i. s. 272.

External Characters.

Its characteristic colour is apple-green, of all degrees

of intensity; it passes into light grass-green, pistachio-green, and olive-green.

It occurs generally massive, and sometimes in plates.

Internally it is dull; glimmering, when it inclines to opal.

Its characteristic fracture is even; some varieties run into small and fine splintery; others, very rarely, into flat conchoidal.

The fragments are angular, and more or less sharp edged.

It is intermediate between translucent and semi-transparent, but always approaches more to the first.

Rather softer than calcedony and flint.

Is rather difficultly frangible.

Rather heavy.

Specific gravity, 2.600, *De la Metherie.* 2.714, *Karsten.* 2.608, *Hoffmann.* 3.250, *Klaproth.*

Chemical Characters.

Before the blowpipe it loses its colour and transparency, and is infusible without addition.

Constituent Parts.

Silica, - -	96.16
Lime, - -	0.83
Magnesia, - -	0.08
Alumina, - -	0.08
Oxide of iron, -	0.08
Oxide of nickel, -	1.00
	100

Klaproth, Beit, t. ii. p. 133.

Geognostic

Geognostic Situation.

It occurs, in plates and cotemporaneous veins, along with quartz, hornstone, common calcedony, semi-opal, asbestus, indurated talc, lithomarge, green-earth, steatite, and pimelite, in primitive serpentine.

Geographic Situation.

It has hitherto been found only in the Principality of Munsterberg, in Lower Silesia, in the vicinity of the towns of Glassendorf, Grochau, and Kosemitz.

Uses.

This beautiful stone is much prized by jewellers. The colour, when deep and pure, is very agreeable to the eye. It is generally cut into a convex form, or what jewellers call *en cabochon*, and is facetted underneath or at the base. It is frequently cut into ring-stones and, when set, green taffeta is used as a foil; and when used in ornamental dress, it is found to harmonise with diamonds and pearls. It is said that its colour fades when kept long in a warm and dry place, or when much exposed to the air: on this account, it is recommended to keep it in moist cotton. As the colouring substance of this mineral is nickel, it is difficult to explain the change of colour just mentioned. It is not easily cut, because it is apt to fly in pieces; and frequently during the facetting it cracks, and sometimes splinters. It is cut on tin and leaden plates, by means of fine emery moistened with water.

The larger and impurer masses are cut into snuff-boxes, seal-stones, and similar articles. Very beautiful masses of chrysoprase are to be seen in the Cathedral Church in Prague.

Observations.

1. Its distinguishishing characters are its apple-green colour, its even and dull fracture, its semi-transparency, and its hardness. Colour, and inferior hardness, distinguish it from *Common Calcedony*; and the dull even fracture, with greater hardness, are the characters by which it is distinguished from *Opal.* Some observers have confounded it with *Prase*, from which it is readily distinguished by colour, the want of lustre, generally even fracture, greater transparency, hardness, and weight.

2. It passes into Opal and Splintery Hornstone.

3. The name Chrysoprase, (Χρυσοπρασος, *Chrysoprasus*). is derived from the Greek, which signifies a mineral, of which the colour is green, passing into yellow. We are ignorant of the stone which the ancients described under this name. Lehman was the first who applied it to the green stone of Kosemitz. Vid. Histoire de l'Academie Royale des Sciences et Belles Lettres, anne 1755; Berlin, 1757, p. 202. Vid. also Meinecke's Monograph, intituled, "Ueber den Chrysopras und die denselben begleitenden fossilien in Schlesien. Ein mineralogischer Versuch von Joh. Ludw. George Meinecke; Erlangen, 1805."

Third

Third Subspecies.

Plasma.

Plasma, *Werner*.

Id. Emm. b. iii. s. 322. *Id. Broch.* t. i. p. 278. *Id. Reuss,* b. 2. s. 286. *Id. Lud. Suck. Bert. Mohs,* b. i. s. 308. *Id. Karst.* Tabel. *Id. Leonhard.* Tabel. *Id. Hab. Id. Kid,* vol. i. p. 205. *Id. Steffens,* b. i. s. 159. *Id. Lenz,* b. i. s. 395.—Quartz-agathe calcedoine vert obscur, *Lucas,* t. ii. p. 110.—Plasma, *Hoff.* b. ii. s. 103.

External Characters.

Its most common colour is a variety intermediate between grass-green and leek-green, and sometimes approaching to pale mountain-green. It frequently occurs greenish-white and ochre-yellow: the latter colour is in dots; the first in spots, or clouded.

Internally its lustre is glistening, inclining to glimmering.

The fracture is perfect, and rather flat conchoidal.

The fragments are angular, and very sharp edged.

It is translucent, inclining to semi-transparent.

It is hard.

It is brittle.

Rather easily frangible.

Rather heavy.

Specific gravity, 2.553, *Klaproth.* 2.445, *Karsten.*

Chemical Characters.

It is infusible before the blowpipe, but loses its colour.

Constituent Parts.

Silica, - -	96.75
Alumina, -	0.25
Iron, - -	0.50
Loss, - -	2.50
	100

Klap. Beit. b. iv. s. 326.

Geognostic and Geographic Situations.

Its geognostic situation is not known: it is associated with common calcedony; and some mineralogists, as Steffens, conjecture that it occurs in beds. Most of the specimens in collections have been found among the ruins of Rome. It occurs also at Prussa, at the foot of Mount Olympus, in Asia Minor, where it is associated with the green calcedony of Klaproth, which, however, is only a variety of plasma.

Use.

It was considered by the Romans as a gem, and was cut into ornaments; and frequently figures were engraved upon it.

Observations.

1. It is distinguished from *Heliotrope* by colour, inferior lustre, and weight, and also by its greater transparency: it is distinguished from *Chrysoprase* by colour, greater lustre, perfect conchoidal fracture, and greater weight, and its fracture distinguishes it from *Common Calcedony.*

2. It has been known for several centuries in Italy under the name *Plasma,* but was first introduced into the system by Werner.

3. It

3. It is the *Prime d'Emeraud* of some authors; and would appear to have been described by Pliny as a variety of his *smaragdus*.

Fourth Subspecies.

Carnelian.

Karneol, *Werner.*

Achates carneolus, *Wall.* t. i. p. 185.—Cornaline, *R. de L.* t. ii. 146.—Blutrothe Kalzedon, *Wid.* p. 318.—Carnelian, *Kirw.* vol. i. p. 300.—Karniol, *Emm.* b. i. s. 157.—Carniola, *Nap.* p. 185.—Agathe Cornaline, *Lam.* t. ii. p. 147.—La Cornaline, *Broch.* vol. i. p. 272.—Quartz-agathe cornaline, *Hauy,* t. ii. p. 425.—Karneol, *Reuss,* b. ii. s. 282. *Id. Suck. Lud. Bert. Mohs,* b. i. s. 298. *Id. Karst.* Tabel. *Id. Leonhard,* Tabel.—Silex cornaline, *Brong.* t. i. p. 296.—Quartz-agathe cornaline, *Lucas,* p. 33. *Id. Brard,* p. 96. *Id. Hauy,* Tabl. p. 26.—Karneol, *Steffens,* b. i. s. 160. *Id. Hoff.* b. ii. s. 118. *Id. Lenz,* b. i. s. 391. *Id. Oken.* b. i. s. 266.

External Characters.

Its principal colour is blood-red, of all degrees of intensity: the dark varieties sometimes incline to reddish-brown; but the paler varieties pass into flesh-red, and reddish-white, and also into a colour intermediate between ochre and wax yellow. It also occurs sometimes milk-white. It generally has but one colour; sometimes, however, it exhibits concentrically striped delineations, or fortification or red dendritic delineations.

It occurs sometimes in rolled pieces, which appear to have been original balls; sometimes in thin layers in agate;

very seldom kidney-shaped. The surface of the rolled pieces is rough, and reddish-brown.

The fracture is perfect conchoidal; in the reniform varieties it is fibrous.

The lustre is glistening, sometimes passing into shining, and is vitreous.

The fibrous varieties occur in concentric lamellar concretions.

It is generally semi-transparent; seldom translucent.

It is hard, but in a lower degree than common calcedony and flint.

Specific gravity, 2.320, 2.581, *Karsten.* 2.606, 2.624, *Brisson.*

Chemical Characters.

It is infusible without addition.

Constituent Parts.

Silica, - -	94
Alumina, -	3.50
Iron, - -	0.75
	100 *Bindheim.*

Geognostic and Geographic Situations.

It frequently occurs as a constituent part of agate, and in general has the same geognostic situation as common calcedony. The flœtz-trap rocks so abundant in Scotland, often contain carnelian, either alone, or in agate. The most beautiful carnelians, viz. those having an uniform blood-red colour, are found in rolled pieces, and are brought to this country from Arabia, India, Surinam, Siberia, and Sardinia: less beautiful varieties are found in in Bohemia, Saxony, and the Palatinate. The fibrous varieties are found in Hungary.

Uses.

It is cut into seal-stones, ring-stones, bracelets, necklaces, broaches, and crosses; and figures are often engraved on it. Artists distinguish three principal kinds of carnelian: the one named *common carnelian*, varies in colour from white. through yellow to red; the second, named *sarde*, (sardoine), displays on its surface an agreeable and rich reddish-brown colour, but appears of a deep blood-red colour when held between the eye and the light; the third, named *sardonyx* is composed of layers of white and red carnelian. In the most esteemed carnelians, the colours are of a uniform tint throughout the mass, without any undulations, and are free from that muddiness to which the European varieties of this stone are so liable. The most highly prized varieties are the white and red striped, or sardonyx, and the blood-red: the next in estimation are the pale red; and the least valuable are the yellow, white, and brown. As it is a softer stone than common calcedony, it is more easily cut, and splinters much less when cutting and polishing; and hence, independent of colour, &c. it has always been preferred by artists to the common calcedony. The finest varieties of carnelian are named by French artists those of the *old rock* (vieille roche), because they are no longer to be found so perfect in colour and transparency. The finest pieces of common carnelian are brought from Arabia, and from Cambay and Surat, in India. The sarde, which is very rare at present, and bears a much higher price than the common carnelian, is procured from the shores of the Red Sea. Formerly carnelians used to

be

be imported from Japan into Holland, and from thence were carried to Oberstein, in the Department of Saare, in France, in order to be exchanged for the agates of that country, which were exported to China.

The carnelian was much esteemed by the ancients. Many fine antique engraved carnelians are preserved in collections; and these have been described by Count Caylus, De Dree, and others. The sardonyx was cut into cameas, and afforded by far the most beautiful articles of this kind. The finest antique camea at present known, is in the French Imperial Museum at Paris: it is cut in a sardonyx, is of an oval shape, and is eleven inches by nine in breadth: it represents the *Apotheosis of Augustus*.

Observations.

1. It is distinguished from *Common Calcedony* by its glistening lustre, and conchoidal fracture. The milk-white variety, which approaches to common calcedony, may be distinguished from it, by conchoidal fracture, and greater transparency.

2. It bears the same relation to Common Calcedony that Conchoidal Hornstone does to Splintery Hornstone.

3. It passes into Hornstone.

4. It was first accurately described by Werner, and united with the sardonyx of the ancients. Werner divided it into two kinds, Conchoidal and Fibrous.

5. Carnelian is named *sarda* by the ancients, according to some, from the city of Sardis in Lydia, in the vicinity of which this stone was found; according to others, from Sardinia, where it was also found; or according to others, from the Greek word σαρδος, which was given to it by reason of its predominating flesh-red colour; lastly, some

some derive *carnelian* from the Latin words *caro* or *carneus*, which may been given to it, owing to its flesh-red colour.

7. Heliotrope.

Heliotrop, *Werner.*

Jaspis variegata, Heliotropius, *Wall.* t. i. p. 315.—Heliotrop *Wid.* s. 316.—Heliotropium, *Kirw.* vol. i. p. 314. *Id. Estner,* b. i. s. 389. *Id. Emm.* b. i. s. 171.—Eliotropio, *Nap.* p. 193.—Jaspe sanguin,, *Lam.* t. ii. p. 166.—L'Heliotrope, *Broch.* t. i. p. 276.—Quartz-jaspe sanguin, *Hauy,* t. ii. p. 436. —Heliotrop, *Reuss,* b. ii. s. 319. *Id. Lud. Suck. Bert. Mohs,* b. i. s. 309. *Id. Karsten,* Tabel. *Id. Leonhard,* Tabel. *Id. Hab.*—Silex Heliotrope, *Brong.* t. i. p. 297.—Quartz-jaspe sanguin, *Lucas,* p. 37. *Id. Brard,* p. 101.—Bloodstone, *Kid,* vol. i. p. 210.—Quartz-agathe ponctue, *Hauy,* Tabl. p. 27.—Heliotrop, *Steffens,* b. i. s. 162. *Id. Hoff.* b. i. s. 105. *Id. Lenz,* b. i. s. 423. *Id. Oken,* b. i. s. 271.

External Characters.

The principal colour is intermediate between celandine-green and leek-green ; sometimes passes into mountain-green, and even into grass and pistachio green. All these colours are dark. Sometimes it is marked with olive-green spots and stripes. The blood and scarlet red and the ochre-yellow dots and spots, are owing to disseminated jasper.

It occurs massive, in angular pieces and rolled pieces.

The

The internal lustre is glistening, and resinous.

The fracture is large and flat, but sometimes imperfect conchoidal.

The fragments are angular, and very sharp edged.

It is generally translucent on the edges; some varieties are translucent.

It is easily frangible.

It is hard; but softer than calcedony.

It is rather heavy.

Specific gravity, 2.623, *Karsten*. 2.700, *Kirwan*. 2.614, *Hoffmann*. 2.633, *Blumenbach*.

Chemical Characters.

It is infusible before the blowpipe.

Constituent Parts.

Silica,	-	84.00
Alumina,		7.50
Iron,	-	5.00

Tromsdorf

Geognostic Situation.

It is found in rocks belonging to the flœtz-trap formation, and probably occurs in the same kind of repository as calcedony.

Geographic Situation.

The ancients procured this mineral from Ethiopia. At present, the most highly esteemed varieties are brought from Bucharia, Great Tartary, and Siberia. It occurs also in Iceland; and it is said also in Transilvania, Sardinia,

dinia, and Bohemia. In Scotland, a variety of mineral nearly resembling heliotrope, occurs in the island of Rum.

Uses.

This mineral was well known to the ancients, who have left us accurate descriptions of it. Figures were seldom cut upon Heliotrope until after the commencement of the Christian era, when representations of religious subjects, were often engraved upon it. There is a fine engraved stone of this kind in the National Library in Paris, representing the head of *Christ flagelle*, so cut that the red spots are made to represent drops of blood. It is also cut into seals, snuff-boxes, and other ornamental articles. The varieties having the greatest degree of transparency, and most numerous red spots, are the most highly prized: these are found in Bucharia. The Siberian varieties are destitute of red spots.

Observations.

1. It is distinguished from *Common Calcedony* by its colour, fracture, lustre, and transparency.

2. It is Calcedony, intimately combined with Green Earth

3. Its fracture and transparency shew that it is a species intermediate between Calcedony and Jasper.

4. The name *Heliotrope* is of Greek extraction, from ἑλιος, *sun*, and τϱοπειν, *to turn*. According to Pliny, it received this name because it was used for solar observations.

8. Siliceous

8. Siliceous Sinter.

Kieselsinter, *Werner*.

This species is divided into three subspecies, viz. Common Siliceous Sinter, Opaline Siliceous Sinter, and Pearly Siliceous Sinter or Pearl-Sinter.

First Subspecies.

Common Siliceous Sinter.

Gemeiner Kieselsinter, *Karsten*.

Kiesel-sinter, *Klaproth & Karsten*, Beit. b. ii. s. 109. *Id. Reuss*, b. ii. s. 241, & 245.—Kieseltuff, *Mohs*, b. i. s. 245. *Id. Leonhard*, Tabel. s. 8.—Gemeiner Kieselschiefer, *Karst.* Tabel. s. 24. —Quartz-agathe concretionne thermogene, *Hauy*, Tabl. p. 27. —Quartz-hyaline concretionne, *Brong.* t. i. p. 274.—Gemeiner Kieselsinter, *Steffens*, b. i. s. 128. *Id. Lenz*, b. i. s. 360. —Tufftripel, *Oken*, b. i. s. 278.

External Characters.

Its colours are greyish-white, and reddish-white, with light red and hair-brown spots and stripes; also smoke-grey and yellowish-grey.

It occurs massive, stalactitic, coralloidal, fine fruc ticose, fine botryoidal, porous; and contains stems of plants.

Externally it is dull; internally, where it is porous, dull; in other forms, glistening and pearly.

The

The fracture is flat conchoidal, also coarse-grained uneven, and sometimes parallel and promiscuous fibrous.

The fragments are angular, and rather blunt edged.

The conchoidal varieties occur in fine granular concretions; the uneven varieties in thin and curved lamellar concretions.

It is more or less translucent on the edges.

It is semi-hard.

Very brittle.

Light.

Specific gravity, 1.807, *Klaproth.* 1.816, *Karsten.*

Chemical Characters.

It is infusible without addition before the blowpipe.

Constituent Parts.

Silica, - -	98.0
Alumina, -	1.5
Iron, - -	0.5
	100

Klaproth, Beit. b. ii. s. 109.

Geognostic and Geographic Situations.

It occurs abundantly around the hot springs in Iceland. It is deposited from the water of these springs, in which it appears to have been held in a state of solution, partly by the alkali the water contains, partly by its high temperature, which is 212° at the surface, but must be greater in the interior of the earth, where the water appears to be subjected to a considerable degree of compression.

Second

Second Subspecies.

Opaline Siliceous Sinter.

Opalartiger Kieselsinter, *Hausman.*

Opalartiger Kieselsinter, *Webers.* Naturkunde, b. ii. s. 111. *Id. Steffens,* b. i. s. 130.

Its colours are yellowish-white and milk-white, with brownish, blackish, or bluish spots; and on the fracture-surface veined and dendritic delineations.

It is massive.

The fracture is imperfect conchoidal, sometimes passing into even.

The lustre is glistening.

The fragments are angular and sharp edged.

Sometimes it occurs in distinct concretions, which are lamellar or granular.

It is translucent on the edges.

It is semi-hard, brittle.

Easily frangible.

Adheres to the tongue.

Specific gravity, 3.0270, *Hausman* *.

Chemical Characters and Constituent Parts.

The same as in the first subspecies.

Geographic Situation.

It occurs at the Hot Springs in Iceland.

Observations.

It bears a striking resemblance to Opal.

Third

* The specific gravity as given by Hausman appears to be too high.

Third Subspecies.

Pearl-Sinter, or Fiorite.

Fiorite, *Thompson*, Bibl. Britan. t. i. Janv. 1790.—Quartz-hyalin concretionne, *Hauy*, t. ii. p. 416.—Perlsinter, *Mohs*, b. i. s. 247.—Quartz-hyalin concretionne, *Brong*. t. i. p. 274. *Id. Lucas*, p. 32. *Id. Brard*, p. 92. *Id. Hauy*, Tabl. p. 25.—Perlsinter, *Steffens*, b. i. s. 131. *Id. Lenz*, b. i. s. 361. *Id. Oken*, b. i. s. 278.

External Characters.

Its colours are milk-white, yellowish-white, greyish-white; also pearl-grey and yellowish-grey.

It occurs coralloidal, stalactitic, tabular, cylindrical, fructicose, botryoidal, reniform, and small globular.

Externally it is sometimes smooth and shining, with a pearly lustre, sometimes rough and dull: internally it is dull, glistening, or shining, with a lustre intermediate between resinous and pearly.

The fracture is fine grained uneven; also flat conchoidal, and fine splintery.

The fragments are angular, and not particularly sharp edged.

It occurs in thin concentric lamellar distinct concretions, which are curved in the direction of the external surface, and incrust massive pearl-sinter, which is in round granular distinct concretions.

It is translucent, often only translucent on the edges; and sometimes is semi-transparent in thin pieces.

It scratches glass, but is not so hard as quartz.

It is brittle, and easily frangible.

Specific gravity, 1.917, *Santi.*

Chemical Characters.

It is infusible before the blowpipe without addition.

Constituent Parts.

Silica,	-	-	94	
Alumina,	-	-	2	
Lime,	-	-	4	
			100	*Santi.*

Geognostic and Geographic Situations.

It was discovered by the late Dr Thompson on the surface of weathered granite at the foot of the hill of Santa Fiora, in the island of Ischia: it has also been found on volcanic tuff and pumice in the Vicentine.

Observations.

According to Dr Thompson it is a volcanic production. The silica, he supposes, was held in solution by means of soda, aided by the high temperature of the vapours which exhale from the bosom of the earth in volcanic countries by natural apertures, named *fumaroli.*

9. Hyalite.

9. Hyalite.

Hyalith, *Werner.*

Hyalite, *Kirw.* vol. i. p. 296. *Id. Broch.* t. i. p. 272. *Id. Reuss,* b. ii. s. 246.—Quartz concretionné, *Hauy,* Tabl. p. 25.—Hyalith, *Steffens,* b. i. s. 132. *Id. Hoff.* b. ii. s. 131. *Id. Lenz,* b. i. s. 365. *Id. Oken,* b. i. s. 273.

External Characters.

Its colours are yellowish and greyish white; also yellowish-grey and light ash-grey *.

It is generally small reniform and small botryoidal, and superimposed on other minerals.

Externally it is smooth and shining; internally it is shining and splendent; and the lustre vitreous, slightly inclining to resinous.

The fragments are angular and sharp edged.

It is translucent, approaching to semi-transparent.

It is intermediate between hard and semi-hard.

It is very easily frangible.

Specific gravity, 2.476, *Karsten.* 2.140, *Kopp.*

Chemical Characters.

It is infusible before the blowpipe without addition.

* M. Verina informs me, that it occurs of a greenish-white colour, which passes into verdigrise-green. This variety occurs at Chemnitz in Hungary.

Constituent Parts.

Silica, - -	92.0
Water, - -	6.33
Trace of Alumina,	
Loss, - -	1.66
	99.99 *Bucholz.*

Geognostic and Geographic Situations.

It has been hitherto found principally near Frankfort on the Mayne, where it occurs in basalt and basaltic greenstone, and generally in fissures or vesicular cavities in these rocks. It is also mentioned as occurring in Hungary, Silesia, in the Vivarais, and at Vicenza in Italy; but these localities are rather of a dubious nature.

Uses.

It is sometimes cut into ringstones, which externally are not unlike those of Topaz, but are easily distinguished from that mineral, partly by inferior hardness, partly by the delicate rents observable in its interior.

Observations.

1. It is distinguished from *Calcedony* by its colour suite, its small reniform and botryoidal shape, its lustre, conchoidal fracture, transparency, and its inferior hardness: it is more nearly allied to *Opal*, but is distinguished from it by its external aspect, greater transparency and hardness, and also by its geognostic situation.

2. It is nearly allied to Pearl-Sinter.

10. Opal.

10. Opal.

Opal, *Werner.*

This Species is divided into seven Subspecies, viz. Precious Opal, Common Opal, Fire Opal, Mother-of-Pearl Opal, or Cacholong, Semi-Opal, Jasper-Opal, and Wood Opal.

First Subspecies.

Precious Opal.

Edler Opal, *Werner.*

Plinius, l. xxxvii. 6. *Wid.* p. 325.—Opal, *Kirw.* t. i. p. 289.—Edler Opal, *Emm.* b. i. s. 341.—Opalo, *Nap.* p. 197.—Opale, *Lam.* t. ii. p. 154.—L'Opale noble, *Broch.* t. i. p. 341. Quartz-resinite opalin, *Hauy,* t. ii. p. 434.—Edler Opal, *Reuss,* b. ii. s. 249. *Id. Lud. Id. Suck. Id. Bert. Id. Mohs,* b. i. s. 341. *Id. Karst.* Tabel. *Id. Leonhard,* Tabel. *Id. Hab.* Silex opale, *Brong.* t. i. s. 300.—Quartz-resinite opalin, *Lucas,* p. 36. *Id. Brard,* p. 100.—Opal, *Kid,* vol. i. p. 227.—Quartz-resinite opalin, *Hauy,* Tabl. p. 27.—Edler Opal, *Steffens,* b. i. s. 135. *Id. Hoff.* b. ii. s. 136. *Id. Lenz,* b. i. s. 397. *Id. Oken,* b. i. s. 275.

External Characters.

The most common colour of precious opal is milk-white, inclining to blue, which, when held between the eye and the light, appears pale wine-yellow. It seldom occurs yellowish-white. Sometimes it is accidentally coloured brown. It almost always displays a beautiful play of colour. The colours it throws out are blue, green, yellow, and red. Generally several of these colours occur in one piece: those specimens are rarer that exhibit but one colour, or where one colour preponderates over the others. The rarest and most beautiful of these colours is the red *.

It occurs massive, disseminated, in plates, and in strings or small veins.

Internally its lustre is generally splendent, seldom passing into shining, and is vitreous.

The fracture is perfect conchoidal.

The fragments are angular and very sharp edged.

It is translucent, and then it exhibits a red and green play of colours; or it passes from translucent into semi-transparent, when it exhibits a most beautiful yellow colour; or it is semi-transparent, approaching to transparent, when the principal colour is azure-blue.

It is semi-hard in a high degree.

It is brittle.

It is uncommonly easily frangible.

Some varieties adhere more or less to the tongue.

Specific gravity, 2.114, *Blumenbach.* 2.073, *Karsten.* 2.110, *Brisson.*

Chemical Characters.

Before the blowpipe it becomes opaque, and milk-white, but is infusible.

Constituent

* The play of colours is caused by numerous minute rents that traverse this mineral: thin layers of air are contained in them, and these have the property of reflecting the prismatic colours. It is a phenomenon analogous to the coloured rings observed by Newton.

Constituent Parts.

Silica, - -	90
Water, - -	10
	100

Opal of Czscherwenitza, according to *Klaproth.*

Geognostic Situation.

It occurs in small cotemporaneous veins in clay-porphyry, and is generally accompanied with semi-opal. It would appear also to occur in flœtz amygdaloid, and in minute portions in veins in gneiss.

Geographic Situation.

It is found in greatest abundance in clay-porphyry at Czscherwenitza, near Kaschau in Upper Hungary; sparingly in flœtz amygdaloid in the Faroe Islands; and in flœtz-trap rocks in the north of Ireland, at Sandy Brae. Formerly small portions of it were found in the mines near Freyberg in Saxony. De Dree mentions that it occurs also in South America.

Uses.

The only opal mines in the world are those of Czscherwenitza in Hungary, which have been worked for a long time: even so early as towards the end of the fourteenth century, about 300 men were employed. This mineral, on account of its beauty and rarity, is considered by jewellers as a gem; and is worked into ring-stones, necklaces, ear-pendants, and other ornaments. It is cut into a convex form, or *en cabochon*, as this form shews its colours to the greatest advantage: as it is soft, it should not be facetted; but if facets are cut on it, these ought to be very flat. The cutting is done on a leaden wheel, with tripoli and water; and then the opal is rubbed with tin ashes, on a piece of chamois leather, by which operation it receives its perfect lustre. When it is deficient in colour, jewellers are in the practice of setting it in a foil of the desired colour; but if it exhibits a beautiful play of colour, it appears to the greatest advantage when set in a black case. At present, the opal is held in great estimation in all countries, but particularly in Hungary, Moldavia and Wallachia, where it forms the chief ornaments in the dress of the oldest and most wealthy families. It is exported to Turkey, and from thence it is frequently imported into Holland, where it is falsely denominated *oriental opal.* Jewels of opal must be very carefully kept, as this mineral easily scratches, and is very apt to crack on sudden changes of temperature. It was much prized by the ancients. Pliny (the only one of the ancient writers who mentions the opal) describes it as uniting the beauties of the carbuncle, amethyst, and emerald; and the Greeks expressed their admiration of this lovely gem, by naming it *pæderos.* Nonius, a Roman Senator, possessed an opal of extraordinary beauty, valued at £160,000; rather than part with which to Mark Antony, he chose to suffer exile. He fled to Egypt with his opal, where it was supposed he secreted it. It is not, after this, mentioned by any ancient writer; and the only other notice published in regard to it, is a story by a French interpreter Roboly, who pretended that he had discovered it amidst the ruins of Alexandria

It

It does not appear that the opal was ever much used for engraving on, and very few engraved stones of this mineral are preserved in collections. The opal is frequently minutely disseminated through the porphyry; and pieces of this kind, when cut and polished, are worked for ornamental articles, as snuff-boxes, &c.

Observations.

1. The pecular play of colour distinguishes this mineral from all others. In all other characters it nearly agrees with Common Opal, differing principally in its higher degree of lustre and transparency.

2. This is one of the few minerals whose name has remained unaltered from the earliest times; but its origin or derivation is imperfectly known. Some derive it from the Greek word οψ or οπος, which signifies *vision*, because it was supposed to possess the power of strengthening the eye. The precious opal was the only kind known to the ancients.

3. The finer varieties are named *oriental opal.* Tavernier, however, informs us, that no precious opal is found in the East, and that those which are sold as oriental are brought from Hungary.

4. Those varieties of precious opal that adhere to the tongue, are only translucent, and scarcely exhibit any of the play of colour which so remarkably distinguishes the common varieties; but when immersed in water, they become more transparent, and acquire a very beautiful play of colour. This property occurs also in some varieties of common and semi-opal; and the older, and some modern mineralogists, considered these varieties as constituting a particular species, to which was given various

rious names, as *Oculus mundi*, *Hydrophane*, or *Changeable opal* *. They are much prized by collectors on account of their rarity, and the property just mentioned. In order to preserve their beauty, we must be careful never to immerse them but in pure water, and to take them out again as soon as they have acquired their full transparency. If we neglect these precautions, the pores will soon become filled with earthy particles deposited from the water, and the hydrophane will cease to exhibit these curious properties, and will always remain more or less opaque. When these changeable opals are well dried, and immersed in melted wax or spermaceti, they absorb a portion of it, and become transparent, but on cooling become opaque again. For some time these prepared opals were imposed on the public as rare and singular minerals, and sold at a very high price, under the name *Pyrophane*.

5. In the Imperial Cabinet at Vienna, there are two pieces of opal from the mines in Hungary, which deserve to be mentioned here. The one is about five inches long and two and a half in diameter, and exhibits a very rich and splendent play of colours; the other, which is of the size and shape of a hen's egg, is also extremely beautiful.

Second

* These hydrophanes were known to the ancients under the name *Pantarbas*. Neuheuser, in his treatise intituled *Coronæ gemmæ nobilissimæ*, published in the sixteenth century, mentions the hydrophane, under the name *Verkehrstein* or *Wunderstein*.

Second Subspecies.

Common Opal.

Gemeiner Opal, *Werner.*

Id. Wid. p. 325.—Semi-opal, *Kirw.* vol. i. p. 290.—Gemeiner Opal, *Emm.* b. i. s. 251.—Opalo, *Nap.* p. 197.—Girasol & Hydrophane, *Lam.* p. 156.—L'Opal commune, *Broch.* t. i. p. 344.—Gemeiner Opal, *Reuss,* b. ii. s. 253. *Id. Lud. Id. Suck. Id. Bert. Id. Mohs,* b. i. s. 344. *Id. Karsten,* Tabel. *Id. Leonhard,* Tabel.—Silex opale, *Brong.* t. i. p. 300.—Quartz-resinite commun, *Hauy,* Tabl. p. 28.—Gemeiner Opal, *Steffens,* b. i. s. 137. *Id. Lenz,* b. i. s. 400. *Id. Oken,* b. i. s. 275.

External Characters.

The principal colour of Common Opal is milk-white; but it occurs also greyish, yellowish, and greenish white. The milk-white passes into bluish-grey: and the greyish-white into smoke-grey: the yellowish-white into yellowish-grey, wax-yellow, honey-yellow, ochre-yellow, hyacinth-red, and an intermediate colour between flesh-red and brick-red: the greenish-white passes into apple-green, pistachio-green, and mountain-green. It never exhibits more than one colour. The milk-white variety, when held opposite the light, appears wine-yellow *.

It occurs, massive, disseminated, in sharp angular pieces, and very rarely vesicular.

Internally

* These varieties are the Opal-resinite Girasol of French mineralogists.

Internally its lustre is generally splendent, sometimes passing into shining; and is vitreous.

The fracture is perfect conchoidal.

The fragments are angular and sharp edged.

It is most commonly semi-transparent; sometimes it approaches to translucent, but seldom to transparent.

It is semi-hard in a high degree.

It is brittle.

It is uncommonly easily frangible.

It sometimes adheres to the tongue.

Specific gravity, 2.015, *Klaproth*. 2.144, *Kirwan*. 2.064, *Haberle*.

Chemical Characters.

Before the blowpipe it is infusible without addition.

Constituent Parts.

	Opal of Kosemutz.	Of Telkobanya.
Silica, -	98.75	93.50
Alumina, -	0.1	
Oxide of iron,	0.1	1.0
Water, - -		5.0
	98.95	99.50

According to *Klaproth*, t. ii. p. 164. & 169.

Geognostic Situation.

It occurs in primitive and flœtz rocks. In primitive country, it occurs in cotemporaneous veins along with precious opal in clay-porphyry; and in cotemporaneous plates or short beds in serpentine. In flœtz rocks it occurs in amygdaloid along with calcedony, either in vesi-

cular

cular cavities, in cotemporaneous veins, or in short and thick beds. It also occurs in metalliferous veins, along with galena and blende, as in Saxony, Island of Elba, and in Bohemia; or in red ironstone veins in Saxony, but never in considerable quantity. These veins traverse granite, gneiss, mica-slate, clay-slate, porphyry, &c.

Geographic Situation.

It is found in Iceland, the Faroe Islands, North of Ireland; in the Electorate of Saxony, as at Freyberg, Hubertsberg, Eibenstock, Johanngeorgenstadt, and Schneeberg; in Bohemia, as at Bleistadt, Fribus, Heinrichsgrun; Brittany in France; Silesia; Poland; Salzburg; at Florence in Italy; and Telkobanya in Hungary.

Uses.

It is cut for ornamental purposes: thus, the green-coloured Silesian variety is sometimes cut into ring-stones; and the yellow variety, which was formerly named *wax-opal* and *pitch-opal*, is also cut and polished by jewellers.

Observations.

This subspecies is characterised by its peculiar milk-white colour, strong lustre, perfect conchoidal fracture, considerable transparency, and its low degree of hardness and weight.

Third

Third Subspecies.

Fire Opal.

Feur Opal, *Karsten.*

Feur Opal, *Karsten & Klaproth,* Beit. iv. s. 156. *Karsten,* Tabel. p. 26. *Id. Steffens,* b. i. s. 138. *Id. Lenz,* b. i. s. 402. *Id. Oken,* b. i. s. 275.

External Characters.

Its principal colour is hyacinth-red, which passes through honey-yellow into wine-yellow; and upon lighter places shews a carmine-red and apple-green iridescence. In its interior, dendritic delineations are sometimes to be observed.

Internally it is splendent, and the lustre is vitreous.

The fracture is perfect conchoidal.

The fragments are angular and sharp edged, or tabular.

It occurs in distinct concretions, which are partly thick and curved lamellar, partly large and coarse granular.

It is completely transparent.

It is hard.

Uncommonly easily frangible.

Specific gravity, 2.120, *Klaproth.*

Chemical Characters.

When exposed to heat, its colour changes into pale flesh-red; it becomes translucent, and traversed with numerous fissures.

Constituent

Constituent Parts.

Silica, - -	92.00
Water, - -	7.75
Iron, - -	0.25
	100.00 *Klaproth.*

Geognostic and Geographic Situations.

It has hitherto been found only in America, at Zimapan in Mexico, where it was first observed by Sonnenschmid and Humboldt. It occurs in a particular variety of hornstone-porphyry, which contains, besides the fire-opal, also imbedded lavender-blue grains the size of a pea, of a substance not unlike porcelain-jasper. In the middle of each grain of this blue substance, there is a whitish siliceous mineral, from which the blue mineral diverges in all directions in a stellular manner.

Fourth Subspecies.

Mother-of-Pearl Opal, or Cacholong.

Perlmutter Opal, *Karsten.*

Achates opalinus, tenax, fractura inæqualis, Cachalonius, *Wall.* gen. 20. sp. 126. p. 285.—Calcedoine alteree, ou Cacholong, *Rome de Lisle,*—Calcedoine blanche opaque, *De Born.*—Cachelonio, *Nap.*—Silex cacholong, *Brong.* t. i. p. 302.—Cacholong, *Kid,* vol. i. p. 225.—Quartz-agathe cacholong, *Hauy,* Tabl. p. 27.—Perlmutter opal, *Steffens,* b. i. s. 139. *Id. Lenz,* b. i. s. 404.—Kacholong, *Oken,* b. ii. s. 274.

External Characters.

It colours are milk-white, yellowish-white, and greyish-white; sometimes dendritic.

It

It occurs massive, disseminated, in blunt angular pieces, in crusts, and sometimes reniform.

Externally it is dull; internally it alternates from dull to glistening and shining, and is pearly.

The fracture is flat conchoidal, but becomes earthy by the action of the atmosphere.

The fragments are angular, and not particularly sharp edged.

It sometimes occurs in coarse granular distinct concretions.

It is opaque.

It is somewhat harder than common opal.

It is brittle, and easily frangible.

Specific gravity, 2.209, *Karsten.* 2.272 Feroe, *Kopp.*

Chemical Character.

It is infusible before the blowpipe.

Geognostic and Geographic Situations.

It occurs, along with calcedony, in trap rocks in the island of Iceland; in the Feroe islands; also in Greenland; and in Bucharia. At Huttenberg in Carinthia, it occurs along with compact and fibrous brown ironstone. It is also mentioned as a production of the Island of Elba, and of Estremadura in Spain.

Uses.

When cut, it is generally *en cabochon;* but it is seldom used for engraving upon, on account of its brittleness. The Valentine III. in the Royal Library at Paris is engraved upon cacholong. Italian artists sometimes use it for mosaic work.

Observations.

Observations.

1. Some mineralogists consider it as a variety of Calcedony; but it is distinguished from that mineral by lustre, fracture, hardness, and specific gravity.

2. The name *Cacholong*, is by some derived from a supposed river in Bucharia, named Cach, where the mineral is said to have been first found; but as there is no river of that name in Bucharia, other mineralogists have derived the name from *cholong*, that is, a stone, and *cach*, which in the language of that country signifies a pebble.

Fifth Subspecies.

Semi Opal.

Halb-Opal, *Werner*.

Id. Wid. s. 325.—Semi-opal, and several of the Pitchstones of *Kirw.* vol. i. p. 290. 292.—Halb-Opal, *Emm.* b. i. s. 256. *Id. Estner,* b. ii. 429.—Semi-opalo, *Nap.* p. 201.—Pissite, *Lam.* t. ii. p. 160.—La Demi-opal, *Broch.* t. i. p. 347.—Quartz-resinite commun, *Hauy,* t. ii. p. 433.—Halb-Opal, *Reuss,* b. ii. s. 257. *Id. Lud.* b. i. s. 97. *Id. Suck.* 1. th. s. 311. *Id. Bert.* s. 266. *Id. Mohs,* b. i. s. 355. *Id. Hab.* s. 8.—Quartz-resinite commun, *Lucas,* p. 36.—Halb-Opal, *Leonhard,* Tabel. s. 13.—Silex Resinite, *Brong* t. i. p. 303.—Halb-Opal, *Karsten,* Tabel, s. 26.—Semi-opal, *Kid,* vol. i. p. 231—Quartz-resinite Hydrophane, *Hauy,* Tabl. p. 27.—Halbopal, *Steffens,* b. i. s. 141. *Id. Hoff.* b. ii. s. 149. *Id. Lenz,* b. i. s. 406. *Id. Oken.* b. i. s. 276.

External Characters.

Its most common colours are white, grey, and brown. Of white, the varieties are yellowish-white and greyish-

 white

white, seldom milk-white and greenish-white. It passes from ash-grey into greyish-black; from yellowish-grey into wax-yellow, into a colour intermediate between ochre and isabella yellow, into yellowish-brown, hair-brown, liver-brown, chesnut-brown, reddish-brown, and nearly into red; and, lastly, from greenish-grey into leek-green, olive-green, and oil-green.

Sometimes several colours occur together, and these are arranged in spotted, concentric striped, clouded, or flamed delineations; but it is most commonly uniform, or of one colour.

It occurs not only massive and disseminated, but also tuberose, small reniform, and small botryoidal, which approaches to the stalactitic.

Externally it is glistening; internally, generally glistening, sometimes approaching to shining, or passing into glimmering.

The fracture is large and flat conchoidal, but less perfect than common opal; and it sometimes inclines to small conchoidal.

The fragments are angular and very sharp edged.

It is more or less translucent, and sometimes passes to translucent on the edges.

It is semi-hard, approaching to hard.

Rather easily frangible.

It adheres to the tongue.

Rather heavy; approaching to light.

Specific gravity,—Yellowish and greenish-grey from Hungary, 2.000, *Hoff.* Yellowish-white from Steinheim, 2.001, *Hoff.* Blackish-brown from Steinheim, 2.059, *Hoff.* Milk-white from Freyberg, 2.167, *Hoff.* From Moravia, 2.167, *Klap.* 2.077, 2.187, *Karsten.*

Chemical

Chemical Characters.

It is infusible before the blowpipe without addition; but with borax it melts, and without intumescing.

Constituent Parts.

Semi-opal from Neu Wieslitz, between Brünn and Kremsier in Moravia:

Silica, - -	85.00
Alumina, -	3.00
Oxide of Iron, -	1.75
Carbon, - -	5.00
Ammoniacal Water,	8.00
Bituminous Oil, -	0.33
	99.08

Klap. Beit. b. v. s. 31

Geognostic Situation.

It occurs, in angular pieces, beds, and veins, in porphyry and amygdaloid; also in metalliferous (most usually silver) veins that traverse granite and gneiss.

Geographic Situation.

It is found in Greenland, Iceland, Feroe Islands, Scotland, in the Isle of Rum, where it occurs in amygdaloid; Electorate of Saxony, Bohemia, Frankfort on the Mayne, Silesia, Lower Austria, Poland, Hungary, Transylvania, Isle of Elba, Piedmont, and Siberia.

Observations.

1. It is distinguished from *Common Opal* by the muddiness of its colours, its particular external shapes, its in-

ferior lustre and transparency, its less perfect conchoïdal fracture, and its greater hardness and weight.

2. It passes into Opal-Jasper, Calcedony, and Conchoidal Hornstone.

3. It has been arranged with Pitchstone by Dolomieu, Fichtel, and other mineralogists; but it is distinguished from that mineral by its vitreous lustre, greater transparency, inferior specific gravity, its want of distinct concretions, and its infusibility.

Sixth Subspecies.

Jasper-Opal.

Opal-Jaspis, *Werner.*

Jaspe-Opale, *Brochant,* t. ii. p. 498.—Opal-Jaspis, *Reuss,* b. ii. s. 317. *Id. Lud.* b. i. s. 95. *Id. Mohs,* b. i. s. 324. *Id. Leonhard,* Tabel. s. 12.—Jasp-Opal, *Karsten,* Tabel. s. 26.—Jasp-opal, *Steffens,* b. i. s. 143.—Opal Jaspis, *Hoff.* b. ii. s. 177.—Opal Jaspis, *Lenz,* b. i. s. 411.—Jasp-opal, *Oken,* b. i. s. 277.

External Characters.

Its colours are scarlet red, light blood-red, brownish-red, ochre-yellow, isabella-yellow, yellowish-grey, and ash-grey. The isabella-yellow passes into yellowish-white; and the blood-red into reddish brown.

The colour is either uniform, or distributed in spotted, veined, and clouded delineations.

It occurs massive.

Internally its lustre is shining, approaching to splendent, and is intermediate between vitreous and resinous.

The

The fracture is perfect conchoidal, and sometimes rather flat conchöidal.

The fragments are angular, and very sharp edged.

It is opaque, and sometimes feebly translucent on the edges.

It is intermediate between hard and semi-hard.

It is easily frangible.

Rather heavy; approaching to light.

Specific gravity,—

Yellow and red striped from Constantinople,	1.863, *Hoff.*
Red from Lauenhayn, - - -	2.053, *Hoff.*
Brownish-red from Telkobanya, -	2.061, *Hoff.*
Reddish-brown from Telkobanya,	2.072, 2.081, *Hoff.*

Chemical Characters.

It is infusible before the blowpipe.

Constituent Parts.

Silica, - -	43.5
Oxide of Iron, -	47.0
Water, -	7.5
	98.0

Klaproth, Beit. b. ii. s. 164.

Geognostic and Geographic Situations.

It is found in large and small pieces in porphyry, near Telkobanya and Tokay in Hungary; near Constantinople; in the Kolyvanian mountains in Siberia; and in veins in the Saxon Erzgebirge.

Observations.

It used to be arranged as a subspecies of Jasper; but its perfect conchoidal fracture, its high lustre, great brittleness, and inferior weight, sufficiently distinguish it from the different subspecies of *Jasper.*

Seventh Subspecies.

Wood-Opal.

Holz-Opal, *Werner.*

Id. Wid. p. 325.—Ligniform opal, *Kirw.* vol. i. p. 295.—Holz-opal, *Emm.* b. i. s. 260.—Semi-opalo, *Nap.* p. 201.—Xilopale, *Lam.* t. ii. p. 102.—Opal ligniforme, *Broch.* t. i. p. 350. —Holz-Opal, *Reuss,* b. ii. s. 267. *Id. Lud.* b. i. s. 98. *Id. Suck.* 1. th. s. 317. *Id. Bert.* s. 267. *Id. Mohs,* b. i. s. 340. *Id. Leonhard,* Tabel. s. 13. *Id. Karst.* Tabel. s. 26.—Quartz-resinite xyloïde, *Hauy,* Tabl. p. 28.—Holz-Opal, *Steffens,* b. i. s. 144. *Id. Hoff.* b. ii. s. 153. *Id. Lenz,* b. i. s. 413. *Id. Oken,* b. i. s. 276.

External Characters.

It occurs most commonly white, grey, or brown, and sometimes also black. The white varieties are milk-white, yellowish-white, and greyish-white: the greyish-white passes into ash-grey, pearl-grey, smoke-grey, and yellowish-grey; which latter passes into ochre-yellow, yellowish-brown, wood-brown, hair-brown; and the greyish-white passes into greyish-black.

The colour is sometimes simple, sometimes in flamed and striped delineations, which are conformable with the original texture of the wood.

It

It occurs in pieces which have the shape of branches, and stems.

Internally its lustre is shining, which sometimes passes on the one hand into splendent, and on the other into glistening, and even glimmering.

The cross fracture is large and flat conchoidal; the longitudinal fracture is sometimes modified by the remaining fibrous woody texture.

The fragments are sometimes sharp edged, sometimes long splintery.

It is more or less translucent; sometimes only translucent on the edges.

It is semi-hard in a high degree.

It is easily frangible.

It is rather heavy; bordering on light.

Specific gravity, 2.080, 2.100, *Kirwan.* 2.048, 2.059, *Hoff.*; also 2.189, *Hoff.*

Geognostic and Geographic Situations.

It is found in alluvial land at Zastravia in Hungary; and is said to occur in flœtz-trap rocks in Transylvania. It has also been found in the neighbourhood of Fain, near Telkobanya in Upper Hungary. Many years ago the trunk of a tree penetrated with opal was found in Hungary, which was so heavy that eight oxen were required to draw it.

Observations.

1. Its woody texture distinguishes it from the other subspecies of *Opal*: and it is distinguished from *Woodstone* by its lighter colours, higher lustre, perfect conchoidal fracture, greater transparency, and inferior hardness and weight.

2. It is wood penetrated with opal, and is intermediate between Common Opal and Semi-opal.

Use.

It is cut into plates, and then used for snuff-boxes, and other ornamental articles.

11. Menilite.

THIS species is by Hoffmann divided into two sub-species, viz. Brown Menilite, and Grey Menilite.

First Subspecies.

Brown Menilite.

Brauner Menilite, *Hoffmann*.

Leberopal, *Reuss*, b. ii. s. 265.—Menilite, *Lud.* b. ii. p. 141.—Leberopal, *Suck.* 1. th. s. 316.—Knollenstein, *Mohs*, b. i. s. 343.—Leberopal, *Hab.* s. 9.—Menilite, *Leonhard*, Tabel. s. 13. Leberopal, *Karst.* Tabel. s. 26.—Silex Menilite, *Brong.* t. i. p. 312.—Menilite, *Kid*, t. i. p. 232.—Quartz-resinite subluisant brunatre, *Hauy*, Tabl. p. 28.—Menilith, *Steffens*, b. i. s. 145.—Brauner Menilite, *Hoff.* b. ii. s. 156.—Leberopal, *Lenz*, b. i. s. 410.—Kalkopál, or Knollenstein, *Oken*, b. i. s. 276.

External Characters.

Its colour is chesnut-brown, which inclines to liver-brown. On the surface, it has sometimes a bluish colour.

It occurs always tuberose, seldom larger than a fist, often smaller.

The

The external surface is rough and dull; internally it is faintly glistening, and the lustre is intermediate between resinous and vitreous.

The fracture is very flat conchoidal.

The fragments are angular and very sharp edged.

It sometimes has a tendency to lamellar distinct concretions.

It is translucent on the edges.

It is semi-hard in a high degree.

Scratches glass.

It is easily frangible.

It is rather heavy; approaching to light.

Specific gravity, 2.185, *Klaproth.* 2.168, *Brisson.* 2.176, *Haberle.* 2.161, 2.169, *Hoff.*

Chemical Characters.

It is infusible before the blowpipe without addition.

Constituent Parts.

Silica, - - - -	85.5
Alumina, - - -	1.0
Lime, - - - -	0.5
Oxide of Iron, - - -	0.5
Water, and Carbonaceous Matter,	11.0
	98.5

Klaproth, Beit. b. ii. s. 169.

Geognostic and Geographic Situations.

It has hitherto been found only at Menil Montant, near Paris, where it occurs imbedded in adhesive-slate, in the same manner as flint is in chalk. It is worthy of remark,

remark, that the direction of the thin lamellar concretions of the menilite correspond with that of the slaty structure of the adhesive-slate in which it is imbedded. This fact shews that the menilite and slate are of cotemporaneous formation.

Observations.

1. This subspecies is distinguished from the following, or the *Grey Menilite*, by its brown colour, its internal lustre, its more perfect conchoidal fracture, its translucency on the edges, its inferior weight, and geognostic situation.

2. It is nearly allied to Semi-opal; but it is distinguished from it by colour, shape, feebler lustre, inferior translucency, greater weight, and geognostic situation.

3. It was at one time arranged along with Pitchstone, under the name *Blue Pitchstone:* more lately it has been described as a member of the Opal species, under the title *Liver opal*, (Leberopal). Werner first accurately described it, and gave it its present place in the system.

Second Subspecies.

Grey Menilite.

Grauer Menilite, *Hoffmann.*

External Characters.

Its colour is yellowish-grey, which sometimes inclines to wood-brown.

It occurs tuberose, but more compressed than in the brown subspecies, and the external surface is smoother.

Internally

Internally it is glimmering or dull.

The fracture is very flat conchoidal, and is sometimes almost even.

The fragments are angular and sharp edged.

It is very feebly translucent on the edges, and sometimes quite opaque.

It is semi-hard in a high degree.

It is easily frangible; and

Rather heavy, but in a low degree.

Specific gravity, 2.286, 2.375, *Hoff.*

Geognostic and Geographic Situations.

It occurs at Argenteuil near Paris, imbedded in a clayey marl; also in gypsum which alternates with this marl. It has also been found at St Ouen, near Paris; and, according to Hauy, on the Maase.

11. Jasper *.

This species is divided into five subspecies, viz. Egyptian Jasper, Striped Jasper, Porcelain-Jasper, Common Jasper, and Agate-Jasper.

First Subspecies.

Egyptian Jasper.

This subspecies is subdivided into two kinds, viz. Red Egyptian Jasper, and Brown Egyptian Jasper.

First

* Etymologists have not been able to ascertain the origin of the word Jasper. We only know that it is of high antiquity, because it occurs in the Hebrew and Grecian languages. We are also ignorant of the particular stone denominated Jasper by the ancients.

First Kind.

Red Egyptian Jasper.

Rother egyptischer Jaspis, *Werner.*

Rother egyptisher Jaspis, *Hoff.* b. ii. s. 162.—Rother Kugel Jaspis, *Steffens,* b. i. s. 181.—Rother egyptisher Jaspis, *Lenz,* b. i. s. 416.

External Characters.

Its colour is intermediate between flesh-red and blood-red; also ochre-yellow, yellowish-brown, and yellowish-grey. These colours form ring-shaped delineations.

It is found in roundish blunt edged rolled pieces.

The external surface is rough; also uneven and dull.

Internally it is dull, or very faintly glimmering.

The fracture is large, and rather flat conchoidal.

It breaks into angular and sharp edged fragments.

It is very feebly translucent on the edges.

It is hard.

Rather easily frangible; and

Rather heavy, but in a low degree.

Specific gravity, 2.632, *Hoff.*

Geognostic and Geographic Situations.

It is found imbedded in red clay-ironstone in Baden.

Use.

It is used, along with agate, for ornamental purposes.

Observation.

Observation.

It sometimes passes into Flint, but is distinguished from that mineral by its opacity, and inferior hardness.

Second Kind.

Brown Egyptian Jasper.

Brauner egyptischer Jaspis, *Werner.*

Silex Ægyptiacus, *Wall.* t. i. p. 276.—Egyptian Pebble, *Kirw.* vol. i. p. 312.—Egyptisher Jaspis, *Emm.* b. i. s. 234.—Caillou d'Egypte, *La Meth.* t. ii. p. 166.—Le Jaspe Egyptien, *Broch.* t. i. p. 332.—Egyptisher Jaspis, *Reuss,* b. ii. s. 302. *Id. Lud.* b. i. s. 93. *Id. Suck.* 1r th. s. 353. *Id. Bert.* s. 227. *Id. Mohs,* b. i. s. 314. *Id. Leonhard,* Tabel. s. 11.—Jaspe Egyptien, *Brong.* t. i. p. 325.—Egyptischer Jaspis, *Karst.* Tabel. s. 38.—Egyptian Jasper, *Kid,* vol. i. p. 208.—Quartz-agathe opaque, *Hauy,* Tabl. p. 27.—Brauner Kugel Jaspis, *Steffens,* b. i. s. 180.—Brauner egyptisher Jaspis, *Hoff.* b. ii. s. 164. *Id. Lenz,* b. i. s. 414.

External Characters.

Its colours are chesnut-brown, yellowish-brown, isabella-yellow, and yellowish-grey. The yellowish-grey, or isabella-yellow, generally form the interior or centre of the pebble; and the brown colours are disposed in concentric stripes, alternating with black stripes. In the brown colour, there sometimes occur black spots, and similar coloured dendritic delineations.

It occurs in roundish blunt edged pieces, and in spheroidal balls.

The surface is uneven or rough.

Externally

Externally it is glimmering, very seldom feebly glistening; internally it is partly glistening, partly glimmering; but the grey is dull.

The fracture is flat and perfect conchoidal.

The fragments are angular and sharp edged.

It is very feebly translucent on the edges, or almost opaque.

It is as hard as hornstone.

Rather heavy, but in a low degree.

Specific gravity, 2.564, *Brisson.* 2.601, 2.624, *Hoff.*

Chemical Characters.

It is infusible without addition before the blowpipe.

Geognostic Situation.

The geognostic situation of this mineral is still imperfectly known. Cordier informs us, that it is found imbedded in a conglomerate rock, which, in his opinion, extends in great beds throughout Egypt to the deserts of Africa; while Mohs, from the resemblance of its colour-delineations to those of agate, supposes that it has been formed by infiltration, in the manner of agate, and therefore, that Egyptian jasper will be found to occur in amygdaloid. In whatever original situation it occurs, it is well known, from the observations of travellers, to occur loose in the sands of Egypt.

Geographic Situation.

It has been hitherto found only in Egypt.

Uses.

As the colours of this mineral are agreeable to the eye, and beautifully disposed, and as it receives a good polish, it it is prized by jewellers as an ornamental stone, and is cut into snuffboxes, and various ornamental articles.

Observations.

Colour, colour-delineation, external shape, and low degree of lustre, are the most distinguishing characters of this mineral.

Second Subspecies.

Striped Jasper.

Band Jaspis, *Werner.*

Striped Jasper, *Kirw.* vol. i. p. 312.—Band Jaspis, *Emm.* b. i. s. 237.—Jaspe rubane, *Lam.* p. 165.—Le Jaspe rubane, *Broch.* t. i. p. 334.—Band Jaspis, *Reuss,* b. ii. s. 305. *Id. Lud.* b. i. s. 94. *Id. Suck.* 1r th. s. 355. *Id. Bert.* s. 228. *Id. Mohs,* b. i. s. 116. *Id. Hab.* s. 13. *Id. Leonhard,* Tabel. s. 11. —Jaspe Rubanne, *Brong.* t. i. p. 324.—Band Jaspis, *Karsten,* Tabel. s. 38.—Riband Jasper, *Kid,* vol. i. p. 207.—Quartz-jaspe Onyx, *Hauy,* Tabl. p. 28.—Band Jaspis, *Steffens,* b. i. s. 182. *Id. Hoff.* b. ii. 166. *Id. Lenz,* b. i. s. 417. *Id. Oken,* b. i. s. 298.

External Characters.

Its colours are grey, green, yellow, and red, and seldom blue. Of grey, it presents the following varieties, pearl-grey, greenish-grey, and yellowish-grey: Of yellow, cream-yellow, which passes into straw-yellow: Of green, mountain-green, which passes into leek-green and greenish-grey: Of red, cherry-red, brownish-red, and flesh-red; the cherry-red passes into plum-blue.

There

There are always several colours together, and these are arranged in striped and flamed, and sometimes in spotted delineations.

It occurs massive, in whole beds.

Internally it is dull, when an admixture of foreign ingredients does not give it a feeble glimmering lustre.

The fracture is large and flat conchoidal, which approaches sometimes to fine earthy, sometimes to even. In the large it sometimes inclines to slaty, and the laminæ are in the direction of the striped delineations.

The fragments are angular, and pretty sharp edged.

It is opaque, or very feebly translucent on the edges.

It is hard, but rather in a lower degree than the Egyptian jasper.

Rather easily frangible.

Rather heavy.

Specific gravity, 2.441, *Haberle*. 2.472, 2.537, 2.553, *Hoff*. 2.491, *Karsten*.

Geognostic and Geographic Situations.

It occurs in flœtz clay-porphyry in the Pentland Hills near Edinburgh; in a similar situation at Gnadenstein and Wolftitz, near Froburg in Saxony. In neither of these countries do we observe the leek-green and brownish-red striped varieties: these latter occur only in the beautiful striped jasper which is found at Orsk, in the district of Orenburg in Siberia. According to Hausmann, it occurs in the Hartz, along with common flinty slate and Lydian stone, in transition mountains.

Use.

This mineral receives an excellent polish, and hence is used like agate for ornamental purposes.

Observations.

Observations.

1. The distinguishing characters of this mineral, are its colour-delineations, its want of lustre, and its very flat conchoidal fracture, which sometimes inclines to earthy, and even to slaty. Its geognostic situation also distinguishes it from all the other subspecies of this species.

2. It is allied to Conchoidal Hornstone and Claystone, and passes into both of these minerals. These transitions are to be observed in the Pentland Hills. It is distinguished from *Conchoidal Hornstone* by its colour, colour-delineations, want of lustre, its more perfect flat conchoidal fracture, and its opacity: from *Claystone* it is distinguished by its greater hardness.

Third Subspecies.

Porcelain-Jasper.

Porzellan Jaspis, *Werner.*

Id. Wid. p. 314.—Porcellanite, *Kirw.* vol. i. p. 313.—Porzellan-Jaspis, *Estner,* b. ii. s. 613. *Id. Emm.* b. i. s. 240.—Diaspro porcellanico, *Nap.* p. 192.—Jaspe porcelaine, *Lam.* t. ii. p. 166. *Id. Broch.* t. i. p. 166.—Thermantide porcellanite, *Hauy,* t. iv. p. 510.—Porzellan Jaspis, *Reuss,* b. ii. s. 307. *Id. Lud.* b. i. s. 94. *Id. Suck.* 1r th. s. 351. *Id. Bert.* s. 226. *Id. Mohs,* b. i. s. 321. *Id. Karsten,* Tabel. s. 38. *Id. Leonhard,* s. 12. *Id. Steffens,* b. i. s. 184. *Id. Hoff.* b. ii. s. 168. *Id. Lenz,* b. i. s. 418. *Id. Oken,* b. i. s. 298.

External Characters.

Its colours are grey, blue, yellow, and seldom black

and red. Of grey, it presents the following varieties smoke, bluish, yellowish, and pearl grey; from pearl-grey it passes into lilac-blue and lavender-blue; also into brick-red, which inclines to yellow; from yellowish-grey it passes into straw-yellow, and ochre-yellow; from smoke-grey into greyish-black, and ash-grey.

It generally exhibits but one colour, and is sometimes marked with dotted, flamed, striped, and clouded delineations.

The grey varieties are generally brick-red in the rifts. It often presents brick-red vegetable impressions; and this is most frequently the case with the lavender-blue varieties.

It occurs most commonly massive, and in angular pieces; and is frequently cracked in all directions.

Internally it is glistening, sometimes approaching to shining, sometimes to glimmering, and even to dull; and the lustre is vitreo-resinous.

The fracture is imperfect conchoidal, and sometimes large and flat, and occasionally small conchoidal.

The fragments are angular and sharp edged.

It is opaque.

Hard in a low degree.

Easily frangible.

Rather heavy in a low degree.

Specific gravity, 2.330, *Kirw.* 2.430, *Karst.* 2.431, 2.432, 2.461, 2.577, 2.595, 2.646, *Hoff.*

Chemical Characters.

The lavender-blue variety, when exposed to a heat of 151° of Wedgwood, according to Kirwan, melts into a spongy yellowish-grey semi-transparent mass. Other varieties,

varieties, according to Link, melt before the blowpipe into a white glass.

Constituent Parts.

Silica,	- -	60.75
Alumina,	- -	27.25
Magnesia,	- -	3.00
Oxide of iron,	-	2.50
Potass,	- -	3.66

According to *Rose.*

Geognostic Situation.

It is always found along with burnt-clay and earth-slags, in places where pseudo-volcanoes have formerly burnt, or where beds of coal are now in a state of inflammation. Hence it follows, that it is a pseudo-volcanic production; and according to Werner, it is slate-clay converted into a kind of porcelain by the action of the heat of the volcano. As the coal wastes, hollows are formed in the bed, and the superincumbent porcelain-jasper breaks in pieces, and falls into them: hence it never occurs in regular beds, but in irregular broken masses, intermixed with burnt-clay, earth-slags, and similar substances.

Geographic Situation.

Porcelain-jasper occurs in Bohemia, principally in the plain betwixt the Erzgebirge and the Mittelgebirge, where immense beds of coal appear. It also occurs at Planitz, near Zwickau in Saxony, and in the neighbourhood of Zittau in Upper Lusatia. Likewise at Erterode, at the Meisner, in the Habichtswald; at Dutweiler, in the department of Saare; and also in Iceland.

 Fourth

Fourth Subspecies.

Common Jasper.

Gemeiner Jaspis, *Werner.*

Jaspis, *Cronst.* § 64, 65. p. 76.—Jaspis, *Wall.* p. 311. & 318. —Gemeiner Jaspis, *Wid.* s. 311.—Common Jasper, *Kirw.* vol. i. p. 310.—Gemeiner Jaspis, *Emm.* b. i. s. 243.—Diaspro commune, *Nap.* p. 189.—Jaspe, *Lam.* p. 164.—Le Jaspe commune, *Broch.* t. i. p. 338.—Quarz-Jaspe, *Hauy,* t. ii. p. 435. —Gemeiner Jaspis, *Reuss,* b. ii. s. 311. *Id. Lud.* b. i. s. 95. *Id. Suck.* 1r th. s. 348. *Id. Bert.* s. 228. *Id. Mohs,* b. i. s. 317. *Id. Hab.* s. 11.—Quartz-Jaspe, *Lucas,* p. 37.—Jaspe commun, *Brong.* t. i. p. 324.—Quartz-Jaspe, *Brard,* p. 101. —Gemeiner Jaspis, *Leonhard,* Tabel. s. 12. *Id. Karst.* Tabel. s. 38.—Jasper *Kid,* vol. i. p. 206.—Quarz-Jaspe, *Hauy,* Tabl. p. 28.—Gemeiner Jaspis, *Steffens,* b. i. s. 185. *Id. Hoff.* b. ii. s. 172. *Id. Lenz,* b. i. s. 420. *Id. Oken,* b. i. s. 298.

External Characters.

The most common colours are red and brown; seldom yellow and black. It occurs brownish-red, cherry-red, blood-red, cochineal-red, scarlet-red, ochre-yellow, yellowish-brown, chesnut-brown, liver-brown, blackish-brown, and pitch-black.

It has generally only one colour; sometimes, however, it occurs with spotted, clouded, flamed, or striped delineations.

It occurs generally massive, sometimes also disseminated, in blunt-cornered rolled pieces, mixed with calcedony in a moss-like manner, and rarely reticulated.

Internally

Internally it varies, according to the fracture, from shining to dull; and the lustre is resino-vitreous.

The fracture of some varieties is more or less perfect and flat conchoidal, and these have a shining or glistening lustre: in others it is even, with a glimmering lustre, or fine earthy and dull.

The fragments are angular, and more or less sharp edged.

It is opaque, or very faintly translucent on the edges.

It is hard in a low degree.

Rather easily frangible.

Specific gravity, 2.554, 2.671, *Haberle*. 2.580, 2.700, *Kirwan*. 2.298, 2.314, 2.349, 2.573, 2.665, *Hoffmann*.

Chemical Characters.

It is infusible without addition before the blowpipe. It retains its colour for a considerable time, and at length becomes white.

Geognostic Situation.

It occurs principally in veins, as a constituent part of agates, or in imbedded cotemporaneous masses in primitive, transition, and flœtz rocks. The veins in which it occurs are either entirely filled with jasper, or they contain also ores of different kinds, as of iron, lead, or silver. It is found more abundantly in ironstone veins than in those of lead and silver; and the iron ores with which it is associated, are red and brown ironstone, accompanied with quartz and iron-flint. The lead ores are lead-glance; and the argentiferous minerals are native silver, and vitreous silver-ore. The most beautiful varieties, and the largest masses, occur in veins entirely filled with jasper, or a mixture of jasper and agate.

 Geographic

Geographic Situation.

Europe.—It occurs in the Pentland Hills, and Moorfoot Hills, near Edinburgh; and in different places along the course of the rivers Tweed and Clyde, where it is contained in transition rocks. It occurs in trap rocks and transition rocks in Ayrshire and Dumfriesshire. To the north of the Frith of Forth it is not unfrequent, both in the form of veins and imbedded portions. In the fine display of rocks described by Colonel Imrie as occurring in the course of the North Esk river in the Mearns, there are cotemporaneous masses and veins of jasper in transition rocks. It occurs also in the Zetland Islands, and in several of the Hebrides. On the Continent of Europe it has been observed in Sweden, Russia, Germany, Hungary, Transylvania, France, Italy, Spain, Portugal; and in the islands of the Mediterranean, particularly Sicily.

Asia.—It is found in great abundance in Siberia.

Uses.

When it occurs in sufficiently large masses, and receives a good polish, it is cut into various ornamental articles, as vases, snuff-boxes, ringstones, &c. The finest varieties are used for engraving on: many beautiful antique engraved stones of common jasper are preserved in collections.

Observations.

1. Colour, lustre, fracture, and geognostic situation combined, distinguish this subspecies of jasper from the others.

2. It passes into Iron-flint and into Clay-Ironstone, and is nearly allied to Hornstone and Claystone.

3. The

3. The Sinopel of some mineralogists is a variety of common jasper.

Fifth Subspecies.

Agate-Jasper.

Agat-Jaspis, *Werner.*

Broch. t. ii. p. 141.—Agath-Jaspis, *Reuss,* b. ii. s. 316. *Id. Lud.* b. i. s. 95. *Id. Mohs,* b. i. s. 322. *Id. Leonhard,* Tabel. s. 12. *Id. Karsten,* Tabel. s. 38. *Id. Steffens,* b. i. s. 187. *Id. Hoff.* b. ii. s. 175. *Id. Lenz,* b. i. s. 422. *Id. Oken,* b. i. s. 298.

External Characters.

Its colours are yellowish-white and reddish-white; the yellowish-white passes into cream and straw yellow, and approaches to ochre-yellow; the reddish-white passes into flesh-red and pale blood-red. Several colours generally occur together, and these are arranged either in clouded, flamed, or striped delineations; of these the striped are either disposed in a circular manner, or fortification-wise.

It occurs massive.

Internally it is dull.

The fracture is small and flat conchoidal, approaching to even.

The fragments are angular, and rather sharp edged.

It frequently occurs in distinct concretions, which are either fortification-wise bent, or concentric lamellar.

It is generally opaque, or slightly translucent on the edges.

 It

It is hard in a low degree.

It sometimes adheres slightly to the tongue.

It is rather light.

Chemical Character.

Before the blowpipe it is affected in the same manner as common jasper.

Geognostic Situation.

It occurs principally in layers, in agate-balls, in amygdaloid; likewise in agate balls and veins in porphyry.

Geographic Situation.

It occurs in the agates of the middle district of Scotland, in Mid Lothian, West Lothian, and East Lothian; also in Saxony, Deuxponts, and Hungary.

Observations.

1. It is distinguished from the other subspecies of Jasper, by its colour-delineations, fracture, hardness, weight, and geognostic situation.

2. It was first accurately examined by Werner, and placed by him in the mineral system.

Agate *.

Agate is not, as some mineralogists maintain, a simple mineral, but is composed of various species of the quartz family, intimately joined together, and the whole mass is so compact and hard, that it receives a high polish. From its compound nature, it ought rather to be considered in the geognostic part of this work; yet as Werner, and other mineralogists, describe it along with the quartz family, we shall not deviate from their plan.

Agate is principally composed of calcedony, with flint, hornstone, carnelian, jasper, cacholong, amethyst, and quartz. Of these minerals sometimes only two, in other instances more than three, occur in the same agate; and these are either massive, disseminated, or in layers. Agates are by Werner divided into different kinds, according to their colour-delineations; and he enumerates the following:—1. *Ribbon* or *Striped Agate*. 2. *Brecciated Agate*. 3. *Fortification-Agate*. 4. *Tubular Agate*. 5. *Landscape-Agate*. 6. *Moss-Agate*. 7. *Jasper-Agate*.

1. *Ribbon or Striped Agate*.

It is composed of layers of calcedony, flint, and amethyst, and also of hornstone, jasper, and quartz, which alternate with each other, vary in breadth, and although sometimes

* The name *Agate*, is derived from the river Achate in Sicily, where it is said this mineral was first found.

sometimes curved, sometimes straight, yet are always parallel. When this agate is cut at right angles to the layers, the *common striped agate* is formed: when the section is made across a reniform elevation, we obtain the *zoned agate;* and when the section is oblique, the *serpentine agate* is formed.

This agate occurs principally in veins. A magnificent vein of this kind occurs at Cunnersdorf and Schlottwitz in Saxony, and another at the Halsbach, near Freyberg, and both are situated in gneiss. Agate veins of this kind occur also in porphyry, and these are of great size, as at Wiederau, near Rochlitz in Saxony; or in numerous small veins, traversing the porphyry in all directions, as at Rothlof, near Chemnitz, also in Saxony.

2. *Brecciated Agate.*

This beautiful agate is composed of fragments of another species, which is usually striped agate, generally connected together by a basis of amethyst. This agate occurs in the middle of the great vein of striped agate at Cunnersdorf.

Ribbon-agate is supposed to have been formed from different solutions which have been successively decomposed in a previously existing rent or fissure. The brecciated agate, which is found in the middle of the vein of ribbon-agate, is conjectured to have owed its origin to a rent or rents taking place in that agate; and the fragments thus formed being afterwards connected together, by a new solution poured into the fissure or fissures.

3. *Fortification*

3. *Fortification-Agate.*

This agate is composed of layers of calcedony, flint, and jasper, which have generally in the middle a nucleus of massive amethyst. These layers are thin, and parallel, and are fortification-wise bent. It generally occurs in irregular balls, which are contained in amygdaloid. When these balls are cut across, their surface sometimes very much resembles a fortification. The largest agates of this description occur at Oberstein on the Rhine; and many beautiful varieties are met with in the amygdaloidal rocks which abound so much in Scotland.

4. *Tubular Agate.*

When the central spaces in stalactitic calcedony are filled with agate, the compound is named Tubular Agate. It is a rare variety.

5. *Landscape-Agate.*

In this the substances are so arranged that the whole may be likened to a landscape. It also is a rare variety.

6. *Moss-Agate.*

In this beautiful kind of agate, jasper of various colours, as brown, yellow, red, appears as it were floating in a basis of calcedony. The jasper resembles moss, and when its arborisations are distinct, it has a very beautiful appearance. All the parts here are evidently of contemporaneous formation.

7. *Jasper-Agate.*

Jasper-agate is a mixture of calcedony, or hornstone, and jasper. The jasper is of a red, yellow, or brown colour, and is the predominating ingredient in the agate. It occurs in veins, which sometimes contain ores of different kinds, as ores of silver and iron.

The following are less important kinds of agate:

1. *Spotted Agate.*

In this beautiful agate, spots of red, yellow, or brown jasper, are dispersed through a calcedonic base. The St Stephen stone, already described under the article Calcedony, may be considered as a spotted agate.

2. *Clouded Agate.*

It is so named from its clouded appearance: the clouded delineations are of jasper.

3. *Star-Agate.*

This is an agate with stellular markings.

4. *Petrifaction-Agate.*

This agate contains petrifactions of marine animal substances, as shells of the turbinites and tubulites tribes.

Geognostic Situation.

Agates, as already mentioned, occur in veins in gneiss, and in porphyry, and in balls in amygdaloid. They also occur in porphyritic transition rocks, and in balls and veins in flœtz porphyry and greenstone; and probably these,

these, as well as all the other repositories of this singular compound mineral, are of cotemporaneous formation with the rocks in which they are contained.

Geographic Situation.

Very beautiful varieties of the different kinds of agate occur in the porphyry, amygdaloid, and greenstone rocks of Scotland. On the Continent there is a great depository of agate at Oberstein on the Rhine: it also occurs in Saxony, Silesia, Bohemia, and Italy; also in the island Sicily. It likewise abounds in Siberia, East Indies, and China.

Uses.

Agate is sometimes cut into snuff-boxes, and ring-stones: the larger masses are hollowed into mortars, or cut into elegant vases. It was much prized by the ancients, who executed several fine works in it. In the Electoral Cabinet at Dresden, and the Ducal Cabinet in Brunswick, there are beautiful vases of agate. At Oberstein on the Rhine, the amygdaloidal rocks are regularly quarried for the agates they contain, and these are cut and polished, and exported to other countries. The cutting, polishing, and selling of the agates (Scotch) of the amygdaloid of this country, is now carried on to a very considerable extent, and is to many a lucrative employment.

VII. PITCH-

VII. PITCHSTONE FAMILY.

This Family contains the following Species: Obsidian, Pitchstone, Pearlstone, and Pumice.

1. Obsidian.

This Species is divided into two subspecies, viz. Translucent Obsidian and Transparent Obsidian.

First Subspecies.

Translucent Obsidian.

Durchscheinender Obsidian, *Hoffmann.*

Achates islandicus, *Wall.* t. ii. p. 378.—Pumex vitreus solidus, Syst. Nat. xii. 3. p. 182. n. 7.—Verre de volcan en masses irreguliers, Pierre obsidienne, Pierre de gallinace, & Agathe noir d'Islande, *Rome de Lisle,* t. ii. p. 635.—Verre ou Laitier de volcan, *Faujas,* des Volcans, p. 308.—Obsidian, *Wid.* s. 348. *Id. Kirwan,* vol. i. p. 265.—Obsidiano, *Nap.* p. 205.—Lave vitreuse obsidienne, *Hauy,* t. iv. p. 494.—Obsidian, *Reuss,* b. ii. s. 355. *Id. Lud.* b. i. s. 85. *Id. Suck.* 1r th. s. 371. *Id. Bert.* s. 270. *Id Mohs,* b. i. s. 349. *Id. Hab.* s. 15.—Lave vitreuse obsidienne, *Lucas,* p. 231.—Obsidienne, *Brong.* t. i. p. 355.—Lava verre noire, *Brard,* p 447.—Obsidian gemeiner, *Leonhard,* Tabel. s. 14.—Obsidian, *Karsten,* Tabel. s. 36.—Obsidian, *Kid,* Appendix, p. 38. *Id. Steffens,* b. i. s. 371.—Durchscheinender Obsidian, *Hoff.* b. ii. s. 191.—Gemeiner Obsidian, *Lenz,* b. i. s. 432.—Obsidian, or Lava-glass, *Oken,* b. i. s. 305.

External Characters.

Its most frequent colour is velvet-black, which sometimes

times passes on the one side into greyish-black, ash-grey, and smoke-grey, and on the other into pitch-black. The colour is generally uniform, seldom spotted or striped. In some varieties (those from South America) a pinchbeck-brown light is to be observed.

It occurs massive, in blunt-cornered pieces, and sometimes in original grains, which are angular or roundish.

The external surface of the blunt-cornered pieces is rough; that of the grains sometimes rough, sometimes smooth.

Internally it is specular splendent, seldom shining, and the lustre is vitreous.

The fracture is perfect, large, and rather flat conchoidal.

It breaks into angular and very sharp-edged fragments, which sometimes incline to the tabular form.

It alternates from translucent to translucent on the edges.

It is hard.

It is very brittle.

It is easily frangible.

Rather heavy, in a low degree.

Specific gravity, Peruvian, 2.348, *Brisson*. Icelandic, 2.382, 2.397, *Hoff*. Hungarian, 2.374, 2.358, *Hoff*.

Chemical Characters.

The black obsidian of Iceland, according to Da Camara, on charcoal, before the blowpipe, melts into a pale ash-grey imperfect vesicular glass. The obsidian of the Island of Candia, before the blowpipe, was changed into a white, light, and uncommonly porous mass. That of Spanish America, before the blowpipe, lost its black colour, became white, spongy, and fibrous, and increased to seven or eight times its original bulk:

bulk: hence, it is conjectured that some gaseous substance escapes; and Von Humboldt is inclined to believe, that the gas evolved during the fusion of obsidian in the interior of the earth, may give rise to the earthquakes that agitate the Cordilleras.

Constituent Parts.

Obsidian of Iceland.		Obsidian of the Serro de las Novagas.	
Silica,	74	Silica,	78.0
Alumina,	2	Alumina,	10.0
Oxide of Iron,	14	Lime,	1.0
		Natron,	1.6
	90	Potass,	6.0
Loss,	10	Oxide of Iron,	1.0
	100		97.6
	Abilgaard.		*Vauquelin.*

Abilgaard is of opinion, that the loss in his analysis was owing to the escape of either potass or soda.

	American.	American.	American.
Silica,	72.0	72.0	71.0
Alumina,	12.5	14.2	13.4
Natron and Potass,	10.0	3.3	5.0
Lime,	0.0	1.2	1.6
Oxide of Iron & Manganese,	2.0	3.0	4.0
	96.5	93.7	95.0
	Collet-Descotils.	*Drappier.*	*Drappier.*

Geognostic Situation.

This mineral occurs both in primitive and flœtz country: the primitive obsidian is said to occur forming mountain masses, in the state of porphyry, (obsidian-porphyry); also in beds alternating with claystone, or felspar-porphyry, or imbedded in pearlstone-porphyry. In

In flœtz country, it appears to be associated with trap rocks, in the form of beds or veins.

Geographic Situation.

Europe.—This singular mineral is found in different parts of Europe. The island of Iceland, so remarkable on account of its volcanoes and hot springs, contains beds of this subspecies of obsidian. According to Shumacher, a bed of obsidian two feet thick, occurs in the Bordafiord Syssel in Iceland; and Sir. George Mackenzie, during his journey through that remote and desolate country, observed a great mass of obsidian, which appeared to him to be part of a stream that had flowed from a volcano. It is also found in the mountains of Tokay in Hungary, imbedded in pearlstone-porphyry; and in the same geognostic situation in Spain. It occurs in several of the islands in the Mediterranean, as Milo, Candia, and the Lipari Islands *.

Africa.—According to Cordier and Humboldt, it occurs at the summit of the Peak of Teneriff; and Dr Forster observed it in great quantity in the Isle of Ascension. It is said also to occur in the Island of Madagascar.

Asia.—Siberia; and near the town Goda, twenty wersts from Teflis in Georgia.

America.—Humboldt and Sonnenschmidt found beds and mountain-masses of obsidian at great heights, both in Peru and Mexico.

Polynesia.—Dr Forster found obsidian in several of the islands in the South Sea, as Easter Island, and Roggewein's Island.

 Uses.

* Vid. Account of the Obsidian of Lipari, by Spallanzani in his Travels, and Colonel Imrie, in the 2d volume of the Memoirs of the Wernerian Society.

Uses.

Although it can be cut and polished, yet its brittleness and frangibility are so great, that it is very apt to fly in pieces during the working: hence it is but seldom used by jewellers. Danish lapidaries cut the obsidian of Iceland into snuff-boxes, ring-stones, and ear pendents. According to Pliny, the ancients are said to have formed it into mirrors, and into ornamental articles. In New Spain and Peru, the natives cut it into mirrors; and formerly they used to manufacture it into knives, and other cutting instruments. Hernandez saw more than 100 of these knives made in an hour. Cortez, in his letter to the Emperor Charles V. relates that he saw at Tenochtitlan, razors made of obsidian; and Von Humboldt examined the mines which afforded the obsidian for these purposes on the Serro de las Novagas, or the *Mountain of Knives*. The natives of Easter and Ascension islands use it in place of cutting instruments; also for pointing their lances and spears, and for striking fire with.

Observations.

1. It was first introduced into the oryctognostic system by Werner. Its name is of great antiquity, being derived from a Roman named Obsidius, who first brought it from Ethiopia to Rome. Pliny speaks of it in the following lines: " In genere vitri et obsidiana numerantur, ad similitudinem lapidis quem in Æthiopia invenit Obsidius, nigerrimi coloris, aliquando et translucidi, crassiori visu, atque in speculis parietum pro imagine umbras reddente," &c.

2. It passes into Pitchstone, Pearlstone, and Pumice.

3. Werner, Hoffmann, Steffens, and Mohs, are of opinion that it is an aquatic production; whereas Faujas St Fond, Von Buch, Cordier, and Hauy, maintain its volcanic origin.

Second

Second Subspecies.

Transparent Obsidian.

Durchsichtiger Obsidian, *Hoffmann.*

Marekanit, *Karst.* Tabel. s. 36.—Obsidienne de Marikan, *Brong.* t. i. p. 342.—Edler Obsidian, *Haus.* s. 87.—Durchsichitiger Obsidian, *Hoff.* b. ii. s. 200.—Marekanit, *Lenz,* b. i. s. 435. *Id. Oken,* b. i. s. 305.

External Characters.

The colours are duck-blue, greyish-white, and clove-brown.

The blue occurs only massive; the white and brown in large and small grains.

The surface of the grains is smooth.

Internally it is splendent.

The fracture is perfect conchoidal.

It breaks into angular and sharp-edged fragments.

It is perfectly transparent.

It is hard.

It is brittle.

Very easily frangible.

Rather heavy, in a very low degree.

Specific gravity, 2.333, 2.360, *Lowitz.* 2.365, *Blumenbach.* 2.366, *Hoff.*

Chemical Characters.

According to Link, it melts more easily than the translucent obsidian, and into a white muddy glass.

Geognostic Situation.

The brown and white varieties of transparent obsidian are found in Siberia, imbedded in pearlstone-porphyry.

Geographic Situation.

The only Siberian locality of this mineral is Marekan, near Ochotsk, where the white and brown varieties are found. The blue variety occurs in the Serro de las Novagas in Mexico.

Observation.

Colour and transparency are highly characteristic of this subspecies.

2. Pitchstone.

Pechstein, *Werner.*

Id. Wid. p. 332.—Pitchstone, *Kirw.* vol. i. p. 292.—Pechstein, *Estner,* b. ii. s 435. *Id. Emm.* b. i. s. 262.—Pietra picea, *Nap.* p. 203.—Pissite, var. h, *Lam.* t. ii. p. 162.—La Pierre de Poix, *Broch.* t. i. p. 353.—Petrosilex resiniforme, *Hauy,* t. iv. p. 386.—Pechstein, *Reuss,* b. ii. s. 345. *Id. Lud.* b. i. s. 98. *Id. Suck.* 1r th. s. 321. *Id. Bert.* s. 225. *Id. Hab.* s. 15.—Resinite, *Brong.* t. i. p. 345.—Pechstein, *Haus.* s. 87. *Id. Leonhard,* Tabel. s. 13. *Id. Karst.* Tabel. s. 36. *Id. Kid,* vol. i. p. 231.—Pechstein, *Steffens,* b. i. s. 375. *Id. Hoff.* b. ii. s. 202. *Id. Lenz,* b. i. s. 436. *Id. Oken,* b. i. s. 304.

External Characters.

The principal colour is green, from which it passes on the one side into black, grey, and blue, and on the other, through

through several varieties of green, into brown, yellow, and red; yellow and blue are the rarest colours. The green colours are blackish-green, mountain-green, leek-green, olive green, and oil-green. From blackish-green it passes into greenish-black, bluish-black. greyish-black, ash-grey, smoke-grey, and a colour intermediate between indigo and Berlin blue: from olive-green and oil-green into liver-brown, reddish-brown, and pale blood-red. The yellowish-grey sometimes approaches to ochre-yellow. These colours are seldom bright, most generally dull and deep. The colour is in general uniform, seldom in veined, clouded, and spotted delineations.

It occurs massive.

Internally it is shining, sometimes passing into glistening, even inclining to glimmering. The red has the feeblest, the bluish and the green the strongest lustre; and the lustre is vitreo-resinous.

The fracture is imperfect conchoidal *, and is sometimes large and flat, sometimes small conchoidal: from the latter it passes into coarse-grained uneven, which sometimes approaches to coarse splintery. The conchoidal has the strongest, the splintery the weakest, lustre.

It breaks into angular and sharp-edged fragments.

It occurs sometimes in coarse, seldom in large and flat granular distinct concretions; sometimes in thick and and wedge-shaped prismatic concretions; and rarely in thick and straight lamellar distinct concretions. The surface of the distinct concretions is generally smooth and shining, and sometimes rather curved.

It is generally feebly translucent; some varieties, particularly the black, are only translucent on the edges.

 It

* Some varieties of the Arran pitchstone have a perfect and large conchoidal fracture: this, with their lamellar concretions, distinguishes them from the more common varieties of this mineral.

It is semi-hard in a high degree.
It is rather easily frangible; and
Rather heavy, but in a very low degree.
Specific gravity, 2.196 to 2.389, *Hoff.* 2.314 to 2.319, *Brisson.*

Chemical Characters.

Before the blowpipe it is fusible without addition. The black variety of Arran, at 21° of Wedgwood's pyrometer, intumesced a little, its colour was slightly altered, the surface glazed, and internally porous; at 31°, intumesced considerably and softened; at 65°, the intumescence was more considerable; at 100°, it was still vesicular, but more compact. The blackish-green variety of Arran becomes black, is much rent, and internally porous at 23°; at 55° formed a porous enamel; at 70° it became perfectly white, and still porous. The pitchstones of Meissen in Saxony, according to Mr Kirwan, appear tc be more infusible than those of Arran. He found some to melt at 130° Wedgwood; others at 152° to 165°; and a red variety remained nearly unaltered at 160°.

Constituent Parts.

Pitchstone of Meissen.

Silica, - -	73.00
Alumina, -	14.50
Lime, - -	1.00
Oxide of Iron, -	1.00
Oxide of Manganese,	0.10
Natron, - -	1.75
Water, - -	8.50
	99.85

Klap. Beit. b. iii. s. 257.

Geognostic

Geognostic Situation.

It occurs in veins that traverse granite; and also in beds in the second porphyry formation. In flœtz country, it appears in beds and veins in the first or old red sandstone formation, and in veins and imbedded portions in flœtz-trap rocks.

Geographic Situation.

Europe.—Pitchstone occurs in considerable abundance in different parts of Scotland. In the Island of Arran it traverses granite in the form of veins; in the red sandstone of that island it appears in beds, and veins of very considerable magnitude; and in small veins in the trap rocks of the Isle of Lamlash. In the islands of Mull, Canna, and Skye, it occurs either imbedded, or in the form of veins, in flœtz-trap rocks. Near Eskdalemuir, in the mountainous part of Dumfriesshire, it rests upon transition rocks, and is associated with flœtz-trap rocks. I am told that it occurs in trap rocks in Ardnamurchan in Argyleshire. Dr Macculloch found it in granite near the summit of Cairngorm; and Mr Murray, Lecturer on Chemistry, several years ago observed black pitchstone amongst trap rocks on the summits of the Cheviot Hills *.

In Ireland, it occurs in a vein traversing granite, in the Townland of Newry †, where it was first observed by Mr Joy of Dublin.

It occurs in the Island of Iceland in flœtz-trap rocks ‡.

 This

* Vid. Jameson's Mineralogical Travels; and Mineralogical Description of Dumfriesshire.

† Fitton's Mineralogy of Dublin, p. 53.

‡ Sir George Mackenzie's Travels in Iceland.

This mineral has never been found in the Scandinavian Peninsula. It occurs in the Electorate of Saxony, near Meissen, along with porphyry and sienite; in a similar formation at Braunsdorf, Spechtshausen and Mohorn, between Dresden and Freyberg, and near Dittersdorf. A newer formation occurs at Planitz, near Zwickau, also in Saxony *. The blue variety of pitchstone occurs near Vicenza in Italy, and probably in flœtz-trap rocks †. Pitchstone occurs in porphyry of the second primitive formation between Schemnitz and Kremnitz, and also near Tokay in Hungary; and in several of the islands in the Mediterranean, it appears to be associated with porphyry.

Asia.—It occurs at Kolyvan in Siberia, and near Mursinsk, in the Uralian mountains.

Ameria.—In Mexico, according to Sonnenschmidt, it occurs in great abundance, associated with clay-porphyry.

In South America it occurs at Pasto, Popayan, and Quito, along with clay-porphyry.

Observations.

1. It has been confounded with *Semi-Opal;* but it is distinguished from it by the following characters: its colour-suite is more extensive, and the colours are duller, and more muddy than those of semi-opal: its lustre is vitreo-resinous; but that of semi-opal is vitreous: its fracture is imperfect conchoidal; that of semi-opal perfect

* German mineralogists have not hitherto ascertained the formation to which the Planitz pitchstone belongs.

† In the Isle of Eigg, there is a blue variety of pitchstone, occurring in flœtz-trap, which resembles that of Vicenza.

fect conchoidal: it is rather heavier than semi-opal, and is differently affected by the blowpipe.

2. It is named *Pitchstone*, from the striking resemblance which some of its varieties have to pitch.

3. It was first discovered, about sixty years ago, by a mineralogist of Dresden, named Schulz; but it was first established as a distinct species by Werner.

4. It appears to pass on the one hand into Obsidian, and on the other into Pearlstone.

3. Pearlstone.

Perlstein, *Werner*.

Le Perlstein, *Broch.* t. i. p. 352.—Lave vitreuse perlée, *Hauy*, t. iv. p. 495.—Perlstein, *Reuss*, b. ii. s. 349. *Id. Lud.* b. i. s. 99. *Id. Suck.* 1r th. s. 367. *Id. Bert.* s. 224. *Id. Mohs*, b. i. s. 353.—Lave vitreuse perlé, *Lucas*, p. 231.—Obsidienne perlé, *Brong.* t. i. p. 340.—Perlstein, *Haus.* s. 88.—Pearlstein, *Leonhard*, Tabel. s. 14. *Id. Karsten*, Tabel. s. 36.—Pearlstone, *Kid*, Appendix, p. 38.—Perlstein, *Steffens*, b. i. s. 378. *Id. Hoff.* b. ii. 208. *Id. Lenz*, b. i. s. 443. *Id. Oken*, b. i. s. 306.

External Characters.

It is generally grey, sometimes also black and red. The varieties of grey are smoke, bluish, ash, yellowish, and pearl grey: from dark ash-grey it passes into greyish-black: from yellowish-grey into a kind of straw-yellow: from pearl-grey into flesh and brick red, and reddish-brown. The colours are sometimes disposed in striped and spotted delineations.

It

It occurs massive, vesicular; and the vesicles are sometimes round, and sometimes so much elongated, that the mass appears fibrous.

Its lustre is shining and pearly.

Its fracture, on account of the thinness of the distinct concretions, is hardly observable, but appears to be small and imperfect conchoidal.

The fragments are angular and blunt-edged.

It occurs in large and coarse angulo-granular distinct concretions, that include small and round granular concretions, which are again composed of very thin concentric lamellar concretions. The surface of the concretions, particularly in the large and coarse granular, is smooth, shining, and pearly, and has a striking resemblance to that of pearl. In the centre of these concretions, we frequently meet with roundish balls of obsidian.

It is translucent on the edges, sometimes even translucent.

It is uncommonly easily frangible.

It is soft, passing into very soft; and

Is rather heavy; approaching to light.

Specific gravity, Mexican, 2.254, *Vauquelin*. Hungarian, 2.340, *Klaproth*. Hungarian, 2.343, *Hoff*,

Chemical Characters.

Before the blowpipe it intumesces very much, and is converted into a white spumaceous glass.

Constituent

Constituent Parts.

Pearlstone of Hungary.		Pearlstone of Mexico.	
Silica, - - - -	75.25	Silica, - -	77.0
Alumina, - -	12.00	Alumina, - -	13.0
Oxide of Iron, -	1.60	Oxide of Iron & Manganese,	2.0
Potass, - -	4.50	Potass, - -	2.0
Lime, - -	0.50	Lime, - - -	1.5
Water, - - - -	4.50	Natron, - - -	0.7
	——	Water, - - -	4.0
	98.35		——
			100.2
Klaproth, Beit. b. iii. s. 326.		*Vauquelin.*	

Geognostic Situation.

It occurs in great beds in clay-porphyry, and both rocks contain imbedded cotemporaneous balls of hornstone. It is frequently intermixed with felspar, mica, and quartz, and thus acquires a porphyritic character. It has also been found in flœtz-trap rocks.

Geographic Situation.

Europe.—It occurs in large beds, in porphyry, near Tokay, Keresztur, and Telkebanya, in Hungary; also at Schemnitz, Glasshütte and Kremnitz, in the same country. It is said to occur at Carbonera, at Cape de Gate in Spain, where it is associated with obsidian. It occurs, along with porphyry, near Sandy Brae in Ireland; and in flœtz-trap rocks in the Island of Iceland *.

Asia.—Beautiful varieties of this mineral occur near Ochotsk.

America.—It occurs, along with porphyry, in Mexico.

Observations.

* Mackenzie's Travels in Iceland.

Observations.

1. The distinct concretions distinguish this species from all the other members of the *Pitchstone* family: it is further characterised by its colour-suite, kind of lustre, great frangibility, and low degree of hardness.

2. Some mineralogists describe it as a variety of Obsidian; others as a Zeolite, under the title *Volcanic Zeolite*; but Werner, with more propriety, views it as a species distinct from either of these, and, from the resemblance of the most characteristic varieties to Pearl in colour, lustre, and form, names it *Pearlstone.*

4. Pumice.

Bimstein, *Werner.*

Porus igneus, *Wall.* t. ii. p. 375—Pumex vulcani, *Rom. de L.* t. ii. p. 629.—Bimstein, *Wid.* 350—Pumice, *Kirw.* vol. i. p. 415—Bimstein, *Emm.* b. i. s. 350.—Pumice, *Nap.* p. 208. Pierre-ponce, *Lam.* t. ii. p. 473.—La Pierre-ponce, *Broch.* t. i. p. 443.—Lave vitreuse pumicée, *Hauy,* t. iv. p. 495.—Bimstein, *Reuss,* b. ii. s. 261. *Id. Lud.* b. i. s. 125. *Id. Suck.* 1r th. s. 374. *Id. Bert.* s. 204. *Id. Mohs,* b. i. s. 356. *Id. Hab.* s. 16.—Lave vitreuse pumicée, *Lucas,* p. 231.—Ponce, *Brong.* t. i. p. 332.—Bimstein, *Haus.* s. 88.—Verre fibreux, *Brard,* p. 448.—Bimstein, *Leonhard,* Tabel. s. 15.—Bimstein, *Karst.* Tabel. s. 36.—Pumice, *Kid,* App. p. 35.—Bimstein, *Steffens,* b. i. s. 379. *Id. Hoff.* b. ii. s. 213. *Id. Lenz,* b. i. s. 439.—Bims, *Oken,* b. i. s. 309.

This species is now divided by Karsten and Werner into three subspecies: these are, Glassy Pumice, Common Pumice, and Porphyritic Pumice.

First

First Subspecies.

Glassy Pumice.

Glasiger Bimstein, *Werner & Karsten.*

External Characters.

Its colour is smoke-grey, of different degrees of intensity; also ash-grey, which sometimes inclines to greyish-white.

It occurs vesicular, and capillary in the vesicular cavities.

Internally the principal fracture is glistening and pearly, the cross fracture shining, and nearly vitreous.

The principal fracture is promiscuous fibrous; the cross fracture small and imperfect conchoidal, inclining to uneven.

The fragments are angular, and blunt-edged.

It is sometimes translucent, sometimes only translucent on the edges.

It is intermediate between hard and semi-hard.

It is very brittle.

It is pretty easily frangible.

It feels very rough, sharp, and meagre; and

Is light or swimming.

Specific gravity, 0.378 to 1.444, *Hoffmann.*

Geognostic and Geographic Situations.

It occurs in beds, along with common pumice and obsidian, in the Lipari islands, and in the islands of Santorini and Milo, in the Grecian Archipelago.

Observations.

Observations.

It is distinguished from *Common* and *Porphyritic Pumice*, by its darker colours, vitreous, shining and conchoidal cross fracture, its greater translucency, and its greater hardness.

Second Subspecies.

Common Pumice.

Gemeiner Bimstein, *Werner & Karsten.*

External Characters.

Its colours are almost always white, and principally greyish and yellowish white, which sometimes approach to yellowish-grey, ash-grey, and smoke-grey.

It occurs vesicular, and the vesicles are much elongated. In the interior of the vesicles there are capillary fibres.

Internally it is sometimes glistening, sometimes glimmering, and the lustre is pearly.

The principal fracture is more or less perfect fibrous, which is curved and parallel; the cross fracture is uneven.

It breaks into blunt-edged fragments, sometimes into splintery fragments.

It is only translucent on the edges.

It is semi-hard, but in a high degree.

It is very brittle.

It is pretty easily frangible.

It feels meagre and rough; and

Is swimming.

Specific gravity, 0.914, *Brisson.* 0.752 and 0.770, *Hoffmann.*

Chemical

Chemical Characters.

At a heat varying from 35° to 40° of Wedgwood, it is so much altered, that its fibrous fracture is no longer distinguishable : at 60°, it melts into a grey-coloured slag

Constituent Parts.

Silica, - - - -	77.50
Alumina, - - -	17.50
Natron and Potass, - -	3.00
Iron, mixed with Manganese, -	1.75
	99.75

Klaproth, Beit. b. iii. s. 265.

Geognostic Situation.

It occurs in beds, along with glassy pumice and obsidian. In the Island of Lipari, according to Spallanzani, there is a whole hill, named Campo Bianco, in which the pumice is distinctly stratified ; and in the same island, it also occurs in globular distinct concretions. It occurs in beds between Andernach and Coblentz ; and a remarkable bed of this mineral is contained in alluvial land, near Neuwied.

Geographic Situation.

Europe.—It occurs in the Island of Iceland, along with obsidian * ; abounds in the Lipari Islands ; is found in the islands of Santorini and Milo, in the Grecian Archipelago ; and, as already mentioned, occurs on the banks of the Rhine.

Africa.—Island of Teneriff.

Asia.

* Mackenzie's Travels in Iceland.

Asia.—Ternate, and the other Molucca Islands.

America.—In Mexico.

Uses.

Common pumice is used for polishing glass, soft stones, and metals; also by parchment-makers, curriers, and hat-makers; and hence it forms a considerable article of trade, and is exported from the Lipari Islands in great quantities to the different countries of Europe. It forms a pernicious ingredient in some teeth-powders: sailors in the Mediterranean use it for shaving; and in the East it is an indispensable article in every bath, for the purpose of removing hairs from the body. On account of its porosity, it is used in Teneriff as a filtering-stone. In Italy it is ground down, and used in place of sand in the making of mortar.

Observation.

It is distinguished from *Glassy Pumice* by its lighter colours, the nature of its lustre, its more perfect fibrous fracture, its lower translucency, and inferior hardness.

Third Subspecies.

Porphyritic Pumice.

Porphyrartiger Bimstein, *Werner & Karsten.*

External Characters.

Its colours are greyish-white, and light ash-grey, and very seldom pale blackish-brown.

It occurs massive, in mountain masses; and internally it is minutely porous.

The

Internally it is glistening or glimmering, and the lustre is pearly.

The fracture is very imperfect curved and parallel fibrous, which sometimes passes into compact, or into splintery and uneven.

The fragments are angular and rather blunt-edged.

It is feebly translucent on the edges.

It is semi-hard.

It is very brittle.

It is rather easily frangible; and

Is light.

Specific gravity, 1.661, *Hoffmann.*

Geognostic and Geographic Situations.

Porphyritic pumice contains crystals of felspar, quartz, and mica, thus forming a kind of porphyry, which is contained in the second porphyry formation, and is generally associated with claystone, obsidian, pearlstone, and pitchstone-porphyry. It occurs in Hungary at Tokay, Keresztur, and Telkebanya; also in the continuation of this tract at Schemnitz, Glashütte, and Kremnitz. It appears also to be associated with porphyry on the northern acclivity of the Carpathians, as at the Green Lake, above Kasemark. It is said to occur also near Rio Mayo, in the province of Quito.

VIII. ZEOLITE FAMILY.

This Family contains the following Species: Prehnite, Zeolite, Apophyllite, Cubicite, Chabasite, Cross-stone, Laumonite, Dipyr, Natrolite, and Wavellite.

1. Prehnite.

Prehnit, *Wern.* Bergm. Journ. for 1790, b. i. s. 110. *Id. Wid.* s. 357. *Id. Emm.* b. i. s. 192.—Prenite, *Nap.* p. 235.—Prehnite, *Lam.* t. ii. p. 311. *Id. Broch.* t. i. p. 295. *Id. Hauy,* t. iii. p. 167. *Id. Reuss,* b. ii. s. 428. *Id. Lud.* b. i. s. 87. *Id. Suck.* 1r th. s. 414. *Id. Bert.* s. 182. *Id. Mohs,* b. i. s. 358. *Id. Lucas,* p. 69. *Id. Brong.* t. i. p. 376. *Id. Haus.* s. 95. *Id. Brard,* p. 171. *Id. Leonhard,* Tabel. s. 15. *Id. Karsten,* Tabel, s. 30. *Id. Kidd,* vol. i p. 250. *Id. Hauy,* Tabl. p. 50. *Id. Steffens,* b. i. s. 382. *Id. Hoff.* b. ii. s. 220. *Id. Lenz,* b. i. s. 444. *Id. Oken,* b. i. s. 356.

This species is divided into two subspecies, viz. Foliated Prehnite, and Fibrous Prehnite.

First Subspecies.

Foliated Prehnite.

Blättriger Prehnit, *Werner.*

External Characters.

The principal colour is apple-green, from which it passes on the one side into leek-green, mountain-green, greenish-grey, and greenish-white, and on the other into grass-green, yellowish-grey, and yellowish-white.

It

It occurs massive, and often also crystallised: the following are its crystallizations.

1. Oblique four-sided table, which is the fundamental figure, (fig. 70.) *. When this table increases in thickness, an oblique four-sided prism is formed.
2. The four-sided table is sometimes truncated, either on all its terminal edges, or only on the acute edges. When the truncations on all the edges increase very much, there is formed
3. An unequiangular eight-sided table, (fig. 71.) †. When the truncations on the acute edges increase considerably, there is formed
4. An unequiangular six-sided table, (fig. 72) ‡. When these truncating planes increase in magnitude, and when the table at the same time becomes thicker, and the obtuse edges are slightly truncated, there is formed
5. A broad rectangular four-sided prism, rather flatly bevelled on the extremities, in which the bevelling planes are set on the smaller lateral planes, and the edge of the bevelment is slightly truncated.
6. The six-sided table sometimes becomes equilateral, and forms, by superposition, an equilateral six-sided prism.

The crystals are small and very small, seldom middle-sized.

They seldom occur single, being generally aggregated, in such a way as to be attached by their lateral planes; sometimes in tabular and manipular groupes, sometimes

 in

* Prehnite rhomboidale, Hauy.

† Prehnite octogonale, Hauy.

‡ Prehnite hexagonale, Hauy.

in cravat and ruff like groupes. All these groupes occur again in druses.

Externally the crystals are almost always shining.

Internally the principal fracture is shining, but in a low degree; the cross fracture is glistening, and the lustre is pearly.

The principal fracture is rather imperfect curved foliated, with a single cleavage, the folia of which are parallel with the lateral planes; sometimes it passes into diverging broad radiated. The cross fracture is fine grained uneven.

The fragments are angular, and not very sharp-edged.

The varieties with a foliated fracture are composed of large, coarse, and fine granular concretions: those with a radiated fracture, on the contrary, are disposed in thick and wedge-shaped columnar concretions. Both kinds of concretions are very much grown together, and not very distinct. It is said also to occur in thick and curved lamellar distinct concretions.

It alternates from translucent, through semi-transparent into transparent.

Hard, but not in a high degree. It scratches glass; but feebly.

It is rather easily frangible; and

Rather heavy, in a middle degree.

Specific gravity, 2.609, 2.696, *Hauy*. 2.924, *Hoff*.

Chemical Characters.

It intumesces before the blowpipe, but does not gelatinate with acids.

Physical Characters.

According to the observations of M. De Dree, it becomes electric by heating.

Constituent

Constituent Parts.

Prehnite of the Cape.		Prehnite of Dauphiny.	
Silica, - -	43.83	Silica, - -	50.0
Alumina, -	30.33	Alumina, - -	20.4
Lime, - -	18.33	Magnesia, - -	0.5
Oxide of Iron, -	5.66	Lime, - -	23.3
Water, - -	1.83	Oxide of Iron, -	4.9
		Water, - -	0.9
	99.89		100

Klaproth, Beobacht. und Endeck der Naturf-Freunde zu, Berlin ; B. ii. s. 211.

Hassenfratz, Journal de Physique, 1780.

Geognostic Situation.

Europe.—It was first found in France in the year 1782, by M. Schreiber, near to Rivoire in Oisans, in a steatitic rock, imbedded in massive hornblende. It is not disseminated through the mass of the rock, but traverses it in the form of cotemporaneous veins, that contain, besides this mineral, also axinite, octahedrite, chlorite, calcareous-spar, and other minerals. In the same country, at St Christophe, it occurs, along with axinite, in cotemporaneous veins that traverse granite. Veins containing foliated prehnite also occur in the Alps of Savoy ; in the Saualpe in Carinthia ; and at Ratschinkes in the district of Sterzing in the Tyrol. It is said to occur massive in amygdaloid, along with calcareous-spar and chlorite, in the Seifer Alp in the Tyrol ; and along with foliated chlorite and adularia, in the valley of Fusch in Salzburg. The yellowish-white variety is found in a slaty rock, with acicular epidote, and delicate fibrous asbestus, in

 Mount

Mount Crelitz, near St Sauveur, in the valley of Bareges in the Department of the High Pyrenees *.

Africa.—Beautiful apple-green massive varieties of this mineral are found in mountains in the country of the Namaquas, in the interior of Southern Africa. These mountains are said to be granitic, and to contain, besides veins of prehnite, also much copper-ore.

America.—It occurs in Greenland, accompanied with calcareous-spar, in minute cotemporaneous veins in sienite.

Observations.

1. This mineral is characterised by its colour, crystallisation, the peculiar grouping of its crystals, its lustre, fracture, transparency, hardness, and weight.

2. It bears some resemblance to Zeolites in lustre, fracture, and in the changes it experiences by exposure to the heat of the blowpipe: but it is distinguished from them by its green colour, its crystallisation, its greater hardness and weight, as also by its chemical characters.

3. It has been confounded with Prase, Chrysolite, Chrysoprase, Emerald, and Felspar. It was Werner who in the year 1783 first established it as a particular species, and named it after its discoverer *Prehn*, at that time Governor of the Cape of Good Hope. He first brought it from the Cape to Europe.

4. The beautiful white-coloured vases sometimes imported from India, and which are said to be of *Jade*, a substance allied to felspar, are, in Count de Bournon's opinion, of the nature of prehnite.

Second

* This variety, from its lightness, was for some time considered as a distinct species, under the name *Loupholite*.

Second Subspecies.

Fibrous Prehnite.

Fasriger Prehnit, *Werner.*

External Characters.

Its colours are siskin-green, oil-green, asparagus-green, mountain-green, and greenish-white.

It occurs massive, reniform, and crystallised in acicular, rectangular, four-sided prisms.

Internally it is glistening, and the lustre is pearly.

The fracture passes from delicate fibrous through coarse fibrous into narrow radiated; and both kinds of fracture are straight and scopiform, or stellular.

It breaks into angular and rather sharp-edged fragments; sometimes into splintery and wedge-shaped fragments.

It occurs in large and coarse angulo-granular distinct concretions.

It is translucent.

It is hard, but not in a high degree; it scratches glass, and gives single sparks with steel.

It is easily frangible; and

Rather heavy, but in a middle degree.

Specific gravity, 2.889, *Hauy.* 2.856, *Hoffmann.*

Chemical Character.

Before the blowpipe it melts into a vesicular enamel.

Physical Character.

It becomes electric by heating.

Constituent Parts.

Silica, - -	42.50
Alumina, -	28.50
Lime, - -	20.40
Natron and Potass,	0.75
Oxide of Iron, -	3.00
Water, - -	2.00
	97.15

Laugier, Annales du Museum, t. xv. p. 205.

Geognostic Situation.

This subspecies appears to be confined to flœtz country; at least it has hitherto been found only in flœtz-trap rocks, as basalt, amygdaloid, basaltic greenstone, and common greenstone. It occurs either in cotemporaneous veins, or in amygdaloidal, and other shaped cavities in those trap rocks.

Geographic Situation.

Europe.—In Scotland, it occurs in veins and cavities in trap rocks near Beith in Ayrshire; at Hartfield, near Paisley; near Frisky Hall, and Loch Humphry in Dunbartonshire; in Salisbury Craig, the Castle Rock, and Arthur Seat near Edinburgh; and in the Island of Mull.

It occurs in small veins, along with zeolite and native copper, in amygdaloid, at Reichenbach, in the department of Saar in France; also near Oberstein. At Fassa in the Tyrol, it occurs in amygdaloid along with zeolite.

America.—It has been found in trap rocks near Boston.

Observations.

1. It is distinguished from *Foliated Prehnite*, by the tendency of its green colours to fall into yellow, its external

ternal shape, low degree of lustre, fracture, and distinct concretions ; and we may add to these, its geognostic situation.

2. It resembles Fibrous and Radiated Zeolites in its lustre, fracture, and easy fusibility; but it is distinguished from these minerals by its green colours, and its greater hardness and weight.

2. Zeolite.

THIS species is divided into four subspecies, viz. Mealy Zeolite, Fibrous Zeolite, Radiated Zeolite, and Foliated Zeolite.

First Subspecies.

Mealy Zeolite.

Mehlzeolith, *Werner.*

Id. Wid. p. 361.—Zeolite, *Kirw.* t. i. p. 278.—Mehl Zeolith, *Estner,* b. ii. s. 481. *Id. Emm.* b. i. s. 199.—Zeolite compatta terrea, *Nap.* p. 235.—La Zeolite farineuse, *Broch.* t. i. p. 298.—Mehl Zeolith, *Reuss,* b. ii. s. 405. *Id. Mohs,* b. i. s. 370.—Erdiger Zeolith, *Haus.* s. 96. *Id. Leonhard,* Tabel. s. 16.—Mehl Zeolith, *Karst.* Tabel. s. 30.—Mesotype altérée aspect terreux, *Hauy,* Tabl. p. 48.—Mehliger Mesotype, *Steffens,* b. i. s. 391.—Mehl Zeolith, *Hoff.* b. ii. s. 232. *Id. Lenz,* b. i. s. 451.—Mehlriger Mesotype, *Oken,* b. i. s. 352.

Externol Characters.

Its colours are yellowish-white, greyish-white, and reddish-

reddish white ; the latter sometimes passes to pale flesh-red, and even approaches to brick-red.

It occurs massive, reniform, coralloidal, and sometimes it forms a crust over the other subspecies of zeolite.

Internally it is dull, or very feebly glimmering.

The fracture is coarse earthy, and sometimes inclines to delicate fibrous.

The fragments are angular and blunt-edged.

It is opaque.

It is very soft.

It is rather sectile.

Uncommonly easily frangible.

It does not adhere to the tongue.

It feels rough and meagre ; and when we draw our finger across it, it emits a grating sound.

It is light, approaching to swimming.

Chemical Characters.

It intumesces before the blowpipe, and forms a jelly with acids.

Constituent Parts.

Silica, - -	60.0
Alumina, - -	15.6
Lime, - - -	8.0
Oxide of Iron, -	1.8
Loss, by exposure to heat,	11.6
	97

Hisinger's Afhandlingar i Fysik, &c. th. 3.

Geognostic Situation.

It occurs in similar repositories with the other subspecies.

Geographic

Geographic Situation.

It is found in Iceland, Faroe Islands, Sweden, and in various parts of Scotland, as in the Islands of Skye, Mull, and Canna, near Tantallon Castle in East Lothian, &c.

Second Subspecies.

Fibrous Zeolite.

Fasriger Zeolith, *Werner.*

This subspecies is divided into two kinds, viz. Common Fibrous Zeolite, and Needle Zeolite.

First Kind.

Common Fibrous Zeolite.

Gemeiner Faser Zeolith, *Werner.*

Faserzeolith, *Karsten,* Tabel. s. 30.—Fasriger Mesotyp, *Steffens,* b. i. s. 387.—Gemeiner Faserzeolith, *Hoff.* b. ii. s. 233.—Fasriger Zeolith, *Lenz,* b. i. s. 454.—Fasriger Mesotyp, *Oken,* b. i. s. 352.

External Characters.

Its colours are generally snow-white, greyish-white, or yellowish-white, seldom reddish-white: from reddish-white it passes into flesh-red, and into a colour intermediate between flesh-red and brick-red; and from yellowish-white into a colour intermediate between yellowish-grey and ochre-yellow, and into pale yellowish-brown.

It

It occurs massive, in blunt angular pieces, in balls, small reniform, and in capillary crystals.

The external surface of the reniform varieties is rough and dull.

Internally it is strongly glimmering, passing into glistening, and the lustre is pearly.

The fracture is straight scopiform, or stellular diverging fibrous, and alternates from delicate to coarse fibrous, and even approaches to narrow radiated.

It breaks into splintery or wedge-shaped fragments.

It occurs in large and coarse granular concretions, which are longish, or angulo-granular, and are very much grown together.

It is faintly translucent.

It is semi-hard in a low degree.

It is rather brittle.

It is easily frangible; and

Is rather heavy, bordering on light.

Specific gravity, 2.158 to 2.197, *Hoffmann.*

Its geognostic and geographic situations are the same with the following subspecies.

Chemical Characters.

It intumesces before the blowpipe, and forms a jelly with acids.

Observations.

1. It is distinguished from the other subspecies of this species, by its inferior lustre, fibrous fracture, low degree of translucency, and hardness; and to these we may add its almost total want of regular crystallisations, and its distinct concretions.

2. It is distinguished from *Calc-sinter*, with which it has been confounded, by its distinct concretions, inferior weight, and its not effervescing with acids.

Second

Second Kind.

Needle Zeolite.

Nadelzeolith, *Werner.*

Mesotype, *Hauy,* t. iii. p. 151.—Prismatischer Zeolith, *Karst.* Tabel. s. 30.—Prismatischer Mesotyp, *Steffens,* b. i. s. 388. —Nadelzeolith, *Hoff.* b. ii. s. 235.—Prismatisher Zeolith, *Lenz,* b. i. s. 455.—Sauliger Mesotyp, *Oken,* b. i. s. 352.

External Characters.

Its colours are greyish or yellowish white, and frequently reddish-white.

It occurs massive ; and crystallized,

1. In acicular rectangular four-sided prisms, very flatly acuminated with four planes, which are set on the lateral plans, (fig. 73. *.)
2. Sometimes two of the acuminating planes disappear, when there is formed an acute bevelment, which is set on somewhat obliquely.
3. The prism is sometimes truncated on the edges, as in fig. 74. †.

The crystals are sometimes scopiformly aggregated, sometimes promiscuously aggregated.

The lateral planes of the crystals are longitudinally streaked, but the acuminating planes are smooth.

Externally

* Mesotype pyramidée, Hauy.

† Mesotype dioctaedre of Hauy.

Externally the crystals are shining, passing into splendent.

Internally it is glistening, and the lustre is vitreous, inclining to pearly.

The principal fracture is very narrow, straight, and scopiform radiated, which sometimes passes into fibrous. The cross fracture is small and fine-grained uneven.

The crystals have a double cleavage.

It breaks into splintery and wedge-shaped fragments.

It occurs in large and coarse granular distinct concretions, and each of these is composed of very thin and straight prismatic concretions, which at their free extremities generally shoot into crystals.

It is translucent: the crystals are semi-transparent and transparent; and it refracts double.

It is semi-hard; it scratches calcareous-spar.

It is brittle; and

Rather heavy, approaching to light.

Specific gravity, 2.179, 2.198, 2.270, *Hoffmann.*

Chemical Characters.

It intumesces before the blowpipe, and forms a jelly with acids.

Physical Characters.

It becomes electric by heating, and retains this property some time after it has cooled. The free extremity of the crystal, with the acumination, shews positive, and the attached end negative electricity.—*Hauy.*

Constituent

Constituent Parts.

Silica, -	50.24	50
Alumina, -	29.30	20
Lime, -	9.46	8
Water, -	10.00	22
	99.00	100
	Vauquelin, Jour. des Mines, N. 44. p. 576.	*Pelletier*, Mem. de Chimie, Paris, 1798, t. i. p. 41.

Geognostic and Geographic Situations.

Europe.—It occurs in flœtz-trap rocks, as in basalt, greenstone, and amygdaloid. In this country it occurs in Dunbartonshire, Ayrshire and Perthshire, and always in trap rocks. It is found in flœtz-trap rocks in the Island of Iceland, and in the Faroe Islands; also in the rocks of the Puy de Marman in Auvergne; and in the Tyrol.

America.—It occurs in flœtz-trap rocks in the Island of Disco, in West Greenland.

Observations.

1. Crystallization, kind of lustre, fracture, distinct concretions, transparency, hardness, and weight, are the most essential characters of this subspecies.

2. It is distinguished from *Radiated Zeolite* by its crystallization, vitreous lustre, prismatic distinct concretions, greater transparency, hardness, and brittleness: it is distinguished from *Common Fibrous Zeolite* by its more frequent and distinct crystallizations, its higher and more vitreous lustre, its prismatic distinct concretions, greater transparency, hardness, and brittleness.

Third

Third Subspecies.

Radiated Zeolite.

Strahl Zeolith, *Werner.*

Id. Wid. p. 363. *Id. Emm.* b. i. s. 202.—Zeolite commune, *Nap.* p. 228.—Zeolite, first variety, *Lam.* t. ii. p. 305.—Zeolithe rayonnée, *Broch.* t. i. p. 301.—Stilbite, *Hauy,* t. iii. p. 161.—Strahl Zeolith, *Mohs,* b. i. s. 372.—Stilbite, *Brong.* t. i. s. 375.—Stilbit, *Haus.* s. 96.—Strahliger Zeolith, *Leonhard,* Tabel. s. 16.—Stilbit, *Karsten,* Tabel. s. 30. —Zeolith, *Steffens,* b. i. s. 393.—Strahlzeolith, *Hoff.* b. ii. s. 237.—Stilbit, *Lenz,* b. i. s. 465. *Id. Oken,* b. i. s. 353.

External Characters.

It occurs almost always yellowish-white and greyish-white, seldom snow-white and reddish-white. The yellowish-white passes into yellowish-grey, into a colour intermediate between ochre and lemon yellow, and into yellowish-brown; and the reddish-white into flesh-red, which sometimes borders on blood-red. The greyish-white sometimes nearly passes into smoke-grey.

It is found massive, in angular pieces, and globular; also frequently crystallised,

1. In broad rectangular four-sided prisms, rather acutely acuminated on both extremities by four planes, which are set on the lateral edges, as in fig. 75 *.
2. The summits of the acuminations are sometimes more or less deeply truncated †. When very deeply

* Stilbite dodecaedre of Hauy.

† Stilbite epointée of Hauy.

deeply truncated, the truncating plane passes into a terminal plane, and the acuminating planes form only truncations on the angles.

3. Sometimes No. 1. is so thin, that it may be considered as a long six-sided table, bevelled on the shorter terminal planes *.

The crystals are sometimes manipularly and scopiformly aggregated, and frequently so grown together that the acuminations only are visible, and project like pyramids.

The crystals are middle-sized, and small.

The broader lateral planes of the crystals are smooth, the smaller longitudinally streaked, and the acuminating planes are smooth, or rough.

The surfaces of the broader lateral planes of the crystals Nos. 1. and 2. are splendent and pearly; the other planes are shining and vitreous: internally, the lustre is more or less shining, and is pearly.

The fracture is generally scopiform, or stellular radiated, alternating from very narrow to very broad radiated, so that it passes on the one side into fibrous, and on the other into foliated. It rarely occurs with a promiscuous radiated fracture.

The fragments are wedge-shaped or splintery.

It occurs in distinct concretions, which are large, coarse, or small granular, generally angular, and sometimes also longish-granular.

The crystals are strongly translucent, sometimes passing into semi-transparent.

It is semi-hard in a low degree; scratches calcareous spar.

It is brittle.

 It

* Stilbite dodecaedre lamelliforme, Haüy.

It is easily frangible; and
Rather heavy, bordering on light.
Specific gravity, 2.132, 2.136, 2.164, *Hoffmann.*

Chemical Characters.

It intumesces and melts before the blowpipe, and during its intumescence emits a phosphoric light. If laid on glowing coal, it becomes white, and then may be easily pulverised. It does not form a jelly with acids.

Constituent Parts.

Silica, - -	40.98
Alumina, -	39.09
Lime, - -	10.95
Water, - -	16.50
	99.52

Meyer, in Beschäftigungen der Berl. Gessellschaft Naturf-Freunde, B. ii. 1776, s. 475.

Its Geognostic and Geographic Situations are the same as in the next subspecies.

Observations.

1. Crystallization, lustre, fracture, degree of transparency, and distinct concretions, are the characteristic marks of this mineral.

2. It is distinguished from *Needle Zeolite* by its crystallizations, pearly lustre, broad radiated fracture, and the want of prismatic concretions.

3. It is distinguished from *Foliated Zeolite* by its crystallizations, and radiated fracture.

Fourth

Fourth Subspecies.

Foliated Zeolite.

Blätter-Zeolith, *Werner.*

Gemeiner Zeolith, *Wid.* p. 363.—Zeolith, *Kirw.* vol. i. p. 278. —Blättriger Zeolith, *Emm.* b. i. s. 204.—Zeolite commune, *Nap.* p. 228.—Zeolite nacrée, *Lam.* t. ii. p. 305.—Zeolite lamelleuse, *Broch.* t. i. p. 302.—Stilbite, *Hauy,* t. iii. p. 161.—Blättriger Zeolith, *Mohs,* b. i. s. 374.—Stilbite, *Brong.* t. i. p. 375.—Stilbit, *Haus.* s. 96.—Blättriger Zeolith, *Leonhard,* Tabel. s. 16.—Stilbit, *Karsten,* Tabel. s. 30.—Stilbit, *Steffens,* b. i. s. 393.—Blätterzeolith, *Hoff.* b. ii. s. 240.—Stilbit, *Lenz,* b. i. s. 465. *Id. Oken,* b. i. s. 353.

External Characters.

Its colours are yellowish and greyish white, seldom milk, snow, and reddish white; from reddish-white it passes into flesh-red and brick-red, even into blood-red. It occurs also yellowish-grey and pinchbeck-brown.

It occurs massive, disseminated, globular, in amygdaloidal-shaped pieces; and crystallised,

1. In low, very oblique, sometimes rather broad, four-sided prisms: these are sometimes truncated on the acute lateral edges, and also on the angles of the obtuse lateral edges, as in fig. 76. *; or all the angles are truncated, as in fig. 77 †.

U 2 When

* Stilbite anamorphique, Hauy.

† Stilbite octoduodecimale, Hauy

When the truncations on the acute lateral edges increase, there is formed

2. A low, equiangular, six-sided prism, as in fig. 77, which is either perfect, or slightly truncated on all the edges. When this prism becomes low, there is formed an

3. Equiangular six-sided table, in which the angles on the two opposite terminal edges are truncated. Sometimes all the lateral edges of the four-sided prism are truncated, and thus there is formed an

4. Eight-sided prism.

The crystals are middle-sized, small, and very small.

The lateral planes of the prisms are transversely streaked, the terminal planes are smooth,

The planes are sometimes shining, sometimes splendent, and the lustre is vitreous.

Internally it alternates from shining to splendent, and the lustre is pearly; the pinchbeck-brown has a semi-metallic lustre.

The fracture is perfect and slightly curved foliated, with a single cleavage, which in the four-sided prism is parallel with the terminal planes, and is parallel with corresponding planes in the other crystallizations. Sometimes a conchoidal cross fracture is to be observed.

The fragments are angular and blunt-edged, and sometimes tabular.

It occurs in distinct concretions, which are large, coarse, and small angulo-granular; seldom in slightly curved lamellar concretions, which are again collected into granular concretions. The lamellar variety resembles straight lamellar heavy-spar.

The massive varieties are strongly translucent: some varieties, particularly the pinchbeck-brown, are only translucent

translucent on the edges; but the crystals are sometimes semi-transparent and transparent. It refracts single.

It is semi-hard, in a low degree; it scratches calcareous-spar.

It is brittle.

It is easily frangible; and

Rather heavy, inclining to light.

Specific gravity, 2.200, *Hoffmann.*

Chemical Character.

The same as in the preceding subspecies.

Constituent Parts.

Silica, -	58.3	52.6
Alumina, -	17.5	17.5
Lime, - -	6.6	9.0
Water, -	17.5	18.5
	100	97.0
	Meyer, in Beschäftigungen der Berl. Gesellschaft. Naturf-Freunde, B. ii. 1776, s. 475.	*Vauquelin*, Jour. de Mines, N. xxxix. p. 164.

Geognostic Situation.

It occurs principally in flœtz amygdaloid, either in drusy cavities, along with calcareous-spar and calcedony, or in cotemporaneous veins. It is also a production of primitive and transition country: there it occurs in metalliferous veins that traverse grey-wacke, as at Andreasberg in the Hartz, where the rectangular four-sided prism is associated with galena; in metalliferous primitive beds at Arendal in Norway, where it is accompanied with

 magnetic

magnetic ironstone, quartz, hornblende, epidote, and augite; at Kongsberg in Norway, where it occurs in metalliferous (apparently cotemporaneous) veins that traverse mica slate and hornblende-slate; and in small cotemporaneous veins in primitive rocks in Dauphiny *.

Geographic Situation.

Europe.—In Scotland it occurs in drusy cavities or veins in the flœtz-trap rocks that abound in the middle division of the country; also in the flœtz-trap rocks of the Hebrides, as of Canna, Skye, Mull, &c. In the north of Ireland it is an inmate of flœtz-trap rocks. It abounds in the trap-rocks of the Faroe Islands, and of Iceland; but it is a rare mineral in the Scandinavian Peninsula. It is found in the trap-rocks of Hessia; in those of Bohemia, of Auvergne, &c.

America.—It occurs in the trap-rocks of Disco in West Greenland; and in those of Zimapan in Mexico.

Asia.—Count de Bournon mentions specimens of this mineral from Kergulen's Island, or the Island of Desolation, which are in his valuable collection †.

OBSERVATIONS

* Lord Webb Seymour found this mineral in drusy cavities in the granite at Garbh coirè du, in the Island of Arran.

† Catalogue de la Collection Mineralogique du Comte de Bournon, p. 101.

OBSERVATIONS ON THE SPECIES IN GENERAL.

1. Zeolite was discovered by Cronstedt in the middle of the last century, and he published an account of it in the Memoirs of the Swedish Academy of Sciences for the year 1756. On account of its intumescing and foaming very much before the blowpipe, he named it *Zeolith*, from the Greek word *ζεω, to foam.* This name was universally adopted by mineralogists until the publication of the system of Hauy. He is of opinion that the zeolite of Cronstedt contains two distinct species; and hence he rejects the name Zeolite altogether, and substitutes in its place the names *Mesotype* and *Stilbite.* Werner, however, still retains the original name, and his own division of the species. It is equally distinct with the method of Hauy, and enables us to avoid the introduction of two new species into the system.

2. This species is well characterised. Its most frequent colour is white: the other colours which it exhibits, viz. yellow, brown, and red occurring but seldom. Its internal lustre is more or less pearly: it is never more than semi-hard; and its specific gravity does not exceed 2.200. But we observe differences in the fracture, degree of lustre, transparency, and crystallization. The differences in the fracture, conjoined with other characters, afford us the distinctions for the four subspecies into which zeolite is subdivided. These subspecies, as Werner observes, pass by almost imperceptible shades into each other; and hence all of them seem to belong to the same species.

 3. Apophyllite.

3. Apophyllite.

Fishaugenstein, *Werner.*

Apophyllite, *Hauy.*

Ichthyophthalm, *Karsten.*

Zeolith von Hallesta, *Rieman,* Vetensk. Acad. Handl. 1784.—Apophyllite, *Lucas,* p. 266. *Id. Brard,* p. 137. *Id. Hauy,* Tabl. p. 36. *Id. Hausmann,* in Weber Beiträge, b. ii. s. 59. *Id. Steffens,* b. i. s. 479.—Ichthyophthalm, *Hoff.* b. ii. s. 357. *Id. Lenz,* b. i. s. 528.—Kalkzeolith, *Oken,* b. i. s. 354.

External Characters.

Its principal colour is greyish-white, which passes into greenish-white, seldom into yellowish or reddish white. The ends of the crystals are sometimes asparagus-green; and the same colour is to be observed in patches or spots throughout the crystals. The fracture-surface is strongly iridescent.

It occurs massive, disseminated; and crystallised in the following figures:

1. Rectangular four-sided prism, which is sometimes so low as to appear tabular. This is the fundamental or primitive figure.
2. The preceding figure truncated on all the angles: when the truncating planes become so large that they touch each other, the terminal planes, in place of the octagonal figure, assume an oblique four-sided form. Frequently only a few of the angles are truncated.
3. The rectangular four-sided prism, in which all the lateral edges are truncated, thus forming an eight

eight-sided prism; sometimes the eight solid angles of this figure are truncated.

4. The rectangular four-sided prism bevelled on all the edges, or only on some of them: sometimes one of the bevelling planes is awanting, when the edge appears to be only obliquely truncated.
5. Slightly oblique four sided prism. It is formed when the truncating planes of No. 3. become so large that the original planes disappear.
6. Rectangular four-sided prism, in which the angles are truncated, and the edges bevelled.
7. Rectangular four sided table, in which the two opposite broader terminal planes are doubly bevelled, and the two smaller planes very flatly acuminated with four planes, of which two are set on the lateral planes, and the other two on the terminal planes, and the terminal edges bevelled.

The crystals are very small, small, middle-sized, and very rarely large.

The surface of the crystals Nos. 1, 2. and 4. is smooth; the surface of Nos. 3. and 5. and the acuminating planes of No. 7. are longitudinally furrowed; the bevelling planes of Nos. 4. 6. and 7. are transversely streaked. All the other planes of the secondary crystals are smooth.

Externally it is splendent; but only the terminal planes of the prism are pearly.

The principal fracture is foliated, with a threefold cleavage: two of the cleavages are parallel with the lateral planes, and one with the terminal planes of a four-sided prism. Traces of other indistinct cleavages are visible: the most distinct cleavage is that parallel with the terminal planes, and which is splendent and pearly. The cross fracture is small and perfect conchoidal, and the lustre is glistening and vitreous.

The

The fragments are tabular, and rather blunt-edged.

The massive varieties occur in distinct concretions, which are straight or curved lamellar, with slightly streaked, splendent and pearly sufaces.

It is semi-transparent, passing on the one side into transparent, on the other into translucent. It refracts single.

It is semi-hard: it scratches calcareous-spar easily, and even fluor-spar, but with difficulty *.

It is brittle, and very easily frangible.

It is rather heavy, but in a low degree.

Specific gravity, 2.417, *Riemann.* 2.467, *Hauy.* 2.430, *Rose.* 2.491, *Karsten.*

Chemical Characters.

It exfoliates very readily before the blowpipe, (it even exfoliates when held in the flame of a candle), and melts with uncommon ease into a white-coloured enamel. It phosphoresces during fusion. When thrown into acids, it exfoliates, and the folia speedily divide into smaller flocculi. When pulverised, and thrown into acids, it gelatinates in the same manner as fibrous zeolite.

Physical Character.

It becomes feebly electric by rubbing.

Constituent

* It separates into folia when we rub it on a hard body.

Constituent Parts.

	Apophyllite of Hællestad.	Apophyllite of Utön.	
Silica, - - -	55.0	50.0	52.00
Alumina, - -	2.3		
Magnesia, - -	0.5		
Lime, - -	24.7	28.0	24.50
Potash, - -	0.0	4.0	8.10
Water, - -	17.0	17.0	15.00
	99.5	99.0	99.60
	Riemann.	*Vauquelin.*	*Rose.*

Rose found the water of the apophyllite to contain a small portion of ammonia; so that the volatile alkali appears, like potash and soda, to form a constituent part of some earthy minerals.

Geognostic and Geographic Situations.

The best known locality of this mineral is the Island of Utön, not far from Stockholm, where it occurs in beds of magnetic-ironstone, along with common felspar, calcareous-spar, and hornblende. It is found also in the great copper-mine of Fahlun; in the mine of Langsoe, at Arendal in Norway; and in ironstone beds at Hällestad in East Gothland. It occurs in Faroe, as well as at Disco in Greenland; and it has lately been found in the Tyrol.

Observations.

1. This mineral, from its close affinity with Zeolite, has been confounded with it. The following characters will assist us in discriminating them. 1. Crystals of Apophyllite have pearly terminal planes, whereas those of Fibrous Zeolite are vitreous: the terminal planes of fibrous

fibrous zeolite are squares, whereas those of apophyllite are rectangles. 2. Apophyllite splits easily in the direction of the terminal planes, but with much difficulty in the direction of the lateral planes; which is precisely the reverse of what takes place in fibrous zeolite. 3. Fibrous zeolite is softer and lighter than apophyllite. 4. In the crystals of radiated zeolite, two of the lateral planes are pearly, but the others are vitreous; whereas in apophyllite, it is the terminal planes which are pearly, not the lateral planes. 5. Apophyllite is also harder and heavier than radiated or foliated zeolites; and it gelatinates with acids, which is not the case with either of these subspecies of zeolite, although it is the case with fibrous zeolite. 6. In apophyllite the fracture-surface is very strongly iridescent, which is not the case with zeolite. 7. Apophyllite exfoliates in a very remarkable manner when exposed even to a very low degree of heat, or when immersed in acids; and will even separate into a number of folia, if struck in a particular direction; and these are characters that do not occur in zeolite.

2. The Portuguese mineralogist D'Andrada, several years ago described a mineral under the name *Ichthyophthalme*, which appears to have been nothing more than a curved foliated pearly felspar. It is therefore a very different substance from the *Ichthyophthalm* of Werner, or *Apophyllite* of Hauy. The name *Ichthyophthalm* given to this mineral by Werner, is derived from the Greek words ιχθυς *fish*, and οφθαλμος *eye*, and was given to it on account of the resemblance of its pearly lustre to that of the eye of a fish. Hauy names it *Apophyllite*, from its great tendency to exfoliate.

4. Cubicite.

4. Cubicite.

Kubizit, *Werner.*

Analcime, *Hauy,* t. iii. p. 180. *Id. Mohs,* b. i. s. 385. *Id. Lucas,* p. 71. *Id. Brong.* t. i. p. 380. *Id. Haus.* s. 94. *Id. Brard,* p. 175. *Id. Leonhard,* Tabel. s. 17. *Id. Karst.* Tabel. s. 30. *Id. Hauy,* Tabl. p. 51. *Id. Steffens,* b. i. s. 401.—Kubizit, *Hoff.* b. ii. s. 251.—Analcim, *Lenz,* b. i. s. 457.—Wurfeliger Cubicit, *Oken,* b. i. s. 349.

External Characters.

Its colours are greyish and yellowish white, but seldom milk and reddish white, which latter approaches to flesh-red.

It occurs seldom massive; generally crystallised, in the following figures:

1. Perfect cube, fig, 78 *.
2. The cube flatly and deeply acuminated on all the angles, with three planes, which are set on the lateral planes, fig. 79. When the acuminating planes become larger, and at length all the planes of the fundamental figure disappear, there is formed
3. An acute double eight-sided pyramid, deeply and somewhat flatly acuminated on both extremities, with four planes, which are set on the alternate lateral edges, fig. 80.

The

* The primitive figure, according to Hauy, is the cube.

The crystals are small and very small, seldom middle-sized, and rarely large; and they are aggregated on one another, or mutually penetrate each other.

The surface of the crystals is smooth, and splendent or shining.

Internally it is intermediate between shining and glistening, and the lustre is vitreous, inclining to pearly.

The fracture is imperfect foliated, with a threefold cleavage, of which the folia are parallel with the sides of the cube. Owing to the imperfection of the foliated fracture, it appears sometimes small or fine grained uneven.

The fragments are generally indeterminate angular, seldom more or less cubical, owing to the imperfection of the foliated fracture.

The massive varieties are disposed in coarse and small angulo-granular concretions, but which are very much grown together, and in general not very distinct.

It is sometimes translucent, sometimes semi-transparent, and in crystals is transparent.

It is semi hard, in a higher degree than zeolite.

It scratches glass sensibly.

It is easily frangible; and

Rather heavy, inclining to light.

Specific gravity 2.244, *Vauquelin.*

Chemical Character.

It melts before the blowpipe into a transparent glass.

Physical Character.

By rubbing, but not by heating, it becomes electric.

Constituent

Constituent Parts.

Cubicite of Montecchio-Maggiore.

Silica, - - - -	58.0
Alumina, - - -	18.0
Lime, - - - -	2.0
Natron, - - -	10.0
Water, - - - -	8.5
	96.5

Vauquelin, Annal. du Mus. d'Hist. Nat. t. ix. p. 241.

Geognostic Situation.

It occurs in primitive and flœtz rocks, but more abundantly in flœtz than primitive country. Thus, it sometimes appears along with magnetic ironstone in gneiss, where it is associated with garnet, augite, hornblende, epidote, and calcareous-spar: in metalliferous veins that traverse clay-slate, where it is accompanied with galena, ores of silver and zinc, and calcareous-spar and quartz; very frequently in amygdaloid, and also in basalt and clinkstone porphyry, in which it occurs either in cotemporaneous veins, or in vesicular cavities.

Geographic Situation.

It occurs in the flœtz greenstone of Salisbury Craigs, and in the porphyritic rock of the Calton Hill, near Edinburgh: in the greenstone it is contained in drusy cavities, where it is associated with calcareous-spar and prehnite; in the porphyry, it is in cavities, and is associated with calcareous-spar. In Dunbartonshire, where it also occurs in flœtz-trap rocks, it is associated with prehnite, needle-zeolite, &c.; and the same is the case with the cubicite

cubicite found near Beith in Ayrshire. The flœtz-trap of Perthshire, also of the islands of Mull, Staffa, Canna, and Skye, contain crystals of this mineral. In general, it occurs in this country more frequently in the leucitic, than in any other form.

It occurs not unfrequently in the trap-rocks of Iceland, and in those of the Faroe Islands, and Disco in Greenland; but it is a rare mineral in Norway, having hitherto been found only in metalliferous beds near Arendal; and its only localities in the North of Germany, are at Andreasberg in the Hartz, where it occurs very rarely in metalliferous veins that traverse primitive clay-slate, and in Bohemia, where it is an inmate of basalt and porphyry-slate.

It occurs in flœtz amygdaloid in the Seiser Alp in the Tyrol; in a similar rock at Montecchio-Maggiore, near Vicenza in Italy; in the Bannat of Temeswar. This species was first discovered by Dolomieu, who found it in the amygdaloid rocks of Etna in Sicily; and a mineral named by the late Dr Thompson of Naples *Sarcolite*, and considered by Hauy as a variety of cubicite, occurs in the rocks of Monte Somma, near Naples *.

Observations.

1. It is distinguished as a species, and also from the nearly allied species *Zeolite*, by its threefold cleavage, its kind of lustre, its degree of hardness, and the form of its crystals. It is distinguished from *Leucite*, by its foliated fracture, inferior hardness, its fusibility, and its occurring superimposed,

* A more particular account of the Sarcolite will be given hereafter.

superimposed, and frequently forming druses, whereas leucite is always singly imbedded.

2. This mineral was formerly known as a subspecies of zeolite, under the name *Cubic Zeolite;* but Hauy ascertained that it was a distinct species, to which he gave the name *Analcime;* that is to say, a body without power, because it is only feebly electricby rubbing. Werner, from its figure, names it *Cubicite.*

3. Count de Bournon observed, in some specimens from Hartfield Moss, a transition from Cubicite into Prehnite.

5. Chabasite.

Schabasit, *Werner.*

Chabasie, *Hauy,* t. iii. p. 176. *Id. Mohs,* b. i. s. 380. *Id. Lucas,* p. 70. *Id. Brong.* t. i. p. 382. *Id. Leonhard,* Tabel. s. 16.—Chabasin, *Karst.* Tabel. s. 30. *Id. Haus.* s. 95.—Chabasie, *Brard,* p. 174. *Id. Hauy,* Tabl. p. 50.—Chabasin, *Steffens,* b. i. s. 399.—Schabasit, *Hoff.* b. ii. s. 257.—Chabasie, *Lenz,* b. i. s. 468.—Rhomboederischer Cubicit, *Oken,* b. i. s. 349.

External Characters.

Its colour is greyish-white, approaching to yellowish-white.

It seldom occurs massive; almost always crystallised, in the form of slightly oblique rhombs, which are,

1. Perfect, fig. 81 *.
2. Truncated on the six obtuse lateral edges.
3. Truncated on the six obtuse lateral edges, and on the six obtuse angles, fig. 82 †.
4. In which each of the original planes of the rhomb is divided into two, fig. 83 ‡.

* Chabasie primitive, Hauy.

† Chabasie tri-rhomboidale, Hauy.

‡ Chabasie disjointé, Hauy.

The crystals are small, middle-sized, and very small, and superimposed and resting on each other.

The lateral planes of the crystals are streaked in a peculiar manner: the streaks shoot from the shorter diagonal, (the dividing edge of the plane), and run parallel with the two adjoining lateral edges of the rhomb. The truncating planes are smooth.

Externally the crystals are splendent: internally glistening, and the lustre is vitreous.

The fracture is imperfect conchoidal, and also small grained uneven.

The fragments are angular.

It is translucent; the crystals sometimes pass into semi-transparent.

It is semi-hard; scratches glass a little.

It is easily frangible; and

Rather heavy, in a middle degree.

Specific gravity, 2.717, *Hauy*.

Chemical Character.

Before the blowpipe it melts into a whitish and vesicular mass.

Constituent Parts.

Chabasite of the Faroe Islands.

Silica, - - -	43.33
Alumina, - -	22.66
Lime, - - -	3.34
Natron, with Potash, -	9.34
Water, - - -	21.00
	99.67

Vauquelin, Annal. de Mus. d'Hist. Nat. t. ix. p. 333.

Geognostic

Geognostic Situation.

It occurs principally in flœtz-trap rocks; most frequently in the cavities of amygdaloid, where it is often associated with agate, calcareous-spar, zeolite, and green-earth. It is said also to occur in a clayey rock, which contains mica and garnet, and in small veins in a rock composed of hornblende and felspar: but we are ignorant of the class to which these rocks belong.

Geographic Situation.

Europe.—The vesicular cavities of the trap-rocks of Mull and Skye afford crystals of chabasite: it occurs in similar rocks in the north of Ireland; and beautiful specimens are found in the amygdaloid of Iceland and the Faroe Islands. The agate balls imbedded in the amygdaloid of Oberstein on the Rhine, sometimes contain beautiful crystals of this mineral; and the clayey, and felspar and hornblende rocks already mentioned, which occur in the Seiser Alp in the Tyrol, afford fine crystals of chabasite. It is said to occur in the basalt of Saxony.

Africa.—It occurs in the trap-rocks of the Isle of Bourbon.

Observations.

1. The principal characters of this species are crystallization, streaking, kind of lustre, fracture, hardness, and weight.

2. It was formerly united with cubicite, to which it is so very nearly allied, that it required the sagacity of Hauy and Werner to establish the marks of difference between them. It is distinguished from *Cubicite* by its

 crystallization,

crystallization, streaking, fracture, and greater specific gravity.

3. In form it is nearly allied both to Calcareous-spar and Axinite: it is distinguished from *Calcareous-spar* by the less obliquity of its rhomb, its fracture, and its remaining unaltered in acids; and its inferior hardness at once distinguishes it from *Axinite*.

4. The name *Chabasite* given to this species, is from *Chabazion*, a stone described by Orpheus in his poems, but unknown to us at present.

6. Cross-Stone.

Kreutzstein, *Werner*.

Hyacinth blanche cruciforme, *R. de L.* t. ii. p. 299.—Staurolite, *Kirw.* vol. i. p. 282. *Id. Estner,* b. ii. s. 499. *Id. Emm.* b. i. s. 209.—Ercinite, *Nap.* p. 239.—Andreolithe, *Lam.* t. ii. p. 285.—Harmotome, *Hauy,* t. iii. p. 191.—Pierre cruciforme, *Broch.* t. i. p. 311.—Kreutzstein, *Reuss,* b. ii. s. 430. *Id. Lud.* b. i. s. 90. *Id. Suck.* 1r th. s. 418. *Id. Bert.* s. 248. *Id. Mohs,* b. i. s. 382. *Id. Hab.* s. 24.—Harmotome, *Lucas,* p. 73.—Kreutzstein, *Leonhard,* Tabel. s. 17.—Harmotome, *Brong.* t. i. p. 385. *Id. Brard,* p. 178. *Id. Haus.* s. 95.—Kreutzstein, *Karst.* Tabel. s. 30.—Staurolite, *Kid,* vol. i. p. 251.—Harmotome, *Hauy,* Tabl. p. 51.—Kreutzstein, *Steffens,* b. i. s. 405. *Id. Hoff.* b. ii. s. 261. *Id. Lenz,* b. i. s. 471. *Id. Oken,* b. i. s. 348.

External Characters.

Its most frequent colour is greyish-white, seldom yellowish and reddish-white: the greyish-white passes into smoke-

smoke-grey; and the yellowish-white into cream-yellow, brick-red, and flesh-red.

It occurs very rarely massive; most frequently crystallised, in the following figures:

1. Generally in broad, seldom in equilateral, rectangular four-sided prisms, rather acutely acuminated on the extremities with four planes, which are set on the lateral edges, fig. 84 *.
2. The preceding figure, in which the edges formed by the meeting of the acuminating planes that rest on the broader lateral planes are truncated, fig. 85 †. When these acuminating planes become so large that the original acuminating planes almost disappear, then the prism appears bevelled on the terminal planes. Very rarely No. 1. becomes very low, when
3. A kind of garnet dodecahedron is formed.
4. Twin crystal, which is formed by two crystals of No. 1. intersecting each other, in such a manner that a common axis and acumination is formed, and the broader lateral planes make four re-entering right angles, fig. 86 ‡.

The crystals are small, middle-sized, and very small, and are singly superimposed.

The surface of the smaller lateral planes is double plumosely streaked, the broader lateral planes transversely streaked, and the acuminating planes streaked parallel with the smaller lateral planes.

 Internally

* Harmotome dodecaedre, Hauy.

† Harmotome partiel, Hauy.

‡ Harmotome cruciforme, Hauy.

The primitive form, according to Hauy, is an octahedron, with isosceles triangular faces.

Internally it is glistening, and the lustre is intermediate between vitreous and pearly.

The fracture is small and imperfect conchoidal, passing into uneven. Probably, also, it possesses an imperfect foliated fracture.

It breaks into angular and pretty sharp-edged fragments.

It is translucent, sometimes passing into semi-transparent.

It is semi-hard in a high degree, and harder than zeolite; scratches glass feebly.

It is easily frangible; and

Rather heavy, in a low degree.

Specific gravity, 2.333, *Hauy*.

Chemical Characters.

According to Link, it becomes opaque, and melts into a white-coloured glass before the blowpipe: others affirm that it is infusible without addition. It does not form a jelly with acids; and, according to Hauy, when pounded and thrown on burning charcoal, emits a yellowish phosphoric light.

Constituent Parts.

Silica,	47.5	44 to 47	49
Alumina,	19.5	20 to 12	16
Barytes,	16.0	25 to 20	18
Water,	13.5	10 to 16	15
Iron,		4	
	96.5	100 99	98
	Tassaert.	*Westrumb.*	*Klap.* Beit. b. ii. s. 83.

Geognostic

Geognostic and Geographic Situations.

It has been hitherto found only in mineral veins and in agate balls. At Andreasberg in the Hartz, it occurs in veins that traverse transition rocks, along with quartz, calcareous-spar, galena or lead-glance, copper pyrites, iron-pyrites, and grey copper ore; and of all the materials of the veins, it is the newest. The mining district of Kongsberg in Norway, which is situated in primitive strata of mica-slate and hornblende-slate, is traversed by numerous metalliferous veins, containing native silver, ores of silver, lead, zinc, arsenic, and iron, and vein-stones of calcareous-spar, heavy-spar, common quartz, and rock-crystal, and sometimes of adularia, zeolite, axinite, chlorite, mountain-cork, fluor-spar, schorl, brown-spar, and *cross-stone*. Strontian in Argyleshire is the only other place where it has been observed in veins. At Oberstein it occurs in single crystals, along with chabasite, in agate balls.

Observations.

1. This is a simple species: its distinguishing characters are, the twin form of its crystals, its kind of lustre, fracture, hardness, and weight.

2. It is allied to Radiated Zeolite; but it is distinguished from it by its twin crystals, vitreous lustre, greater hardness and weight, insolubility in acids, and infusibility before the blowpipe.

7. Laumonite.

Lomonit, *Werner*.

Zeolithe efflorescente, *Hauy*, t. iv. p. 410.—Lomonit, *Haus*. s. 95. *Id. Karsten*, Tabel. s. 32.—Laumonite, *Hauy*, Tabl. p. 49. *Id. Lucas*, t. ii. p. 188. *Id. Steffens*, b. i. s. 409. *Id. Hoff*. b. ii. s. 267. *Id. Lenz*, b. i. s. 470.—Spathiger Dolomit, *Oken*, b. i. s. 393.

External Characters.

Its colours are yellowish-white, snow-white, and greysh-white.

It occurs massive, and sometimes crystallised, in slightly oblique four-sided prisms, in which the lateral edges are rounded off, so that the crystals acquire a reed-like aspect, and are bevelled on the extremities, the bevelling planes set on the obtuser lateral edges.

The crystals are small, superimposed, and form druses.

Internally it is sometimes shining, sometimes glistening, and the lustre is pearly.

The fracture is foliated, with a twofold cleavage; and the folia are delicately longitudinally streaked.

The fragments are angular and blunt-edged.

It occurs in distinct concretions, which are large and small longish granular.

When in a fresh state it is transparent, but on exposure to the atmosphere, it very soon becomes opaque.

It is so hard as to scratch glass when fresh; but on exposure to the atmosphere, it soon becomes so soft as to yield to the mere pressure of the finger.

Is uncommonly easily frangible; and

Rather heavy, approaching to light.

Specific gravity, 2.234, *Bournon*.

Chemical

Chemical Characters.

It forms a jelly with acids. According to Vogel, it dissolves with effervescence in cold muriatic and nitric acids, and the solution immediately forms a transparent jelly: it dissolves in sulphuric acid slightly heated, and forms with it a white-coloured opaque jelly. Before the blowpipe it intumesces, and is changed into a pearly shining compact mass.

Constituent Parts.

Silica,	49.0	
Alumina,	22.0	
Lime,	9.0	
Water,	17.5	
Carbonic Acid,	2.5	
	100	*Vogel.*

Geognostic and Geographic Situations.

Europe.—This mineral was first found, in the year 1785, in the lead-mines of Huelgoet in Brittanny, by M. Gillet Laumont, a distinguished French mineralogist. Since that period, it has been discovered in other parts of the world. It is found, along with cubicite, in amygdaloid, near Paisley in Renfrewshire, and in a similar rock in the counties of Fife and Perth. At Portrush in Ireland, it is an inmate of trap-rocks, along with crystals of foliated zeolite and cubicite; and in amygdaloid in the Faroe Islands. It has been brought from Dupapiatra, near Zalathna in Transyslvania; and it is contained in the amygdaloid of the Vicentine; it likewise accompanies the beautiful apatite of St Gothard.

Asia.—It is said to occur in China, along with prehnite.

Observations,

Observations.

1. Laumonite is allied to Zeolite, but is distinguished from it by its readily falling into powder on exposure to the air. It disintegrates so readily, that if we wish to preserve our specimens unaltered, they must be kept in well closed glass vessels, or their surface must be covered with gum or varnish; and it is said that they will not disintegrate if immersed in distilled water.

2. It was named by Werner in honour of M. Gillet Laumont, its first discoverer.

3. The most complete and satisfactory account of this mineral hitherto published, is that of Count de Bournon in the first volume of the Memoirs of the Geological Society.

8. Dipyre.

Schmelzstein, *Werner.*

Leucolite de Mauleon, *Lam.* t. ii. p. 275.—Dipyre, *Hauy,* t. iii. p. 242. *Id. Broch.* t. ii. p. 508.—Schmelzstein, *Reuss,* b. iv. s. 154.—Dipyre, *Lucas,* p. 79.—Schmelzstein, *Leonhard,* Tabel. s. 17.—Dipyre, *Brong.* t. i. p. 384. *Id. Brard,* p. 191. *Id. Haus.* s. 95. *Id. Karsten,* Tabel, s. 32. *Id. Hauy,* Tabl. p. 55.—Schmelzstein, *Steffens,* b. i. s. 411. *Id. Hoff.* b. ii. s. 270.—Dipyre, *Lenz,* b. i. s. 461. *Id. Oken,* b. i. s. 350.

External Characters.

Its colour is light pearl-grey, which passes into greyish-white and reddish-white.

It occurs massive, and in minute indistinct disseminated crystals.

Internally it is shining on the longitudinal fracture, and glistening on the cross fracture, and the lustre is intermediate between vitreous and pearly

The

The longitudinal fracture has not been ascertained; but the cross fracture is small and fine-grained uneven.

The fragments are angular.

It occurs in very thin and straight prismatic concretions, and these are longitudinally streaked and shining.

It is translucent.

It is intermediate between hard and semi-hard; it scratches glass.

It is uncommonly easily frangible; and

Rather heavy.

Specific gravity, 2.630, *Hauy*.

Chemical Characters.

When pounded, and thrown on burning coals, it emits a weak phosphoric light in the dark.

It melts, and intumesces very much before the blowpipe.

Constituent Parts.

Silica,	60
Alumina,	24
Lime,	10
Water,	2
	96

Geognostic and Geographic Situations.

Brongniart says that it occurs along with iron-pyrites in steatite, and has hitherto been found only near Mauleon, in the Western Pyrenees.

Observations.

1. It is particularly characterised by its prismatic distinct concretions, its colour, great frangibility, its degree

gree of hardness, and its chemical relations. At first sight it might be confounded with Schorlite; but it is distinguished from it by the thinness of its distinct concretions, and the want of cross rents, which indicate the foliated cross fracture in schorlite. It is further distinguished from Schorlite by its degree of hardness, specific gravity, and its chemical relations. It cannot readily be confounded with Radiated Zeolite, because it differs from that mineral in colour and distinct concretions; and it is also harder and heavier than it. It is distinguished from *Needle Zeolite* by colour, fracture, greater hardness, easy frangibility, and weight.

2. It is named *Schmelzstein* by Werner, on account of its being very easily fused before the blowpipe: the name *Dipyre* given to this mineral by Hauy, has a reference to the double action of fire on it, as it melts and phosphoresces at the same time, when exposed to the action of heat.

9. Natrolite.

Natrolith, *Werner.*

Natrolith, *Reuss,* b. iv. s. 153. *Id. Mohs,* b. i. s. 364. *Id. Brong.* t. i. p. 370. *Id. Brard,* p. 415. *Id. Haus.* s. 96. *Id. Leonhard,* Tabel. s. 15. *Id. Karsten,* Tabel. s. 36. *Id. Hauy,* Tabl. p. 64. *Id. Steffens,* b. i. s. 412. *Id. Hoff.* b. ii. s. 273. *Id. Lenz,* b. ii. s. 945. *Id. Oken,* b. i. s. 356.

External Characters.

Its colour is intermediate between cream-yellow and ochre-yellow, sometimes approaching to pale yellowish-brown,

brown, or yellowish-white. The colours are generally arranged in narrow striped delineations, which are parallel with the reniform external shape.

It occurs massive, in plates, reniform, and in delicate capillary crystals.

Internally it is glistening, passing into glimmering, and the lustre is pearly.

The fracture is fibrous, and is straight, and either scopiform or stellular fibrous.

It breaks in angular and wedge-shaped pieces.

The massive and reniform varieties are composed of large and coarse granular distinct concretions, which are intersected with curved lamellar concretions. The surfaces of the concretions are streaked.

It is translucent on the edges.

It is semi-hard in a high degree: it is harder than zeolite.

It is easily frangible; and

Is rather heavy, approaching to light.

Specific gravity, 2.200, *Klaproth*.

Chemical Characters.

Before the blowpipe, it becomes first black, then red, intumesces, and forms a white compact glass.

Constituent Parts.

Silica, - -	48.00
Alumina, -	24.25
Natron, -	16.50
Oxide of Iron, -	1.75
Water, - -	9.00
	99.50

Klaproth, Beit. b. v. s. 44.

Geognostic

Geognostic and Geographic Situations.

It occurs in small cotemporaneous veins in clinkstone porphyry, in the hills of Hohentwiel, Stauffen, Hohenkrahen, and Magdeberg in Wurtemberg. Also in Scotland, as in the trap-tuff hill named the Bin, behind Burntisland, and in the trap rocks of the islands of Mull and Canna.

Observations.

1. The colour, and in particular the circular colour-delineation, the reniform external shape, the fracture, and distinct concretions, are the principal specific characters. It is distinguished from *Fibrous Zeolite*, with which it has been confounded, by its colour, external shape, distinct concretions, and hardness.

2. It was first analysed by Klaproth, who gave it its name on account of the great quantity of Natron or mineral alkali which it contains.

10. Wavellite.

Hydrargillite, *Davy*, Nicholson s Jour. xi. p. 153. *Gregor. Id.* xiii. p. 247.—Wavelit, *Karsten & Klaproth,* in Magazin der Gesellschaft Naturf Freunde zu Berlin, b. ii. s. 2. *Id. Karst.* Tabel. s. 48. *Id. Kid,* vol. i. p. 136. *Id. Haus.* s. 85.—Diaspore, *Hauy,* Tabl. p. 59.—Wavellite, *Lucas,* t. ii. p. 240. *Id. Brong.* t. i. p. 434.

External Characters.

Its colours are greyish-white, greenish-white, ash-grey, asparagus-green, and sometimes spotted brown.

It occurs botryoidal, globular, stalactitic; and also crystallised in the following figures:

1.

1. Very oblique four-sided prism, flatly bevelled on the extremities, the bevelling planes set on the obtuse lateral edges.
2. The preceding figure, very deeply truncated on the obtuse lateral edges.

Externally it is shining: internally shining, passing into splendent; and the lustre is pearly.

The fracture is narrow radiated, and is scopiform or stellular diverging.

The fragments are wedge-shaped.

It occurs in distinct concretions, which are large and coarse granular, and sometimes concentric lamellar.

It is translucent.

It is so hard as to scratch quartz.

It is brittle; and

Rather heavy.

Specific gravity, 2.270, *Lucas*. 2.22, 2.253, *Gregor*. 2.7, *Davy*.

Chemical Characters.

It becomes opaque and soft by the action of the blowpipe, but neither decrepitates nor fuses. By the aid of heat it is soluble in the mineral acids and fixed alkalies, with effervescence, and leaves very little residue.

Constituent Parts.

	Barnstaple Wavellite.		Cornish Wavellite.	Hualgayoc Wavellite.
Alumina,	71.50	70.0	58.70	68.00
Oxide of Iron,	0.50		0.19	1.00
Lime,	-	1.4	0.37	
Silica,	-		6.12	4.50
Water,	28.0	26.2	30.75	26.50
Loss,	-	-	3.87	
	100	97.6	100	100
	Klap. Beit. b. iv. s. 110.	*Davy*, Nichol. Jour. xi. 157.	*Gregor.*	*Klap.* Beit. b. v. s. 111.

It is said also to contain a small portion of Fluoric Acid.

Geognostic

Geognostic and Geographic Situations.

This mineral occurs in veins, along with fluor-spar, quartz, tinstone, and copper-pyrites, in granite, at St Austle in Cornwall. At Barnstaple in Devonshire, where it was first found by Dr Wavell, it traverses slate-clay in the form of small cotemporaneous veins *. The Secretary of the Wernerian Society, Mr Neill, found it in a similar situation in Corrivelan, one of the Shiant Isles in the Hebrides. Dr Fitton informs us, that it has been found at Spring Hill, about ten miles south-eastward from the city of Cork; and Captain Laskey collected specimens of it from rocks of slate-clay near Loch Humphry in Dunbartonshire. Humboldt brought specimens of it from the mines of Hualgayoc in South America, where it is associated with grey copper-ore; and Mr Mawe found it in Brazil.

Observations.

1. This beautiful mineral was found many years ago by Dr Wavell in a quarry near Barnstaple in Devonshire. Dr Babington examined it, and from its characters concluded that it was a particular species, to which he gave the name *Wavellite*, from the discoverer.

2. Daubuisson, Bournon, and others, are of opinion, that this mineral is a variety of the Diaspore of Hauy.

3. I have been induced to place the Wavellite in the Zeolite Family, from its general resemblance in external characters to the members of this division of the system. In this I follow Dr Thomson, who, in his System of Chemistry, places Wavellite in the Zeolite Family.

IX. AZURE-

* I satisfied myself of the true nature of these veins, by the examination of a beautiful and interesting collection of this mineral, which I owe to the politeness of Dr Wavell.

IX. AZURESTONE FAMILY.

This Family contains the following species: Azurestone, Azurite, Hauyne, and Blue-Spar.

1. Azurestone.

Lasurstein, *Werner*.

Zeolithes particulis, &c. Lapis lazzuli, *Wall.* t. ii. p. 326.—Lapis lazzuli, *R. de L.* t. ii. p. 49.—Lazurstein, *Wid.* s. 371.—Lapis lazuli, *Kirw.* vol. i. p. 283.—Lapis lazzoli, *Nap.* p. 241.—Lazulite, *Lam.* t. ii. p. 185.—La pierre d'azur, *Broch.* t. i. p. 313.—Lazulite, *Hauy,* t. iii. p. 145.—Lasurstein, *Reuss,* b. ii. s. 436. *Id. Lud.* b. i. s. 91. *Id. Suck.* 1r th. s. 423. *Id. Bert.* s. 169. *Id. Mohs,* b. i. s. 387. *Id. Hab.* s. 25.—Lazulite, *Lucas,* p. 66.—Lasurstein, *Leonhard,* Tabel. s. 16.—Lazulite, *Brong.* t. i. p. 367. *Id. Brard,* p. 164.—Lasurstein, *Haus.* s. 94. *Id. Karst.* Tabel. s. 44.—Lapis lazuli, *Kid,* vol. i. p. 244.—Lazulite, *Hauy,* Tabl. p. 47.—Lasurstein, *Steffens,* b. i. s. 414. *Id. Hoff.* b. ii. s. 276. *Id. Lenz,* b. i. s. 475. *Id. Oken,* b. i. s. 355.

External Characters.

Its colour is azure-blue, of all degrees of intensity: the lighter varieties pass into Berlin-blue and smalt-blue; and the darker into blackish-blue. The white spots it sometimes contains, are probably owing to an intermixed mineral.

It is found massive, disseminated, in rolled pieces; and crystallised, in rhomboidal dodecahedrons.

Internally

Internally it is either glistening or glimmering.

The fracture is small and fine-grained uneven.

The fragments are angular, and rather blunt-edged.

It is feebly translucent on the edges.

It is intermediate between hard and semi-hard: it scratches glass, and in some places gives a few sparks with steel.

It is easily frangible.

It is rather heavy, but in a middling degree.

Specific gravity, 2.771, *Blumenbach*. 2.767 to 2.945, *Hauy*. 2.896, *Kirwan*. 2.761, *Brisson*. 2.959, *Karsten*.

Chemical Characters.

It retains its colour in a low degree of heat: in a higher heat, it melts into a blackish mass; and in a very high heat, it melts into a white enamel. When pounded and calcined, it forms a jelly with acids.

It is deprived of its colour by all the mineral acids: with great rapidity by nitrous acid; less rapidly by muriatic acid; and slowest by means of sulphuric acid.

Constituent Parts.

Silica, - -	46.00
Alumina, - -	14.50
Carbonate of Lime,	28.00
Sulphate of Lime,	6.50
Oxide of Iron, -	3.00
Water, - -	2.00
	100

Klaproth, b. i. s. 196 *.

Geognostic

* The older chemists were of opinion, that the beautiful colour of this mineral was owing to copper; but it is now known that iron is the only colouring principle it contains.

Geognostic Situation.

Its geognostic situation is still imperfectly known. It appears sometimes to occur in primitive limestone, along with iron pyrites, in Persia, Tartary and China; in veins that traverse granite, along with quartz, mica, and iron-pyrites in the Altain mountains; and at the southern end of the Lake Baikal in Siberia, in a vein, associated with garnets, mica, felspar, and iron-pyrites.

Geographic Situation.

It is found in Persia, Bucharia, China, Great Tartary, and Siberia. Mr Pennant, in his Outlines of the Globe, informs us, that it is found in considerable quantities in the Island of Hainan in the Chinese sea, from whence it is sent to Canton, where it is employed in china painting.

Uses.

On account of its beautiful blue colour, and the fine polish it is capable of receiving, it is much prized by lapidaries, and is cut as ring-stones, seal-stones, vases, snuff-boxes, and other ornamental articles of the same nature: it is also used in mosaic and Florentine work. It is highly valued by painters, on account of the fine ultramarine blue colour obtained from it.

The whole art in preparing this colour, consists in freeing the azurestone from all impurities, and reducing it to an extremely fine powder. This is done in the following manner: The azurestone is first reduced to a coarse powder, and then exposed for an hour in a crucible to a pretty strong heat. Vinegar is then poured on it, and the whole is allowed to stand for some days: at the end of this time, the vinegar is poured off, and the powder is still further comminuted, by rubbing in a glass mortar. The roasting or calcination of the azurestone must be repeated one or more times, if the first heating has not rendered it so friable as to allow of its being reduced to a sufficiently fine powder. The powder is now to be repeatedly washed with water, in order to free it from the vinegar with which it is combined, and then to be ground on a stone of porphyry or agate, until it is rendered completely impalpable. It is next to be thrown into a melted mixture of pitch, wax, and linseed oil, and carefully mixed with it, and then allowed to cool. Tepid water is next to be poured on this mixture, and the whole is to be well triturated by means of a pestle: the water becomes muddy, and is to be poured off; fresh water is to be added, which very soon assumes a beautiful blue colour. When this water is sufficiently saturated, it is poured off, fresh water is added to the mixture, and soon assumes a blue colour, but of a paler tint than the former, and this process is repeated, until the water becomes only of a dirty grey colour. A powder is deposited from each of these ablutions, and the beauty of its colour depends on the purity of the azurestone and the ablution itself, the first always affording the finest and richest colour. The foreign parts remain combined with the cement. It was formerly an article of the materia medica, and was therefore kept in apothecaries shops; but very often its place was supplied by azure copper-ore, mixed with limestone, which was named *Armenian Stone.*

Observations.

1. This mineral is distinguished by its colour, low degree of lustre, fracture, its low degree of translucency on the

the edges, its hardness, and geognostic situation. It has been confounded with Azure Copper-Ore; but it differs from that mineral in lustre, fracture, hardness, and geognostic situation.

2. Azurestone was well known to the Greeks and Romans, under the name of *Sapphire:* when it contained much disseminated iron-pyrites, it was then called *Sapphirus regius*, because the pyrites was supposed to be gold.

3. It is generally known under the name *Lapis lazuli:* Lazulus is derived from the Arabian word *azul*, the heaven, and refers to the fine blue colour of this mineral. The name *Ultramarine*, given to the fine pigment obtained from azurestone, is said to have been given it on account of its having been brought into Europe from beyond the sea.

2. Azurite.

Lazulit, *Werner.*

Le Lazulithe, *Broch.* t. i. p. 315. *Id. Hauy,* t. iii. p. 145.—Unachter Lazurstein, *Reuss,* b. ii. s. 440.—Lazulit, *Lud.* b. i. s. 86. *Id. Suck.* 1r th. s. 319. *Id. Bert.* b. i. s. 263. *Id. Mohs,* b. i. s. 185. *Id. Lucas,* p. 277.—Lazulite de Klaproth, *Brong.* t. i. p. 369.—Gemeiner Lazulit, *Karst.* Tabel. s. 46.—Lazulit de Verner, *Hauy,* Tabl. p. 62.—Lazulith, *Steffens,* b. i. s. 418. *Id. Hoff* b. ii. s. 285.—Gemeiner Lazulit, *Lenz,* b. i. s. 481. *Id. Oken,* b. i. s. 336.

External Characters.

The most frequent colour of this mineral is indigo-blue, which sometimes inclines to sky-blue, sometimes to smalt blue.

It occurs in small massive portions, disseminated, and crystallised in very oblique four-sided prisms, which are rather flatly acuminated on the extremities, with four planes set on the lateral edges.

The cross fracture is glistening, the principal fracture shining, and the lustre vitreous.

The longitudinal fracture is imperfect foliated, the cross fracture small and fine-grained uneven.

The fragments are angular, and rather sharp-edged.

It is opaque, or very feebly translucent on the edges.

It is intermediate between hard and semi-hard; it scratches glass.

It is easily frangible.

It is rather heavy.

Chemical Characters.

It is infusible without addition before the blowpipe; but with borax, it forms a clear pale wine-yellow vitreous pearl.

Constituent Parts.

Alumina, - -	66
Silica, - -	10
Magnesia, - -	18
Lime, - -	2
Oxide of Iron, -	2½
	98½ *Tromsdorf.*

Geognostic Situation.

It occurs imbedded in small portions in quartz; also in fissures in clay-slate, along with sparry-ironstone, heavy-spar, and quartz.

Geographic Situation.

It occurs principally in the district of Vorau in Stiria; also in the neighbourhood of Wienerisch-Neustadt in Austria, and near Schwatz in the Tyrol. In all these places it is imbedded in quartz. It occurs in transition clay-slate in the Pinzgau, and near Werfen in Salzburg.

Observations.

1. The foliated fracture distinguishes it from *Azure-stone.*

2. It is named *Azurite,* from its resemblance to azure-stone in general appearance.

3. Lenz subdivides this species into two subspecies, viz. Common Lazulite and Siderite; but it does not appear that this subdivision was necessary. The name *Siderite* is applied by some mineralogists to the beautiful blue quartz of Golling in Salzburg.

3. Hauyne.

Häuyn, *Karsten.*

Latialite, *Gismondi & Hauy,* Tabl. p. 62.—Saphirin, *Nose,* in Mineral Studien, p. 162.—Häuyn, *Steffens,* b. i. s. 416. *Id. Lenz,* b. i. s. 479. *Id. Oken,* b. i. s. 355.

External Characters.

Its colour is sky-blue, passing on the one side into pale Berlin-blue, on the other into celandine-green. When held between the eye and the light, all the colours appear somewhat muddy.

It occurs disseminated, in angular imbedded grains, and crystallised in rhomboidal dodecahedrons like garnet.

Internally it is shining and vitreous.

The fracture is imperfect conchoidal, passing into splintery, or inclining to imperfect foliated.

The fragments are angular and sharp-edged.

It is semi-transparent and transparent.

It scratches glass; also felspar and quartz very feebly.

It is easily frangible.

Specific gravity, 3.333, *Gismondi.* 3.100, *Neergaard.*

Chemical Characters.

It is infusible without addition before the blowpipe, and its colour remains unaltered; but with borax forms yellowish-green glass. With acids it forms a translucent jelly.

Constituent Parts.

Silica, - -	30.0	
Alumina, -	15.0	
Sulphate of Lime,	20.5	
Lime, - - -	5.0	
Potash, - -	11.0	
Iron, - -	1.0	
Trace of Sulphureted Hydrogen.		
Loss, - -	17.5	
	100	*Vauquelin.*

The

The great loss in this analysis, is conjectured by Vauquelin to be owing to the escape of water, which appears to be an essential ingredient in minerals that form a jelly with acids.

Geognostic and Geographic Situations.

It occurs imbedded in the basalt rocks of Albano and Frascati, along with mica, augite, leucite, and vesuvian; also in the basalt of Andernach.

Observations.

1. It is distinguished from the other minerals of the Azurestone Family by its fracture, lustre, and hardness, and particularly its geognostic situation.

2. It was first discovered by the Abbé Gismondi, who named it *Latialite*, from Latium, the ancient name of the country where it occurs: the German mineralogist Nose, who observed it in the trap-rocks of Andernach, considered it as allied to Sapphire, and has described it under the name *Saphirin*: Ferber names it *Blue-schorl* of Andernach: Cordier arranged it with Spinel: Bruun Neergaard, who has given the fullest account of it, has placed it in the system under the name *Hauyne*; and Steffens, in his System of Mineralogy, places it between Azurestone and Azurite.

4. Blue-Spar.

Blauspath, *Werner*.

Wid. Bergm. Journ. 1791, p. 345.—Felsite, *Kirw.* vol. i. p. 326. —Dichter Feldspath, *Emm.* b. i. s. 271.—Le Feldspath compacte, *Broch.* t. i. p. 367.—Feld-spath bleu? *Hauy,* t. ii. p. 605.—Dichter Feldspath, *Reuss,* b. ii. s. 46. *Id. Lud.* b. i. s. 100. *Id. Suck.* 1r th. s. 420. *Id. Bert.* s. 238. *Id. Mohs,* b. i. s. 420. *Id. Leonhard,* Tabel. s. 19.—Feldspath Bleu, *Brong.* t. i. p. 360.—Splittriger Lazulit, *Karsten,* Tabel. s. 46. —Blue Felspar, *Kid,* vol. i. p. 160.—Feldspath Bleu, *Hauy,* Tabl. p. 60.—Blauspath, *Steffens,* b. i. s. 420. *Id. Hoff.* b. ii. s. 287. *Id. Lenz,* b. i. s. 479. *Id. Oken,* b. i. s. 337.

External Characters.

Its colour is pale smalt-blue, which sometimes passes into sky-blue, and occasionally into milk-white.

It occurs massive and disseminated.

Internally it is glistening, which rarely approaches to shining.

The fracture is imperfect foliated, inclining to splintery.

It is translucent in a low degree.

It is hard, inclining to semi-hard; scratches glass.

It is rather difficultly frangible.

Yields a greyish white coloured streak.

It is rather heavy, but in a low degree.

Specific gravity, 3.046, *Klaproth.* 3.060, *Karsten.*

Chemical Characters.

Before the blowpipe it becomes white and opaque; and affords a black-coloured glass with borax.

Constituent

Constituent Parts.

Silica, - - - -	14.00
Alumina, - - -	71.00
Magnesia, - - -	5.00
Lime, - - - -	3.00
Potash, - - -	0.25
Oxide of Iron, - -	0.75
Water, - - - -	5.00
	99.00

Klaproth, Beit. b. iv. s. 285.

Geognostic and Geographic Situations.

It occurs along with quartz, mica, and garnets, and probably either in the form of a bed or a mountain-mass. It is found in the valley of Murz, near Krieglach in Stiria.

Observations.

1. This species is not very extensive, and, as far as we know at present, does not appear to be of great importance. Its essential characters are its blue colour, low degree of lustre, very imperfect foliated fracture, inconsiderable translucency, its greyish-white coloured streak, and hardness, which is not higher than hard, approach- to semi-hard, and a specific gravity of 3.

2. It is allied on the one hand to Azurite, and on the other to Compact Felspar, with which it was long confounded. It is distinguished from *Azurite* by its paler colour, inferior lustre, more imperfect foliated fracture, which even inclines to splintery, rather inferior hardness, greater specific gravity, its chemical relations, and composition: It is distinguished from *Compact Felspar* by its blue

blue colour, higher lustre, more distinct foliated fracture, slightly inferior hardness, but greater weight, and also by its chemical characters, and composition.

X. FELSPAR FAMILY.

THIS Family contains the following Species: Andalusite, Saussurite, Chiastolite, Indianite, Felspar, Spodumene, Bergmannite, Scapolite, Elaolite, Sodalite, Meionite, Nepheline, and Ice-Spar.

1. Andalusite.

Andalusit, *Werner*.

Spath adamantin d'un rouge violet, *Bournon*, Journ. de Phys. 1789, p. 453.—Andalusit, *La Meth.* Id. An 6. p. 386. *Id. Reuss*, b. iv. s. 135.—Foretzer Feldspath, *Suck.* 1r Abh. s. 396.—Andalusit, *Mohs*, b. i. s. 423. *Id. Karsten*, Tabel. s. 46. *Id. Leonhard*, Tabel. s. 19. *Id. Brong.* t. i. p. 363.—Feld-spath apyre, *Hauy*, Tabl. p. 60.—Andalusit, *Steffens*, b. i. s. 455. *Id. Hoff.* b. ii. s. 291. *Id. Lenz*, b. i. s. 484. *Id. Oken*, b. i. s. 317.

External Characters.

Its colour is flesh-red, which sometimes inclines to pearl-grey.

It occurs massive, and crystallised in rectangular four-sided prisms, in which the terminal angles and lateral edges are sometimes truncated.

The crystals are seldom large, generally middle-sized or small, and almost always imbedded.

The

The principal fracture is shining, in a low degree; the cross fracture glistening, and the lustre is vitreous.

The principal fracture is rather imperfect foliated, apparently with a twofold rectangular cleavage, in which the folia are parallel with the lateral planes of the prism; the cross fracture is uneven.

The fragments are indeterminate angular.

It is feebly translucent.

It is hard; it scratches quartz, and sometimes even spinel *.

It is rather easily frangible.

It is rather heavy, in a middle degree.

Specific gravity, 3.050, *Romé de Lisle.* 3.074, *Guyton.* 3.165, *Hauy.* 3.060, 3.127, *Von Voith.*

Chemical Characters.

It becomes white before the blowpipe, but does not melt. The andalusite of Herzogau was exposed by Bucholz for an hour and half to a temperature equal to that of melting silver, when its colour was changed, its lustre almost destroyed, but it appeared to have increased in hardness and brittleness.

Constituent Parts.

Silica, -	32	29.12
Alumina, -	52	51.07
Potash, -	8	
Oxide of Iron,	2	7.83
	94	88.02
	Vauquelin.	*Guyton.*

Geognostic

* Herr Von Voith says that he found the schorl of Hörlberge, the steatite in gneiss near Werneberg, and the andalusite of Herzogau, so soft in their orignal repositories, that he could flatten them between the fingers, and cut them with a knife, but that they became very hard on exposure to the air.

Geognostic Situation.

It occurs in gneiss and mica-slate; also in veins that traverse granite, or gneiss, either along with felspar, or with felspar, quartz, mica, and schorl.

Geographic Situation.

It occurs in gneiss in Aberdeenshire; in mica-slate in the Island of Unst; also in mica-slate on the north-east side of Douce mountain, in the county of Wicklow; and at Killeny, in the county of Dublin, where it was first noticed by my friend Dr Blake *.

On the Continent it was first found in the province of Andalusia, or Castile; afterwards by Count de Bournon at Imbert, near Montbrison in Forez, in a vein of felspar traversing granite. It also occurs in veins in gneiss near Herzogau, in the Upper Palatinate; and in gneiss near Bodenmais in Bavaria. In the Fichtelgebirge, and at Braunsdorf, near Freyberg in Saxony, it occurs imbedded in mica-slate.

Observations.

It is distinguished by its colour, external shape, fracture, hardness, and weight. It has been confounded with Felspar; but it is distinguished from that mineral by its greater hardness, and weight, and its infusibility. Its double rectangular cleavage, and its inferior specific gravity, distinguish it from *Corundum*.

2. Saussurite.

---

* Fitton's Mineralogy of Dublin, p. 47.

2. Saussurite.

Magerer nephrite, *Reuss*, b. ii. s. 192.—Jade, *Saussure*, Voyages.—Jade tenace, *Hauy*, t. iv. p. 368.—Jade de Saussure, *Brong.* t. i. p. 348.—Saussurite, *Karst.* Tabel. s. 34.—Feldspath tenace, *Hauy*, Tabl. p. 36.—Saussurite, *Theodore de Saussure*, Journal des Mines, n. cxi. p. 205. *Id. Steffens*, b. i. s. 451.—Variolit, *Hoff.* b. ii. s. 338.—Saussurit, *Lenz*, b. i. s. 507. *Id. Oken*, b. i. s. 332.

External Characters.

Its colours are white, grey, and green: it passes from greyish-white into greenish-white, greenish-grey, bluish-grey, and mountain-green; and sometimes it occurs smoke-grey and pearl-grey.

It occurs massive, disseminated, and in rolled pieces.

Internally it is dull, or feebly glimmering.

The fracture is splintery; and sometimes we can discern an imperfect foliated fracture, with a double rectangular cleavage.

It breaks into very sharp-edged pieces.

It is faintly translucent on the edges.

It is very difficultly frangible.

It is hard; according to Saussure it scratches quartz.

It is meagre to the feel; and

Rather heavy.

Specific gravity, 3.200, *Klaproth*. 3.310, 3.319, *Saussure*.

Chemical Characters.

Before the blowpipe it melts on the edges and angles, but is not entirely melted.

Constituent

Constituent Parts.

Silica, -	44.00	49.00
Alumina. -	30.00	24.00
Lime, - -	4.00	10.50
Magnesia, - -		3.75
Natron, -	6.00	5.50
Potash, - -	0.25	
Iron, - -	12.50	6.50
Manganese, -	0.05	
	96.80	99.25
	Saussure, Journ. des Mines, n. cxi. p. 217.	*Klaproth*, Beit. b. iv. s. 278,

Geognostic and Geographic Situations.

It occurs at the foot of Mount Rose. Rolled pieces are found at the mouth of the Reuss; and large blocks, containing diallage, abound in the Pays de Vaud. Rolled masses are scattered on the shores of the Lake of Geneva; and the well known rock from Corsica, named *Verde di Corsica*, is a compound of diallage and saussurite. It is also found in Norway, Finland, Italy, France, and Savoy.

Observations.

The older Saussure, who has particularly described this mineral, was of opinion that it belonged to the magnesian class, and arranges it in the system under the name *Jade*. Saussure the Younger, after a careful examination and analysis, found that it could not be arranged with jade, but formed a distinct species, nearly allied to felspar, to which he gave the name *Saussurite*. Its affinity

affinity with felspar is so great, that Hauy describes it under the title *Feldspath tenace*; and Werner places it in his system as a kind of compact felspar, under the name *Variolite*: but its high specific gravity, great hardness, difficult frangibility, and chemical characters and composition, induce me, with Saussure the Younger, and Steffens, to retain it in the system as a distinct species.

3. Chiastolite.

Hohlspath, *Werner*.

Röbien, in Nouv. idées sur la Format. des Foss. p. 108.—Pierre de croix, *Romé de L.* t. ii. p. 440.—Crucite, *Lam.* t. ii. p. 292.—Macle, *Broch.* t. ii. p. 514. *Id. Hauy*, t. iii. p. 267.—Chiastolith, *Reuss*, b. ii. s. 47. *Id. Lud.* b. i. s. 149. *Id. Suck.* 1r th. s. 476. *Id. Bert.* s. 201. *Id. Mohs*, b. i. s. 539. *Id. Hab.* s. 35.—Macle, *Lucas*, p. 85.—Chiastolith, *Leonhard*, Tabel. s. 20.—Macle, *Brong.* t. i. p. 498. *Id. Brard*, p. 200.—Chiastolith, *Haus.* s. 88. *Id. Karsten*, Tabel. s. 34.—Macle, *Hauy*, Tabl. p. 56.—Chiastolith, *Steffens*, b. i. s. 447.—Hohlspath, *Hoff.* b. ii. s. 330.—Chiastolith, *Lenz*, b. i. s. 503.—Hohlspath, *Oken*, b. i. s. 324.

External Characters.

Its colours are white and grey: the white colours are yellowish-white, greenish white, greyish-white, and reddish-white: the grey colours are pearl-grey, greenish-grey, and yellowish-grey.

It occurs always crystallised, in the following figures:

1. Four-sided prism, which is nearly rectangular *.

* Macle prismatique, Hauy.

2. Four-sided prism, in which the lateral edges are rounded *.

3. Four prisms arranged in the form of a cross †.

These crystals always appear as if they had been at one time hollow, and these hollows filled up with clay-slate, the position of which varies in regard to the crystals, and gives rise to the following varieties:

a. In the centre of the crystal there is a small prism of clay-slate, the lateral planes of which are parallel with those of the crystal, and from the angles of this prism black lines run to each angle of the crystal ‡, fig. 87.

b. In this variety there is, in addition to the central prism and black lines of the former, smaller black-coloured prisms of clay-slate, one on each angle of the crystal, and their lateral planes are parallel with those of the crystal ||, fig. 88.

c. In this variety the terminal planes of the crystal are marked with black lines, which run from each of the lateral planes, parallel with the adjacent planes, to the black diagonal lines §, fig. 89.

d. A black prism, in which the lateral planes are covered with a thick or thin crust of the hollow-spar ¶.

The black or clay-slate mass is often thickest in the middle, and becomes thinner towards the extremities of the crystal: in other instances it is thinnest in the middle, and becomes gradually thicker towards the extremities of

---

* Macle cylindroide, Hauy.

† Macle quaternée, Hauy.

‡ Macle tetragrammé, Hauy.

|| Macle pentarhombique, Hauy.

§ Macle polygrammé, Hauy.

¶ Macle circonscrite, Hauy.

of the crystal; and frequently the clay-slate mass is of equal thickness throughout.

The crystals are large, middle-sized, and small; sometimes also acicular, and always imbedded.

The lustre of the principal fracture is glistening, that of the cross fracture glimmering.

The fracture is foliated, with a double cleavage, the folia of which are parallel with the lateral planes of the prism; sometimes it has also a splintery appearance.

It is translucent.

It is hard; it scratches glass.

It is rather difficultly frangible; and

Rather heavy.

Specific gravity, 2.944, *Hauy*. 2.923, *Karsten*.

Chemical Characters.

It is infusible before the blowpipe, and becomes white and nearly opaque.

Its constituent parts have not hitherto been ascertained.

Geognostic and Geographic Situations.

It occurs in small acicular crystals in clay-slate in Cumberland; also at Aghavanagh, and Baltinglass-hill, in the county of Wicklow *. The largest and most beautiful crystals are found in clay-slate near to St Brieux in Brittany: smaller crystals occur in the clay-slate of St Jago di Compostella in Gallicia; the variety *d* is found in the valley of Barreges in the Pyrenees; and the variety 3. in the plain of Thourmouse, in the High Pyrenees.

 It

* Fitton's Mineralogy of Dublin, p. 51. & 52.

It has been observed in micaceous clay-slate in the Serra de Marao in Portugal; and in very small acicular crystals in clay slate near Gefrees in Bareuth.

Observations.

1. Some mineralogists, as Hauy and Mohs, maintain that chiastolite is a talcaceous mineral: others, as Bernardi, Dr Fitton, and the late Mr Stephens of Dublin, are of opinion, that it is the same substance with Andalusite; whilst Werner and Hoffmann consider it as nearly allied to Common Felspar.

I agree with Steffens in considering it as a distinct species, and, from its affinity with Andalusite, place it in the system beside it.

2. It agrees with Andalusite in form, fracture, and transparency, but differs from it in hardness and structure.

3. It is evident from the arrangement of the clay-slate in the chiastolite, that it cannot be considered as having filled previously existing hollows in it, but is to be viewed as of cotemporaneous formation with the spar.

4. Indianite.

Indianite, *Bournon.*

External Characters.

Its colours are white and grey.

It occurs massive.

Its lustre is shining.

The fracture is foliated.

It occurs in granular distinct concretions, which alternate from large to very small granular.

It

It is translucent, sometimes inclining to transparent.
It is hard; scratches glass, but is scratched by felspar.
It is rather heavy.
Specific gravity, 2.7420.

Chemical Characters.

It is infusible before the blowpipe: some of the varieties form a kind of jelly with nitrous acid.

Constituent Parts.

Silica,	42.5	
Alumina,	37.5	
Lime,	15.0	
Iron,	3.0	
Trace of Manganese.		
	100.0	*Chenevix.*

Geognostic and Geographic Situations.

It has hitherto been found only in the Carnatic, where it is associated with hornblende, and contains imbedded crystals of corundum.

Observations.

The preceding description of this species is drawn up from the account of it by the discoverer Count de Bournon, contained in his Catalogue Mineralogique.

 5. Felspar.

5. Felspar.

This Species is divided into the following subspecies: Adularia, *a*. Adularia, *b*. Glassy Felspar; Labrador-stone; Common Felspar, *a*. Fresh, *b*. Disintegrated; and Compact Felspar.

First Subspecies.

This subspecies is subdivided into two kinds, Adularia, and Glassy Felspar.

First Kind.

Adularia.

Adular, *Werner*.

Moonstone, *Kirw*. vol. i. p. 322.—Adular, *Estner*, b. i. s. 525. *Id. Emm.* b. i. s. 277.—Adularia, *Nap*. p. 218.—Adulaire, *La Meth*. t. ii. p. 194. *Id. Broch.* t i. p. 371.—Feldspath nacré, *Hauy*, t. ii. p. 600.—Adular, *Reuss*, b. ii. s. 379. *Id. Lud.* b. i. s. 101.—Opalisirender Feldtein, *Bert*. s. 242.—Adular, *Suck*. 1r th. s. 389. *Id. Mohs*, b. i. s. 394.—Adularischer Feldspath, *Hab*. s. 21.—Feldspath nacré, *Lucas*, p. 50.—Opalisirender Feldspath, *Leonhard*, Tabel. s. 17.—Feldspath Adulaire, *Brong*. t. i. p. 358.—Feldspath limpide, *Brard*, p. 134.—Opalisirender Feldspath, *Karsten*, Tabel, s. 34.—Adularia, *Kid*, vol. i. p. 158.—Feldspath nacré, *Hauy*, Tabl. p. 36—Adular, *Steffens*, b. i. s. 422. *Id. Hoff*. b. i. s. 296. *Id. Lenz*, b. i. s. 486.—Opalisirender Feldspath, *Oken*, b. i. s. 375.

External Characters.

The principal colour is greenish-white, which sometimes

times passes into greyish-white and milk-white, and even inclines to asparagus-green. It is frequently iridescent; and the milk-white varieties, in thin plates, when held between the eye and the light, sometimes appear pale flesh-red.

It occurs massive; and frequently crystallised.

The fundamental crystallization is,

1. A very oblique four-sided prism, flatly bevelled on the extremities, and the bevelling planes set on the obtuse lateral edges.

Sometimes two diagonally opposite bevelling planes are smaller, and at length disappear, when there is formed

2. An oblique four-sided prism, in which the terminal planes are set on obliquely.
3. The fundamental figure is sometimes truncated on the acute lateral edges. When these truncating planes become larger, there is at length formed
4. A broad rectangular six-sided prism, flatly bevelled on both extremities, and the bevelling planes are set on the lateral edges, which are formed by the smaller lateral planes.

Sometimes the prism becomes so broad and thin, that it may be described as a

5. Six-sided table, in which the smaller lateral planes of the preceding figure form bevelments on the terminal planes.
6. Rectangular four-sided prism, in which the terminal planes are obliquely bevelled.

Sometimes twin-crystals occur: one variety is the same as that afterwards to be described as occurring in common felspar; the other is formed by two tabular crystals of the variety 5 growing together by their broader lateral planes.

The crystals are generally middle-sized and large, sometimes very large, but seldom small. They are always superimposed, and either single or variously aggregated.

The lateral planes of the prism are longitudinally streaked.

Externally it is splendent; internally the principal fracture is splendent, and the cross fracture shining and glistening. The lustre is intermediate between vitreous and pearly.

The principal fracture is foliated, with a double rectangular cleavage. In the rectangular four-sided prism, the two cleavages are parallel with the lateral planes; consequently in the six-sided prism they are parallel with the broader lateral planes, and the two diagonally opposite bevelling planes; and in the fundamental figure with the acute lateral edges, and the two diagonally opposite bevelling planes. These two cleavages are obliquely traversed by a third cleavage, which is parallel with one of the lateral planes of the fundamental figure: it is, however, but imperfect, and seldom to be seen, the mineral in this, as in all other directions across the two rectangular cleavages, having a small and imperfect conchoidal fracture.

It generally breaks into sharp angular fragments, and occasionally into rhomboidal fragments.

The massive varieties occur in large granular concretions, and sometimes in straight lamellar concretions, which are imperfect and thick.

It is semi-transparent, sometimes inclining to transparent, or is translucent.

The translucent varieties, when viewed in a certain direction,

rection, sometimes exhibit a silvery or pearly light *. It refracts double.

It is hard in a low degree; but is harder than common felspar, as it affords more sparks with steel than that mineral.

It is easily frangible.

It is rather heavy, in a low degree.

Specific gravity, 2.564, *Brisson.* 2.531 & 2.560, *Hoffmann.*

Chemical Character.

It melts before the blowpipe, without addition, into a white-coloured transparent glass.

Constituent Parts.

Silica,	64
Alumina,	20
Lime,	2
Potash,	14
	100 *Vauquelin.*

Geognostic Situation.

It occurs in cotemporaneous veins or drusy cavities in granite and gneiss: in these repositories it is associated with rock-crystal, calcareous-spar, epidote, amianthus, but principally with chlorite and felspar.

Geographic Situation.

Europe.—It occurs in the granite of the Island of Arran; and in the granite and gneiss rocks of Switzerland, France,

* This beautiful pearly light is generally seen when the specimen is viewed in the direction of the imperfect or third cleavage.

France, and Germany. The largest and most beautiful crystals are found in the mountain of Stella, a part of St Gothard.

Asia.—Rolled pieces, having a most beautiful pearly light, are collected in the Island of Ceylon.

Uses.

The variety which exhibits the pearly light is valued by jewellers: it is cut into a semi-globular form, and is sold under the name of *Moonstone.* It is usually worn as a ring-stone, and, when set round with diamonds, its pearly lustre forms a striking and agreeable contrast with that gem.

Observations.

1. This mineral is known by its white colour, iridescence, pearly light, splendent external and internal lustre, conchoidal cross fracture, high degree of transparency, specific gravity, and considerable hardness.

2. It was first discovered by an Italian mineralogist, Professor Pini of Milan, in the mountain of Stella, belonging to the St Gothard groupe. He named it *Adularia Felspar*, in the belief that the mountain on which he had collected it was named *Adula;* but the truth is, the mountain of Adula does not occur near St Gothard; it is situated in the Grisons.

3. Werner is of opinion that the moonstone of Ceylon is the ὑαλοειδης of Theophrastes, the Asteria, Astrios, and Androdamas of Pliny, and the Girasole of the Italians. Before Werner ascertained its true nature, it was by some arranged with the Opal, and by others with the Cat's-eye.

Second

Second Kind.

Glassy Felspar.

Glasiger Feldspath, *Werner.*

Nose, Orthographische Briefe, 1. s. 128.—Nöggerath Studien: s. 27.—*Reuss,* Mineralogishe Briefe, 1. n. 2. a. a. o.—Glasiger Feldspath, *Karsten,* Tabel. s. 34. *Id. Haus.* s. 88. *Id. Steffens,* b. i. s. 441. *Id. Hoff.* b. ii. s. 328. *Id. Lenz,* b. ii. s. 502. *Id. Oken,* b. i. s. 375.

External Characters.

Its colour is greyish-white, sometimes passing into grey.

It occurs always crystallised, in broad rectangular four-sided prisms, bevelled on the extremities. These crystals are often very much cracked; they are generally small, seldom midddle-sized, and always imbedded.

Internally it is splendent, and the lustre is vitreous.

The principal fracture is foliated, with the same cleavage as in adularia: the cross fracture is uneven, or small and imperfect conchoidal.

It is transparent.

In all its other characters it agrees with adularia.

Specific gravity, 2.575, *Klap.* 2.518, 2.589, *Stucke.*

Chemical Characters.

Before the blowpipe, it melts without addition into a grey semi-transparent glass.

Constituent.

Constituent Parts.

Silica, - -	68.0
Alumina, -	15.0
Potash, - -	14.5
Oxide of Iron, -	0.5
	98.0

Klaproth, Beit. b. v. s. 18.

Geognostic and Geographic Situations.

It occurs imbedded in a porphyritic rock in the Siebengebirge; also in a rock composed of white felspar, and very small blackish-brown scales of mica, and fine disseminated magnetic ironstone, in the Drachenfels on the the Rhine. It is an inmate of the flœtz-trap rocks of the Bohemian Mittelgebirge; and has been noticed in the porphyritic pumice of Hungary; and the flœtz-trap and pitchstone rocks of Scotland. It is also said to occur in veins in Dauphiny, along with axinite and epidote; and in the lava of Solfatara.

Observations.

Glassy Felspar is distinguished from the other minerals of the felspar species, by its white colour, splendent vitreous lustre, transparency, and the frequent rents or fissures with which it is traversed.

Second

Second Subspecies.

Labrador Felspar.

Labradorstein, *Werner.*

Pierre de Labrador, *R. de L.* t. ii. p. 497.—Feldspath, var. *Wid.* p. 335.—Labradore-stone, *Kirw.* vol. i. p. 324.—Labradorstein, *Emm.* b. i. s. 273.—Feldspato commune, var. *Nap.* p. 213.—Labradorite, *Lam.* t. ii. p. 197.—Feldspath opalin, *Hauy,* t. ii. p. 607.—La Pierre de Labrador, *Broch.* t. i. p. 369.—Labradorstein, *Reuss,* b. ii. s. 387. *Id. Lud.* b. i. s. 100. *Id. Suck.* 1r th. s. 380.—Gemeiner Feldstein, *Bert.* s. 238.—Labradorstein, *Mohs,* b. i. s. 407.—Labradorischer Feldspath, *Hab.* s. 22.—Feldspath opalin, *Lucas,* p. 50.—Labradorischer Feldspath, *Leonhard,* Tabel. s. 18.—Feldspath opalin, *Brong.* t. i. p. 359.—Feldspath opalin, *Brard,* p. 134. Farbenspielender Feldspath, *Haus.* s. 88.—Labrador Feldspath, *Karst.* Tabel. s. 34.—Opaline Felspar, *Kid,* vol. i. p. 160.—Feldspath opalin, *Hauy,* Tabl. p. 36.—Labradorstein, *Steffens,* b. i. s. 432. *Id. Hoff.* b. ii. s. 304.—Labrador Feldspath, *Lenz,* b. i. s. 490.—Labradorstein, *Oken,* b. i. s. 376.

External Characters.

Its most frequent colours are light and dark ash-grey, and smoke-grey, seldom yellowish-grey. When light falls on it in determinate directions, it exhibits a great variety of colours: of these the most frequent are blue and green, more seldom yellow and red, and the rarest variety is pearl-grey. The blue varieties are indigo, berlin, azure, violet, smalt, and sky blue: this latter colour passes into verdigris-green; from this variety through celandine, mountain, leek, emerald, grass, pistachio, olive, and oil, into siskin green; the siskin-green passes into sulphurphur-yellow, and through brass, gold, lemon, honey, and orange yellow, into yellowish and reddish brown, copper-red, brick-red, flesh-red, brownish-red; and lastly, into pearl-grey and bluish-grey. The same specimen exhibits different colours, which run imperceptibly into each other, and are disposed in large patches or stripes.

It occurs massive, or in rolled pieces.

The principal fracture is splendent, the cross fracture glistening, and the lustre is intermediate between vitreous and pearly.

The fracture is the same as in common felspar.

It breaks into rhomboidal or sharp-edged fragments.

It generally occurs in distinct concretions, which are large and coarse, seldom small granular, and very seldom straight lamellar.

It is translucent, but in a low degree.

It is hard in a low degree.

It is rather more difficultly frangible than common felspar.

Is rather heavy, but in a low degree.

Specific gravity, 2.692, *Brisson.* 2.590, *Hoffmann.*

Chemical Characters.

According to Mr Kirwan, it is more infusible than common felspar, and is difficultly melted before the blowpipe, without addition.

We have no accurate analysis of this mineral.

Geognostic and Geographic Situations.

It occurs in rolled masses of syenite? associated with common hornblende, hypersthene, and magnetic ironstone, in the Island of St Paul, on the coast of Labrador, where it was first discovered, upwards of thirty years ago, by the Moravian Missionaries settled in that remote and dreary

dreary region. Some years afterwards, several varieties of it were found imbedded in a granitic rock in Ingermannland; but the colours of these were neither so vivid nor numerous as in the Labrador felspar of St Paul's. In the interesting country around Laurwig in Norway, Labrador felspar occurs as a constituent part of the zircon-syenite; its colours are brighter than in the Ingermannland varieties, but not so vivid as those of St Paul. Blue is the principal colour of the Norwegian felspar, but it sometimes also exhibits a beautiful bluish mother-of-pearl opalescence, like that observed in adularia. A variety of this mineral is said to occur in the Hartz. Rolled pieces of it have been brought from West Greenland; and it has been found on the banks of Lake Champlain in North America *.

Uses.

On account of its beautiful colours, it is valued as an ornamental stone, and is cut into ring-stones, snuff-boxes, and other similar articles. It receives a good polish; but the streaks caused by the edges of the folia are frequently so prominent as to injure the appearance of the stone.

Observations.

1. This mineral is distinguished by its grey colours, and its changeability of colours.

2. The beautiful changeability of colours which Labrador felspar exhibits, appears to be caused by small rents, that run parallel with the folia of the cleavage, in this differing from the play of colour observed in the precious opal, which is owing to rents that run in every direction.

Third

---

* I have specimens in my possession, said to have been found in Aberdeenshire.

Third Subspecies.

Common Felspar.

This Species is divided into two kinds, viz. Fresh Common Felspar, and Disintegrated Common Felspar.

First Kind.

Fresh Common Felspar.

Frischer gemeiner Feldspath, *Werner*.

Spathum scintillans, *Wall.* t. i. p. 214.—Feldspath, *Wid.* p. 335. *Id. R. de L.* t. ii. p. 445.—Common Felspar, *Kirw.* vol. i. p. 316.—Blättrig Feldstein, *Estner*, b. i. s. 513.—Gemeiner Feldspath, *Emm.* b. i. s. 266.—Feldispato commune, *Nap.* p. 213.—Feldspath, *Lam.* t. ii. p. 187. *Id. Hauy*, t. ii. p. 590. —Le Feldspath commun, *Broch.* t. i. p. 362.—Gemeiner Feldspath, *Reuss*, b. ii. s. 369. *Id. Lud.* b. i. s. 100. *Id. Suck.* 1r th. s. 380.—Gemeiner Feldstein, *Bert.* s. 238.—Gemeiner Feldspath, *Mohs*, b. i. s. 407. *Id. Hab.* s. 20.—Feldspath, *Lucas*, p. 50.—Gemeiner frischer Feldspath, *Leonhard*, Tabel. s. 18.—Feldspath commun, *Brong.* t. i. p. 367.—Feldspath, *Brard*, p. 131.—Gemeiner Feldspath, *Haus.* s. 88. *Id. Karst.* Tabel. s. 34.—Felspar, *Kid*, vol. i. p. 157.—Feldspath, *Hauy*, Tabl. p. 35.—Frisher gemeiner Feldspath, *Steffens*, b. i. s. 436. *Id. Hoff.* b. ii. s. 309. *Id. Lenz*, b. i. s. 494. *Id. Oken*, b. i. s. 374.

External Characters.

Its most frequent colours are white and red, seldom grey, and rarely green and blue. The white varieties are greenish-white, milk-white, yellowish-white, greyish-white,

white, snow-white, and reddish-white: from reddish-white it passes into flesh-red, and into a colour intermediate between flesh-red and blood-red: from greenish-white it passes into apple-green, asparagus-green, grass-green, emerald-green, leek-green, mountain-green, verdigris-green; and from this latter into sky-blue: from milk-white it passes into bluish-grey, smoke-grey, and yellowish-grey. The grey varieties are generally spotted.

It occurs most frequently massive and disseminated, seldom in blunt angular rolled pieces and grains; sometimes crystallised, in the following figures:

1. Very oblique four-sided prism, flatly bevelled on both extremities, and the bevelling planes set on the obtuse lateral edges, fig. 90 *. This is the fundamental figure.
2. The preceding crystallization, in which two diagonally opposite bevelling planes are smaller than the two others †. Sometimes the latter entirely disappear, when there is formed
3. A perfect and very oblique four-sided prism, in which the terminal planes are set on obliquely; or when the prism becomes shorter, and all the planes diminish in an equal proportion, there is formed
4. An acute rhombus ‡, fig. 91.

When the prism of the fundamental figure becomes shorter, and the bevelling planes become much larger than the lateral planes, we can view the former

* Feldspath ditetraedre, Hauy.

† The bevelling planes mentioned above, are those that form the greatest angle with the lateral edges.

‡ Feldspath binaire, Hauy.

former as lateral planes, and the latter as bevelling planes, and thus there is formed

5. A very oblique four-sided prism, acutely bevelled on the extremities, and the bevelling planes set on the acute lateral edges. This figure sometimes passes into a kind of
6. Elongated octahedron.
7. The fundamental figure, truncated on the acute lateral edges.
8. The variety N° 3. truncated on the acute edges, in the same manner as the variety N° 7 *, fig. 92.

When the truncating planes of the variety N° 7. become larger than the lateral planes, there is formed

9. A broad equiangular six-sided prism, flatly bevelled on the extremities, and the bevelling planes set on those lateral edges which are formed by the meeting of the smaller lateral planes, fig. 93 †.
10. The preceding figure, in which the edges formed by the meeting of the larger and smaller lateral planes are truncated, fig. 94 ‡.
11. The crystallization N° 9. in which the angles formed by the meeting of the smaller bevelling planes and the lateral edges on which they are set, are more or less deeply truncated, fig. 95 ||.
12. The preceding variety, in which the edges formed by the smaller bevelling planes and the broader lateral planes are truncated, fig. 96 **

13. The

* Feldspath prismatique, Hauy.

† Feldspath bibinaire, Hauy.

‡ Feldspath quadridecimal, Hauy.

|| Feldspath dihexaedre, Hauy.

** Feldspath sexdecimal, Hauy,

13. The preceding variety, in which the edges formed by the meeting of the other bevelling planes with the broader lateral planes, are truncated *.
14. In all the preceding varieties from N° 9. the proper edge of the bevelment is sometimes truncated †.
15. The smaller bevelling planes in N° 11. sometimes disappear, whilst the truncating planes on the angles become larger, and form with the larger bevelling plane a new and much more acute bevelment, fig. 97 ‡.

When two bevelling planes in variety 9. become very large, as in N° 2. whilst the prism becomes very broad and short, so that these two large bevelling planes approach near to each other, and increase in equal proportion with the broader lateral planes with which they meet under a right angle, they form with these

16. A rectangular four-sided prism, in which the smaller lateral planes of the six-sided prism form a kind of oblique bevelment on the terminal planes, and which is variously modified by the remains of the smaller bevelling planes of the fundamental figure, and the other planes of alteration ||.
17. The preceding figure truncated on the lateral edges.

 These

* These truncating planes, along with some others, occur in Hauy's Feldspath synoptique and Feldspath decidodecaedre.

† This transition is to be seen in Hauy's Feldspath apophane and Feldspath synoptique.

‡ As in Hauy's Feldspath decidodecaedre.

|| Vid. Hauy, fig. 91. and 92. Romé de Lisle assumed this as the fundamental form of felspar.

These truncating planes correspond with those of the 13th crystallization.

Sometimes the planes at the extremities of the figure N° 16. almost totally disappear, and there remains only the two truncating planes of the 11th crystallization, and then the crystallization becomes

18. A nearly perfect rectangular four-sided prism, in which the terminal planes are set on obliquely, fig. 98 *.

Besides these simple crystallizations, twin crystals also occur, of which the following are the principal varieties :

19. Twin crystal, which we may suppose to have been formed by two prisms of N°s. 9. or 15. being pushed into each other in the direction of their thickness, in such a manner that their axes are either parallel to each other, or form a more or less obtuse angle. The lateral planes, and also some of those at the extremities of the crystals, form re-entering angles.

20. Twin crystal, which we can conceive to be formed when N°s. 16. & 17. are divided longitudinally from one extremity to the other, in the direction of the two opposite lateral planes, (the broader lateral planes of the six-sided prism), and the one-half turned completely around and applied to the other. In this way a rectangular four-sided prism is formed, in which the diagonally opposite planes of the two extremities of the single crystal will be placed together. This is the *hemitrope* crystal of Hauy †.

The

* Feldspath unitaire, Hauy.

† Romé de Lisle, t. ii. p. 478.—492. var. 10.—16. Pl. 3. fig. 94—106.

The crystals are generally small and middle-sized, seldom very small, large, and very large. They are generally imbedded, sometimes also superimposed, and variously aggregated, forming druses.

Internally the principal fracture is shining, and sometimes splendent; the cross fracture is glistening, and frequently not more than feebly glistening. The lustre is intermediate between vitreous and pearly, but inclining rather more to the former than the latter.

The principal fracture is perfect foliated, with a double rectangular cleavage: there is also a third cleavage, which is imperfect and intersects the other two under an oblique angle; it is often very difficultly discoverable, and frequently the fracture in that direction, in place of being foliated, is uneven, or splintery. The foliated fracture is occasionally curved and floriform, which latter passes into broad and scopiform radiated.

The fragments are rhomboidal, and have only four splendent shining faces.

It generally occurs in granular distinct concretions, of all degrees of magnitude.

It is sometimes translucent, sometimes only translucent on the edges.

It is hard, but in a lower degree than quartz.

It is very easily frangible; and

Is rather heavy, but in a low degree.

Specific gravity, 2.594, *Brisson.* 2.551, 2.567, *Hoff.*

Chemical Characters.

Before the blowpipe, it is fusible without addition into a grey semitransparent glass.

Constituent Parts.

	Siberian Green Felspar.	Flesh-red Felspar.	Felspar from Passau.
Silica,	62.83	66.75	60.25
Alumina,	17.02	17.50	22.00
Lime,	3.00	1.25	0.75
Potash,	13.00	12.00	14.00
Oxide of Iron,	1.00	0.75	a trace.
Water,			1.00
	96.85	98.25	98.00
	Vauquelin, Jour. des Mines, n. 49. p. 23.	*Rose*, in Scherer's Jour. der Chimie, b. 7. s. 244.	*Bucholz*, in Von Moll's Neue Jahrb der Berg und Hüttenkunde, b. 2. s. 361.

Geognostic Situation.

This is one of the most abundant minerals in nature; it forms fully two-thirds of the mass of granite and gneiss, two of the most widely distributed and oldest rocks hitherto discovered. It occurs as an accidental mixed part in mica-slate and clay-slate. It is a constituent part of white-stone, and syenite: in white-stone it is associated with garnet, mica, and hornblende: in syenite always with a subordinate portion of hornblende. It forms the basis of certain porphyries, and then it occurs in fine granular concretions; and in the form of imbedded crystals it occurs in all the different kinds of porphyry. Greenstone, a rock so abundant in Primitive country, is a compound of common felspar and hornblende, but in which the hornblende predominates. Frequently the felspar is tinged of a green colour, owing to an intermixture

ture of hornblende, and in this state it is heavier than the pure varieties of this mineral. But it occurs not only as a constituent part, and accidentally mixed with primitive mountain rocks, but we find it also in beds alternating with these, in nests and kidneys contained in them, and in veins traversing them. The beds occurring in granite or gneiss, are sometimes entirely composed of felspar, with the addition of very little mica and quartz; or in them it is associated with hornblende, garnet, actynolite, epidote, and copper and iron ores, as in Sweden and Norway. The kindneys and nests vary from a few inches to fathoms in extent, and are contained in granite or gneiss. The veins appear to be of cotemporaneous formation with the granite and gneiss rocks in which they are contained: they are sometimes entirely composed of felspar; in other instances of felspar, with a little quartz and mica, or of felspar, with rock-crystal, mica, chlorite, epidote, schorl, beryl, and rutile. It is in these veins that the greater number of the crystallizations of felspar occur. The most beautiful crystallizations occur in the Alps of Switzerland, in Lombardy, France, and Siberia. The green felspar analysed by Vauquelin, is said to occur in a vein in granite, in the government of Ubinsky, in the Uralian mountains in Siberia; also in cotemporaneous masses in the granite of Onega.

Felspar is not confined to primitive rocks; it occurs abundantly in Transition mountains, and also in those of the Flœtz class. In transition mountains, it forms an essential constituent part of granite, syenite, porphyry, greenstone, and grey-wacke; and occurs accidentally intermixed in other rocks of this class. In flœtz rocks, it occurs in many sandstones, in porphyry, greenstone, clinkstone-porphyry, and basalt.

 Geographic

Geographic Situation.

As granite, gneiss, mica-slate, porphyry, syenite, greenstone, grey-wacke, sandstone, basalt, and other rocks in which common felspar occurs, are found in almost every great tract of country, it would be superfluous to attempt detailing the individual geographic localities of a mineral so widely distributed.

Uses.

It is one of the ingredients in the finer kinds of earthen-ware, and is the substance used by the Chinese under the name *Petunse* or *Petuntze*, in the manufacture of their porcelain. The green varieties of felspar, which are rare, are considered as ornamental stones, and are cut and polished, and made into snuff-boxes, and other similar articles. When the green varieties are spotted with white, they are named *Aventurine Felspar*, and are prized by collectors. Other two varieties, having the same name, and much esteemed by collectors, are found in Russia: the one is a red felspar, with white spots, from the coast of the White Sea; the other a yellow felspar, with shining yellow spots, from the Island of Cedlowatoi, near Archangel. The green felspar from South America, which is cut and polished, and sold under the name *Amazon Stone*, is found in small rolled pieces, on the banks of the river of Amazons.

Observations.

1. It is distinguished from the other subspecies of this species by its more extensive colour-suite, its want of changeability of colour, its distinct concretions, passing into

into fine granular, easy frangibility, and inferior translucency.

2. It has been confounded with Corundum, but it is distinguished from that mineral by its fracture, inferior specific gravity, and inferior hardness. It is distinguished from *Chrysoberyl* by its fracture, inferior hardness, and inferior weight. The green-coloured felspar is distinguished from *Green Diallage* by its superior hardness, and its double cleavage.

3. The German name *Felspar* was given to this mineral on account of its sparry or foliated texture, and from the circumstance of its frequently occuring as a constituent part of those loose blocks of stone we observe scattered over the country, *(Feldern)*. Hence it appears that the name *Felspar*, used by the English, and sometimes by French authors, is not quite correct.

Second Kind.

Disintegrated Common Felspar.

Aufgelöster gemeiner Feldspath, *Werner*.

External Characters.

Its colours are greyish-white, yellowish-white, and reddish-white, all of which incline very much to grey.

It generally occurs massive, and disseminated, and sometimes in imbedded crystals, which agree in form with those of fresh common felspar.

Internally it is sometimes glistening, sometimes glimmering, or even dull, according to the kind of fracture.

The

The fracture is sometimes imperfect foliated, sometimes coarse and small grained uneven, which approaches to earthy.

It breaks into blunt angular pieces.

It occurs in granular distinct concretions.

It is either translucent on the edges, or opaque.

It is soft, sometimes inclining to very soft, or to semihard.

It is sectile, and easily frangible.

It is rather heavy, but in a low degree.

The chemical characters and composition of this substance have not been ascertained.

Geognostic and Geographic Situations.

It has hitherto been found only in granite, as in that of Cornwall, Saxony, and other countries.

Observations.

1. This mineral seems in some instances to be felspar in a state of decomposition: in others, as in Cornwall, to be an unaltered substance, very nearly of the nature of common felspar. The *Growan* of Cornwall appears to contain principally the disintegrated felspar.

2. The *Feld-spath décomposé* of Hauy, which he considers as identical with the Disintegrated Felspar of Werner, appears rather to be a variety of Porcelain Earth.

Fourth

Fourth Subspecies.

Compact Felspar.

Dichter Feldspath, *Werner*.

Continuous Felspar, *Kirw.* vol. i. p. 323.—Dichter Feldstein, *Estner,* b. ii. s. 511. *Id. Emm.* b. i. s. 271.—Felspato compatto, *Nap.* p. 218.—Le Feldspath Compacte, *Broch.* t. i. p. 367.—Dichter Feldspath, *Reuss,* b. ii. s. 366. *Id. Lud.* b. i. s. 10. *Id. Suck.* 1r th. s. 393.—Dichter Feldstein, *Bert.* s. 238.—Dichter Feldspath, *Mohs,* b. i. s. 420. *Id. Hab.* s. 19.—Feldspath compacte, *Lucas,* p. 50.—Dichter Feldspath, *Leonhard,* Tabel. s. 19.—Petrosilex, *Brong.* t. i. p. 351. —Feldspath Compacte ceroide, *Brard,* p. 133.—Dichter Feldspath, *Haus.* s. 88. *Id. Karst.* Tabel. s. 34.—Feldspath Compacte ceroide, *Hauy,* Tabl. p. 35.—Dichter Feldspath, *Steffens,* b. i. s. 442. *Id. Hoff.* b. ii. s. 334. *Id. Lenz,* b. i. s. 506. *Id. Oken,* b. i. s. 299.

External Characters.

Its colours are white, grey, green, and red: it passes from greyish-white through greenish-white into apple-green, oil-green, inclining to olive-green, mountain-green, greenish-grey, smoke-grey, pearl-grey, flesh-red, and brick-red.

It occurs massive, disseminated, in blunt angular rolled pieces, and crystallised in rectangular four-sided prisms.

The crystals are either middle sized, or small, and imbedded.

Internally it is sometimes glistening, sometimes glimmering.

The

The fracture of the glistening varieties is imperfect foliated; but in the glimmering varieties it approaches to even and splintery.

It breaks into fragments which are rather sharp-edged.

It is feebly translucent, sometimes only translucent on the edges.

It is hard in a low degree; or rather softer than common felspar.

When pure, it is rather easily frangible.

It is rather heavy, but in a low degree.

Specific gravity, 2.609, *Kirwan*. 2.666, *La Metherie*. 2.659, *Saussure*.

Chemical Characters.

Before the blowpipe, it melts with difficulty into a whitish enamel.

Constituent Parts.

Compact Felspar of Salberg in Sweden.		Compact Felspar of the Pentland Hills, near Edinburgh.	
Silica,	68.0	Silica,	71.17
Alumina.	19.0	Alumina,	13.60
Lime,	1.0	Lime,	0.40
Potash,	5.5	Potash,	3.19
Oxide of Iron,	4.0	Oxide of Iron,	1.40
Water,	2.5	Manganese,	0.10
		Volatile Matter,	3.50
	100		93 36
		Loss,	6.64
			100

Godon de St Memin, Journ. de Physique, t. lxiii. p. 60.

Mackenzie, Mem. Wern. Soc. vol. i. p. 618.

Geognostic

Geognostic Situation.

This mineral occurs in mountain-masses, beds and veins, either pure, or intermixed with other minerals, in primitive, transition, and flœtz rocks. In primitive mountains, it is associated with hornblende in greenstone, and greenstone-slate; and it forms the basis of several felspar-porphyries. Beds of it in a pure state occur in gneiss, and other primitive rocks. In transition mountains, it occurs in beds, as a constituent part of porphyry and greenstone; and beds of it occur either pure, or in porphyry, or greenstone, in flœtz mountains.

Geographic Situation.

The Pentland Hills contain beds of compact felspar, associated with claystone, old red sandstone, and conglomerate. It occurs in a similar situation on the hill of Tinto, described by Dr Macknight in the 2d volume of the Memoirs of the Wernerian Society. Mr Mackenzie found it along with flœtz rocks in the Ochil Hills *; and Mr Fleming observed it associated with rocks of the same nature in the Island of Papa Stour, one of the Zetland group †. Beds of it, which are sometimes porphyritic, occur in the transition rocks of Dumfriesshire and Galloway; and in rocks of the same class to the north of the Frith of Forth, as in Perthshire, and the Mearns ‡. In the primitive rocks to the north of the Frith of Forth, it occurs in beds and veins, either pure, or in the state of porphyry. Examples of both occur in Perthshire, in the course

* Memoirs of the Wernerian Society, vol. ii. p. 20.

† Memoirs of the Wernerian Society, vol. i. p. 170.

‡ Imrie, in Transactions of Royal Society of Edinburgh, vol. vi.

course of the Gairy and the Tilt; in the country around Castletown, in the upper part of Aberdeenshire; and in other places, as will be mentioned in the 3d volume of this work *.

It has been observed in Scandinavia. A well known instance is at Sala in Sweden, where the reddish translucent splintery variety, so often mistaken for hornstone, occurs in beds. Grey and green varieties occur in the greenstone-slate at Siebenlehn and Gersdorf, at the foot of the Saxon Erzgebirge. A beautiful greenish-grey and mountain-green compact felspar was found by Saussure the Father near Pissevache in the Vallais.

Observations.

1. The principal characteristic distinctions of this mineral are colour, lustre, fracture, translucency, hardness, and weight.

2. It has been frequently confounded with Splintery Hornstone; but is distinguished from it by colour, and foliated fracture, but principally by its lustre, inferior hardness, easier frangibility, fusibility before the blowpipe, and its being frequently intermixed with hornblende and mica.

6. Spodumene.

* Dr Macknight mentions several localities of this mineral in his elegant and interesting sketch of the scenery and mineralogy of the Highlands, in the 1st volume of the Memoirs of the Wernerian Society.

6. Spodumene.

Spodumene, *D'Andrada.*

Spodumene, *D'Andrada,* Scherer's Journ. b. iv. 19. s. 30. *Id. Reuss,* b. ii. s. 495. *Id. Lud.* b. ii. s. 162. *Id. Suck.* 1r th. s. 725. *Id. Bert.* s. 174.—Triphane, *Lucas,* p. 209.—Spodumene, *Leonhard,* Tabel. s. 19.—Triphane, *Brong,* t. i. p. 388. *Id. Brard,* p. 417. *Id. Haus.* s. 88.—Spodumen, *Karsten,* Tabel. s. 34.—Triphane, *Hauy,* Tabl. p. 37.—Spodumene, *Steffens,* b. i. s. 474. *Id. Hoff.* b. ii. s. 341.—Triphan, *Lenz,* b. i. s. 525. *Id. Oken,* b. i. s. 372.

External Characters.

Its colour is intermediate between greenish-white and mountain-grey, and sometimes passes into oil-green.

It occurs massive and disseminated.

The principal fracture is shining, the cross fracture glistening, and the lustre is pearly.

The principal fracture is foliated, with a threefold very oblique angular cleavage, of which two of the cleavages are parallel with the sides of a rhomboidal prism, and the third with the smaller diagonal of the basis of the prism. The cross fracture is fine-grained uneven.

It sometimes breaks into very oblique rhomboidal fragments, but more frequently into such as are tabular and indeterminately angular.

It occurs in distinct concretions, which are large or coarse granular.

It is translucent.

It is hard, but in a low degree; it scratches glass.

It

It is uncommonly easily frangible; and
Rather heavy.
Specific gravity, 3.192, *Hauy*. 3.218, *D'Andrada*.

Chemical Characters.

Before the blowpipe, it first separates into small gold-yellow coloured folia; and if the heat is continued, they melt into a greenish-white coloured glass. D'Andrada says, that it first separates into golden-coloured scales, and then into a kind of powder or ash: hence the name *Spodumene* (from *σποδύω*, *I change into ash*, or *σποδος*, *ashes*), given by him to this mineral.

Constituent Parts.

Silica, -	56.5	64.4
Alumina, -	24.0	24.4
Lime, -	5.0	3.0
Potash, -		5.0
Oxide of Iron,	5.0	2.2
	90.5	99.0
	Vauquelin.	*Vauquelin*, Hauy's Tabl. p. 168.

Geognostic and Geographic Situations.

Spodumene has hitherto been found only in the Island of Uton, in Sudermanland in Sweden, where it is associated with red felspar and quartz *.

7. Scapolite.

---

* Count de Bournon, in his *Catalogue*, p. 96. mentions specimens of Spodumene in his possession from Ireland.

7. Scapolite.

Skapolith, *Werner.*

THIS species is subdivided into three subspecies, viz. Radiated Scapolite, Foliated Scapolite, and Compact Scapolite.

First Subspecies.

Radiated Scapolite.

Strahliger & Nadelförmiger Skapolith, *Karsten,* Tabel. s. 34.—Glasartiger Scapolith, *Haus.* s. 189.—Strahliger & Glasartiger Skapolith, *Steffens,* b. i. s. 461. & 464.—Stangensteinartiger Scapolit, *Schumacher,* Verzeichniss, s. 97.—Strahliger grauer Skapolith, *Hoff.* b. ii. s. 346.—Paranthine dioctaedre, aciculaire & cylindroide, *Hauy,* Tabl. p. 46.

External Characters.

Its most frequent colour is grey, seldomer white and green: it occurs greyish-white, yellowish-white, greenish-white, yellowish-grey, greenish-grey, mountain-green, olive-green, and asparagus-green.

It seldom occurs massive, generally crystallised, and in the following figures:

1. Slightly oblique four-sided prism, flatly acuminated on the extremities with four planes, which are set on the lateral planes.
2. The preceding figure, in which the lateral edges are truncated.

The crystals vary very much in length as well as thickness; for we meet with them from the acicular form to the thickness of a finger, and from very long to short. Sometimes the long prisms are curved, and are traversed with rents.

The crystals are frequently columnarly aggregated, or intersect one another.

The lateral planes of the crystals are deeply longitudinally streaked, and shining.

Internally it is intermediate between shining and glistening, and the lustre is intermediate between resinous and pearly.

The fracture is torn, narrow, and scopiform diverging radiated, and the rays have a double cleavage; and both cleavages are in the direction of the diagonals of the prism, and consequently parallel with the truncating planes on the lateral edges. The fracture-surface is streaked. The cross fracture is fine-grained uneven.

The fragments are angular, and not very sharp-edged.

It occurs in thick and wedge-shaped distinct concretions.

It is translucent, and semi-transparent in crystals.

It is hard in a low degree, and that only when quite fresh and undecomposed; the grey and semi-transparent crystals scratch glass, the others only calcareous-spar.

It is rather easily frangible; and

Rather heavy, in a high degree.

Specific gravity, 3.680, 3.708, *D'Andrada*. 2.740, *Laugier*. 2.691, 2.773, *Simon*. 2.400, 2.857, *Schumacher*. 3.500, *Steffens*. 2.660, 2.743, 3.269, *Hausmann*.

These specific gravities differ so much from one another, that either some of them must be erroneous, or different species of mineral must have been weighed.

Chemical

Chemical Characters.

Green scapolite, before the blowpipe, becomes white, and melts into a white glass.

Constituent Parts.

Silica, - -	53.50	Silica, -	45.0
Alumina, -	15.00	Alumina, -	33.0
Magnesia, -	7.00	Lime, -	17.6
Lime, -	13.75	Natron, -	1.5
Natron, -	3.50	Potash, -	0.5
Iron, - -	2.00	Iron and Manganese,	1.0
Manganese, -	4.00		——
Water, - -	0.50		98.6
	——		
	99.24		

Simon, Chem. Journ. b. iv, s. 411.

Laugier, Annales du Museum d'Hist. Nat. cah. Ix. p. 472.

Geognostic and Geographic Situations.

It has hitherto been found only in Scandinavia, and in primitive rocks: thus, it occurs in the neighbourhood of Arendal in Norway, where it is associated with magnetic ironstone, felspar, quartz, mica, garnet, augite, hornblende, actynolite, and calcareous-spar.

The magnetic-ironstone occurs in gneiss, in the form of beds, that vary in thickness from four to sixty feet. In these beds, the scapolite and other accompanying minerals already mentioned, are either contained in cotemporaneous veins, or are irregularly disseminated throughout the beds. M. Hausmann has also observed it in beds of specular iron-ore or iron-glance, in the Swedish province of Wermeland, where it is associated with calcareous-spar and garnet; and the same excellent mineralogist found it at Malsjo in Wermeland, in a bed of limestone; and at Garpenberg in Dalecarlia, in beds of copper-pyrites.

Observations.

All the subspecies decay very readily on exposure to the weather, a circumstance which has induced Hauy to name this species *Paranthine:* the name *Scapolite* given it by Werner, is derived from the Greek word *σκαπος, a rod,* and refers to the columnar form and mode of aggregation of its crystals.

Second Subspecies.

Foliated Scapolite.

Micarell, *Abilgaard.*—Talkartiger Scapolit, Blättriger Scapolit, Pinitartriger Scapolit, *Schumacher,* Verzeichniss, s. 98.–100. —Wernerit, *Karsten,* Tabel. s. 34.—Arcticit, *Werner.*—Gemeiner Skapolith & Glimmeriger Skapolith, *Steffens,* b. i. s. 462. 464.—Blättriger grauer Skapolit, *Hoff.* b. ii. s. 353. Sodait, *Eckeberg.*

External Characters.

Its most frequent colours are grey and green; it occurs also white, blue, and brown: from greenish-grey it passes on the one side into greenish-white, greyish-white, and snow-white; on the other into mountain-green, which on the one side passes into celandine-green, verdigris-green, sky-blue, smalt-blue, indigo-blue, and duck-blue, on the other into pistachio-green, asparagus green, olive-green, oil-green, liver-brown, reddish-brown, yellowish-brown, and muddy honey-yellow. It occurs also greyish-black and pitch-black. The colours are seldom pure, generally pale and muddy, and sometimes two colours occur

cur in the same specimen; and greenish-grey coloured crystals sometimes appear sky-blue externally.

It occurs massive, disseminated; and crystallised in the following figures:

1. Slightly oblique four-sided prism, rather flatly acuminated with four planes, which are set on the lateral planes.
2. The preceding figure, in which the lateral edges are deeply truncated.

When these truncating planes become so large that the original ones disappear, there is formed

3. A four-sided prism, somewhat more oblique than the former, acuminated on the extremities with four planes, which are set on the lateral edges.
4. The preceding figure, in which the summits of the acuminations are deeply truncated. When these truncations increase very much, there is formed
5. A simple rather oblique four-sided prism.
6. When the truncating and lateral planes of N° 2. become equal in magnitude, there is formed
7. An eight-sided prism, acuminated with four planes, the acuminating planes set on the alternate lateral planes; and the acuminations are generally deeply truncated.
8. Perfect eight-sided prism.
9. Eight-sided prism, truncated on all its angles.
10. The crystallization N° 3. slightly truncated on the lateral edges, and on the extremities of the acuminations.

The crystals are sometimes middle-sized, seldom large, and not often very small.

The crystals are generally superimposed, seldom imbedded.

Externally the crystals are shining or splendent, and vitreous.

 Internally

Internally the principal fracture is shining, the cross fracture glistening, and the lustre intermediate between resinous and pearly.

The principal fracture is imperfect and torn foliated, with a threefold rather oblique angular cleavage; the cross fracture small and fine grained uneven, or small conchoidal.

The fragments generally angular and sharp edged, sometimes rhomboidal.

It occurs in distinct concretions, which are large and coarse, sometimes longish granular.

It is generally translucent, and passes sometimes into transparent, sometimes to translucent on the edges.

It is hard, but in a low degree; and

Yields a white streak.

It is brittle.

It is very easily frangible; and

Rather heavy in a high degree.

Specific gravity, 3.600, *Werner*.

Geognostic and Geographic Situations.

Whitestone, a species of primitive rock, principally composed of felspar, occurs in considerable quantity on the north-western acclivity of the Saxon Erzgebirge, and includes cotemporaneous masses of granite, that vary in magnitude from a few feet to some miles in extent. In these granitic masses, various minerals have been observed, as schorl, tourmaline, lepidolite, and *Foliated Scapolite* *. It occurs also in Scandinavia, along with the radiated subspecies.

Third

* Vid. Pusch, uber Granit. Leonhard, Tachenbuch 1812, p. 137.

Third Subspecies.

Compact Scapolite.

Dichter Scapolith.

This subspecies is divided into two kinds, viz. Compact Green Scapolite, and Compact Red Scapolite.

First Kind.

Compact Green Scapolite.

Dichter Skapolith, *Hausmann,* Magazin Naturf. Freunde, b. iii. s. 220. *Id. Karsten,* Tabel. s. 34. *Id. Steffens,* b. i. s. 465. —Wernerit, *Hauy.*—Fuscit, & Gabbronit, *Schumacher.*

External Characters.

Its colours are greyish-white, greenish-grey, olive-green, pistachio-green, and leek-green.

It occurs massive; and crystallised in the following figures:

1. Rectangular four-sided prism, acuminated with four planes, which are set on the lateral edges.
2. The preceding figure, truncated on the lateral edges, and the acuminating planes set on the truncating planes.

The prisms are short and thick; and the crystals are very small, small, and middle-sized.

Externally it is more or less shining and pearly.

Internally it is glistening, or nearly dull.

The fracture is uneven and splintery, and sometimes

a very faint trace of a twofold longitudinal cleavage is to be observed.

The fragments are angular.

It is translucent on the edges, or opaque.

It gives sparks with steel, and scratches glass feebly.

It is difficultly frangible.

Specific gravity, 3.600, *Hauy*.

Chemical Characters.

When thrown on glowing coal, it phosphoresces in the dark. It intumesces before the blowpipe, and melts into white enamel.

Geognostic and Geographic Situations.

This subspecies agrees with the other subspecies, both in Geognostic and Geopraphic Situations.

Second Kind.

Compact Red Scapolite.

External Characters.

Its colour is dark brick-red, passing into pale blood-red.

It seldom occurs massive, more frequently crystallised, in long, frequently acicular, four-sided prisms, which are often curved, and are without terminal crystallizations.

Externally the crystals are rough and dull.

Internally it is very feebly glistening, almost glimmering.

The fracture is fine-grained uneven, approaching to splintery.

The

The fragments are angular, and sharp edged.
It is opaque, or very faintly translucent on the edges.
It is hard in a low degree.
It is easily frangible ; and
Rather heavy.

Geognostic and Geographic Situations.

It occurs along with the other subspecies, in metalliferous beds at Arendal in Norway.

Observations.

This mineral is characterised by its red colour, low degree of lustre, compact fracture, and nearly complete opacity.

8. Bergmannite.

Bergmannite, *Schumacher,* Verzeichniss. *Id. Steffens,* b. i. s. 471. Fasriger Wernerit, *Hausmann,* Magazin der Naturf. Freunde, b. iii. s. 221.

External Characters.

Its colours are greenish and greyish white, greenish and yellowish grey, and muddy flesh-red.

It occurs massive.

Externally it is glistening, and the lustre is intermediate between pearly and resinous.

Internally it is glistening, and the lustre the same as externally.

The fracture is very delicate, curved, scopiform or stellular fibrous, which passes into fine grained uneven, and dull.

The

The fragments are angular, and not very sharp edged.

It is faintly translucent on the edges.

It scratches glass, and even felspar.

Chemical Characters.

It melts before the blowpipe, without intumescing, into a white semi-transparent enamel.

Geognostic and Geographic Situations.

It occurs along with grey and red coloured quartz, in a bed at Friedichswärn in Norway.

Observations.

This species was first pointed out by Schumacher. The celebrated naturalist Steffens, places it in the system immediately after Scapolite, and before Elaolite, an arrangement which is here followed.

9. Elaolite.

Eläolith, *Klaproth.*

Fettstein, *Werner.*

Dichter Wernerit, *Hausmann.*

External Characters.

The colours of this mineral are duck-blue, which inclines more or less to green, and flesh-red, which falls more or less into grey, sometimes even inclines to brown.

It occurs massive.

Internally it is shining or glistening, and the lustre is resinous.

The

The fracture, principally in the red variety, is flat and imperfect conchoidal; but the blue variety is imperfect foliated, with a double cleavage *: the cross fracture small splintery.

The fragments are angular, and not very sharp edged.

It is translucent, in a low degree. The blue variety, when cut in a particular direction, displays a peculiar opalescence, not unlike that observed in the cat's-eye.

It is hard; it scratches glass.

It is rather easily frangible.

It is rather heavy, but in a low degree.

Specific gravity, 2.613, *Hauy*. From 2.588 to 2.618, *Hoffmann*.

Chemical Characters.

When pounded, and thrown into acids, it gelatinates. Before the blowpipe it melts into a milke-white enamel.

Constituent Parts.

Silica, -	46.50	Silica, -	44.00
Alumina, -	30.25	Alumina,	34.00
Lime, -	0.75	Lime, -	0.12
Potash, -	18.00	Potash & Soda,	16.50
Oxide of Iron,	1.00	Oxide of Iron,	4.00
Water, -	2.00		
	98.50		98.62

Klaproth, in Magazin für die Neuesten Endeckungen in der Naturkunde, &c. 3ter Jahrg, s. 45. Also *Klaproth*, Beit. b. v. s. 178.

Vauquelin, in Hauy's Tabl. Comparative, p. 178.

Geognostic

* Hauy mentions a fourfold cleavage, of which two of the folia only are distinct, and are those mentioned above.

Geognostic and Geographic Situations.

The blue variety is found at Laurwig, and the red at Stavern and Friedrichswärn, both in the rock named *zircon-syenite.*

Observations.

1. It is named *Elaolite* by Klaproth, and *Fettstein* by Werner, on account of its resinous lustre.

2. It was at one time considered to be a variety of Foliated Scapolite or Arcticite; but it is distinguished from that mineral by colour, lustre, fracture, weight, and chemical composition. Hausmann names it *Compact Wernerite;* and Werner places it in the system between Jasper and Cat's-eye.

10. Sodalite.

Sodalite, *Thomson.*

Transactions of Royal Society of Edinburgh, vol. vi. p. 390.

External Characters.

Its colour is intermediate between celandine and mountain green.

It occurs massive, and crystallised in rhomboidal or garnet dodecahedrons.

Externally it is smooth, and shining or glistening: internally the longitudinal fracture is vitreous, and the cross fracture resinous. The longitudinal fracture is foliated, with a twofold cleavage; the cross fracture small conchoidal.

The fragments are angular, and sharp-edged.

It

It is translucent.
It is as hard as felspar.
It is brittle, and easily frangible.
Specific gravity, 2.378.

Chemical Characters.

When heated to redness, it does not decrepitate, nor fall to powder, but becomes dark grey ; and is infusible before the blowpipe.

Constituent Parts.

Silica, - -	38.52	36.00
Alumina, - -	27.48	32.00
Lime, - -	2.70	
Oxide of Iron, -	1.00	0.25
Soda, - -	25.50	25.00
Muriatic Acid, -	3.00	6.75
Volatile Matter, -	2.10	
Loss, - -	1.70	
	100.00	100.00
	Thomson, in Tr. R. S. of Ed. vol. vi. p. 394.	*Ekeberg*, in Tr. R. S. of Ed. vol. vi. p. 395.

Geognostic and Geographic Situations.

It was discovered at Kanerdluarsuk, a narrow tongue of land, upwards of three miles in length, in lat. 61°, in West Greenland, by Mr Giesecké. It is found in a bed from six to twelve feet thick, in mica-slate, and is associated with sahlite, augite, hornblende, and garnet *.

Observations.

* Allan, in Thomson's Annals of Philosophy, vol. ii. p. 390.

Observations.

Sodalite was first established as a distinct species, and analysed by Dr Thomson. I have given it its present place in the system, on account of its external characters.

11. Meionite.

Meionite, *Hauy & Werner.*

Hyacinthe blanche de la Somma, *Romé de Lisle,* t. ii. p. 290.—Meionite, *Hauy,* t. ii. p. 586. *Id. Broch.* t. ii. p. 519, 520. *Id. Lucas,* p. 49. *Id. Leonhard,* Tabel. s. 17. *Id. Brong.* t. i. p. 583. *Id. Brard,* p. 130. *Id. Haus.* s. 95. *Id. Karsten,* Tabel. s. 34. *Id. Hauy,* Tabl. p. 34. *Id. Steffens,* b. i. s. 458. *Id. Hoff.* b. ii. s. 361. *Id. Lenz,* b. i. s. 512. *Id. Oken,* b. i. s. 351.

External Characters.

Its colour is greyish-white.

It occurs sometimes massive, but more frequently crystallised, and in the following figures:

1. Rectangular four-sided prism, flatly acuminated with four planes, which are set on the lateral edges *.
2. The preceding figure, truncated on the lateral edges, fig. 99 †.

Sometimes

---

* The primitive form of Meionite, according to Hauy, is a prism, whose bases are squares.

† Meionite dioctaedre, Hauy.

Sometimes one of the acuminating planes becomes so large that the others disappear, when there is formed

3. A four-sided prism, in which the terminal planes are set on obliquely.
4. N° 1. bevelled on the lateral edges, and the edges of the bevelment truncated; and the edges between the acuminating planes and the lateral planes also truncated, fig. 100 *.

The crystals are small, seldom middle-sized; they are superimposed, and form druses.

Externally the crystals are smooth and splendent, internally splendent and vitreous.

The fracture is foliated, with a double rectangular cleavage, in which the folia are parallel with the lateral planes of the prism.

The fragments are indeterminate angular.

It is generally transparent, or semi-transparent, seldom translucent.

It is hard in a low degree; scratches glass.

It is easily frangible; and

Is rather heavy, but in a low degree.

Specific gravity, 2.612, *Mohs*.

Chemical Characters.

It is easily fusible before the blowpipe; intumesces during fusion, and is converted into a white vesicular glass.

It has not hitherto been analysed.

Geognostic

---

* Meionite soustractive, Hauy.

Geognostic and Geographic Situations.

It occurs, along with ceylanite and nepheline, in granular limestone, at Monte Somma, near Naples. It is said also to occur in basalt, along with augite and leucite, at Capo di Bove, near Rome.

Observations.

1. This species is characterised by its white colour, simple crystallizations, splendent vitreous lustre, cleavage, transparency, inconsiderable hardness, low specific gravity, and the changes it experiences before the blowpipe.

2. It is distinguished from *Adularia* by its crystallizations, its cleavage, and the changes it undergoes before the blowpipe: its crystallizations, foliated fracture, and easy frangibility, distinguish it from *Nepheline:* it is readily distinguished from *Cross-stone*, by the flatness of its acuminations, the equality of its lateral planes, and its never occurring in twin crystals; it is further discriminated by its stronger lustre, foliated fracture, and fusibility: it was formerly confounded with *Hyacinth* or *Zircon*, but is distinguished from that mineral by colour-suite, the flatness of its acuminations, perfect vitreous lustre, double cleavage, inferior hardness and weight, and infusibility before the blowpipe.

3. It was Romé de Lisle who first attended to the crystallization of this mineral: it was more particularly examined by Hauy, who established it as a distinct species, under the name *Meionite*, from the Greek word μειων, *smaller*, *shorter*, because the acumination of its principal crystallization is flatter, and also lower, than in similar crystals in other minerals.

12. Nepheline.

12. Nepheline.

Nepheline, *Hauy & Werner*.

Sommite, *La Metherie*, t. ii. p. 271.—Nepheline, *Broch.* t. ii. p. 522. *Id. Hauy*, t. iii. p. 186. *Id. Lucas*, p. 72.—Sommit, *Leonhard*, Tabl. s. 16.—Nepheline, *Brong.* t. i. p. 387. *Id. Brard*, p. 176. *Id. Haus.* s. 94.—Sommit, *Karst.* Tabel. s. 32.—Nepheline, *Hauy*, Tabl. p. 51. *Id. Steffens*, b. i. s. 476. *Id. Hoff.* b. ii. s. 365.—Sommit, *Lenz*, b. i. s. 513. Weicker Smaragd, *Oken*, b. i. s. 319.

External Characters.

The colours are snow-white, greyish-white, yellowish-white, and greenish-white, which latter sometimes passes into greenish-grey.

It occurs massive; and crystallised in the following figures:

1. Perfect equiangular six-sided prism, fig. 101 *.
2. The preceding figure, truncated on the terminal edges, fig. 102 †.

When the prism becomes shorter, there is formed,

3. A thick six-sided table, in which the lateral edges are truncated.

The crystals are small and very small, always superimposed, and forming druses.

Externally the crystals are splendent: internally shining, and the lustre is vitreous.

A fourfold cleavage is to be observed: three of the cleavages are parallel with the lateral planes, and one with

* Nepheline primitive, Hauy.

† Nepheline annulaire, Hauy.

with the terminal planes of the prism; the cross fracture is conchoidal.

The fragments are sharp-edged.

It is strongly translucent, passing into transparent.

It is hard in a low degree; its sharp corners scratch glass.

It is rather heavy, but in a middling degree.

Specific gravity, 2.274, *Hauy*.

Chemical Characters.

It is not soluble in nitrous acid; but transparent pieces, when immersed in that acid, become cloudy in the interior: hence the name *Nepheline*, from νεφελη, *a cloud*, given to this species by Hauy. It melts with difficulty before the blowpipe into a dark glass.

Constituent Parts.

Silica, - - -	46
Alumina, - -	49
Lime, - - -	2
Oxide of Iron, - -	1
	98 *Vauquelin.*

Geognostic and Geographic Situations.

It occurs in drusy cavities in granular limestone, along with ceylanite, vesuvian, and meionite, at Monte Somma, near Naples; also in fissures of basalt at Capo di Bove, near Rome. It is mentioned also as a production of the Isle of Bourbon.

Observations.

1. This species is characterised by its white colours, which sometimes incline to green, its crystallizations, vitreous

vitreous lustre, conchoidal fracture, high degree of translucency, inferior hardness, and specific gravity.

2. It is distinguished from *Meionite* by its crystallizations, fourfold cleavage, conchoidal cross fracture, inferior specific gravity, and its appearance when exposed to heat: its conchoidal cross fracture, and superior hardness, distinguish it from *Appatite*: it is readily distinguished from *Felspar* by its crystallizations, fracture, inferior hardness, and greater specific gravity: and its colour, and inferior hardness, distinguish it from *Emerald* and *Beryl*.

3. It is described by early writers under the name *White Schorl*. La Metherie named it *Sommite*, from the place where it was first found; and Hauy denominates it *Nepheline*.

4. The small acicular crystals of this species found near Rome, are described by Fleuriau Bellevue, under the name *Pseudo-Nepheline*, and are considered as belonging to a distinct species. Judging from the accounts of this pseudo-nepheline published by authors, we are still inclined to consider it but as a variety of nepheline.

13. Ice Spar.

Eis-spath, *Werner*.

Eis-spath, *Chierici*, Moll's Ephem. 5. 1. s. 126. *Id. Steffens*, b. i. s. 478. *Id. Hoff*. b. ii. s. 369. *Id. Lenz*, b. i. s. 515.

External Characters.

Its colour is greyish-white, which inclines sometimes to yellowish-white, sometimes to greenish-white.

It occurs massive, cellular, porous, and crystallised, in thin longish six-sided tables, in which the shorter terminal

minal planes meet under an obtuse angle, and the terminal planes are bevelled.

The crystals are small.

The lateral planes of the prism are longitudinally streaked.

Externally the crystals are shining, and sometimes splendent: internally shining, and the lustre is vitreous.

The fracture is imperfect foliated, and probably with several cleavages. The massive varieties are sometimes radiated.

The fragments are sharp-edged.

It occurs in large granular distinct concretions, which are again composed of concretions which are thin and straight lamellar.

It is strongly translucent; the crystals are transparent.

It is hard in a low degree.

It is very easily frangible; and

Rather heavy, but in a low degree.

Geognostic and Geographic Situations.

It occurs, along with nepheline, meionite, mica, and hornblende, at Monte Somma, near Naples.

Observations.

1. It is a very simple, but well characterised species. It is characterised by its white colour, tabular crystallizations, pretty strong vitreous lustre, foliated fracture, distinct concretions, low degree of hardness, and weight.

2. It was first established as a distinct species by Werner, who named it *Ice-Spar*, on account of its icy appearance, and foliated or sparry fracture.

XI. CLAY

XI. CLAY FAMILY.

This Family contains the following species: 1. Aluminite; 2. Alum-stone; 3. Porcelain-Earth; 4. Common Clay; 5. Claystone; 6. Adhesive-Slate; 7. Polier-Slate; 8. Tripoli; 9. Floatstone.

1. Aluminite.

Reine Thonerde, *Werner*.

Reine Thonerde, *Wid.* s. 385. *Id. Wern.* Cronst. s. 176.—Native Argil, *Kirw.* vol. i. p. 175.—Argilla pura, *Nap.* p. 246.—L'Alumine pure, *Broch.* t. i. p. 318.—Reine Thonerde, *Reuss,* b. ii. s. 102. *Id. Lud.* b. i. s. 104. *Id. Suck.* b. i. s. 471. *Id. Bert.* s. 277. *Id, Mohs,* b. i. s. 434. *Id. Leonhard,* Tabel. s. 20.—Argile native, *Brong.* t. i. p. 515.—Aluminit, *Haus.* s. 85. *Id. Karsten,* Tabel. s. 48.—Alumine pure, *Hauy,* Tabl. p. 58.—Alument, *Steffens,* b. i. s. 194. *Id. Lenz,* b. i. s. 541.—Verwitterter Alaunstein, *Oken,* b. i. s. 368.

External Characters.

Its colour is snow-white, which verges on yellowish-white.

It occurs in small reniform pieces.

It has no lustre.

The fracture is fine earthy: its consistence is intermediate between friable and solid.

 It

It is opaque.
It soils slightly.
It affords a glistening streak.
It adheres feebly to the tongue.
It passes from very soft into friable.
It feels fine, but meagre.
Specific gravity, 1.669, *Schreber.*

Chemical Character.

It is very difficultly fusible.

Constituent Parts.

Alumina, -	32.50	31.0
Water, -	47.00	45.0
Sulphuric Acid,	19,25	21.5
Silica, -	0.45	
Lime, -	0.35	2.0
Iron, -	0.45	
		99.5
	Simon, in Allgem. Journ. der Chemie, 5. Jahrg. s. 137.	*Bucholz.*

Geognostic and Geographic Situations.

It occurs, along with selenite, in calcareous loam, which rests on brown coal, in the alluvial strata around Halle in Saxony; and it is said to occur at Newhaven, near Brightelmstone in England. The white crusts sometimes observed in the clay ironstone of Scotland, appear to be aluminite.

Observations.

Steffens and Keferstein are of opinion, that this mineral, and the selenite with which is accompanied, is formed ed by the decomposion of iron-pyrites: the sulphuric acid thus formed is supposed to unite with the lime and alumina; with the lime it forms sulphate of lime or selenite, and with the alumina an alum, with a superabundance of alumina.

2. Alum-Stone.

Alaunstein, *Werner*.

Calcareus aluminaris albus, *Wall.* t. ii. p. 34.—Alumen marmoris, *Breislac,* Sagg. di Observ. sulla Tolfa, &c. Rom. 1796, 8vo.—Alaunstein, *Wid.* s. 399.—Pietra d'Allume, *Nap.* p. 266.—Alumenilite, *La Meth.* t. ii. p. 113.—La pierre alumineuse, *Broch.* t. i. p. 381.—Alaunstein, *Reuss,* b. ii. 2. s. 139. *Id. Lud.* b. i. s. 109. *Id. Suck.* b. i. s. 526. *Id. Bert.* s. 279. *Id. Mohs,* b. i. s. 445. *Id. Leonhard,* Tabel. s. 22. *Id. Karst.* Tabel. s. 36. *Id. Steffens,* b. i. s. 143. *Id. Lenz,* b. ii. s. 567. *Id. Oken,* b. i. s. 366.

External Characters.

Its colours are greyish-white, snow-white, reddish-white, seldom yellowish-white, and spotted brown, and peach-blossom-red.

It occurs massive, sometimes porous, and the pores have a crystalline drusy appearance.

Internally it is feebly glimmering, passing into dull.

The fracture is small-grained uneven, sometimes passing into splintery, sometimes into earthy.

The fragments are sharp-edged.

It is semi-hard in a low degree.

It is feebly translucent on the edges.

It is brittle, and easily frangible.

Specific gravity, 2.587, *H.* 2.633, *Karsten.*

 Chemical

Chemical Character.

It is difficultly fusible.

Constituent Parts.

		Alumstone from Tolfa.	Hungarian Alumstone.
Alumina, -	43.92	19.00	17.50
Silica, -	24.00	56.50	62.25
Sulphuric Acid,	25.00	16.50	12.50
Potash, -	3.08	4.00	1.00
Water, -	4.00	3.00	5.00
		99.00	98.25
	Vauquelin.	*Klaproth*, Beit. b. iv. s. 252.	*Klaproth*, Beit. b. iv. s. 256.

Geognostic and Geographic Situations.

It occurs at Tolfa, near Civita Vecchia, in nests, kidneys, and small veins, in a flœtz rock, which is said by Dolomieu to be volcanic. The Hungarian varieties are found in beds at Beregszaz and Nagy-Begamy, in the county of Beregher, in Upper Hungary.

Uses.

Alum is obtained from this mineral, by repeatedly roasting it, then lixiviating it, and crystallising the solution thus obtained. The art of preparing alum is an eastern discovery; and the most ancient known alum-work is that of Rocca, the present Edessa, in Syria. In the middle ages, all the alum of commerce was prepared in the Levant; but in the fifteenth century, some Ge-noese,

noese, skilled in the Levant art of alum-making, discovered alumstone in Italy, and immediately began to extract alum from it; and this new source of wealth soon became very considerable, by an edict of the Pope Pius I. who prohibited the use of Levant alum.

3. Porcelain-Earth, or Kaolin.

Porcellanerde, *Werner*.

Porcelain Clay, *Kirw*. vol. i. p. 178.—Argilla da Porcellana, *Nap*. p. 248.—La terre à Porcelaine, *Broch*. t. i. p. 320.—Feldspath argiliforme, *Hauy*, t. ii. p. 616.—Porcellanerde, *Reuss*, b. ii. s. 107. *Id. Lud*. b. i. s. 105. *Id. Suck*. 1r th. s. 492. *Id. Bert*. s. 213. *Id. Mohs*, b. i. s. 431. *Id. Hab*. s. 38. *Id. Leonhard*, Tabel. s. 21.—Argil Kaolin, *Brong*. t. i. p. 516.—Kaolin, *Haus*. s. 85. *Id. Karst*. Tabel. s. 36.—Porcelain Clay, *Kid*, vol. i. p. 165.—Feldspath decomposé, *Hauy*, Tabl. p. 36.—Porcellanerde, *Steffens*, b. i. s. 445.—Kaolin, *Lenz*, b. ii. s. 546.—Kieskaolin, *Oken*, b. i. s. 371.

External Characters.

Its most frequent colour is reddish-white, of various degrees of intensity; also snow-white, and yellowish-white.

It occurs massive, and disseminated.

It is composed of dull dusty particles, which are feebly cohering.

It soils strongly.

It feels fine and soft, but meagre.

It adheres slightly to the tongue.

Specific gravity, 2.216, *Karsten*.

Chemical

Chemical Characters.

It is infusible before the blowpipe.

Constituent Parts.

	Porcelain-Earth from Aue in Saxony.
Silica, - - -	52.00
Alumina, - -	47.00
Iron, - -	6.33
	99.33 *Rose.*

Geognostic Situation.

It generally occurs in granite and gneiss countries, either in beds contained in the granite, when it appears to be an original deposite, or on the sides and bottom of granite and gneiss hills, when it is certainly formed by the decomposition of the felspar of these rocks.

Geographic Situation.

Europe.—It occurs in small quantity in the different granite districts in Scotland, and also in England and Ireland. The finest porcelain-earth of Saxony, that which is used in the porcelain manufactory at Meissen, is brought from Aue, near Schneeberg, where it occurs in a bed contained in granite: a similar bed of porcelain-earth occurs in granite, in the valley of Gatach, above Haussach in Wirtemberg. The Austrian porcelain is made from a fine porcelain-earth which is dug near Passau. At St Yrieux la Perche, near Limoges in France, there is a bed or vein of porcelain-earth, in granite; and it has been discovered in granite near to Bayonne.

Asia.—Very valuable varieties of this mineral are found China and Japan, where they are denominated *Kaolin.*

Uses.

Uses.

This mineral forms a principal ingredient in the different kinds of porcelain. It is not used in the state in which it is found in the earth, but is previously repeatedly washed, in order to free it from impurities. After the process of washing, only fifteen parts of pure white clay remain, which is the kaolin of the Chinese. This clay, mixed in proper proportions with quartz, flint, gypsum, steatite, and other substances, forms the composition of porcelain; and this mixture is sifted several times through hair-sieves. The mixture is afterwards moistened with rain-water, in order to form a paste, which is put into covered casks. This paste is called by the workmen the *mass*. A fermentation soon takes place, which changes its smell, colour, and consistence. Sulphuretted hydrogen gas is evolved: the colour passes from white into dark grey; and the matter becomes tougher and softer. The older the mass is, the better it succeeds. It must be carefully moistened from time to time, to prevent it from drying. The preparation of the mixture, and the art of rightly managing the mass, are secrets in most manufactories. The second operation is to give the paste the form we wish; and this is done by first kneading it with the hands, in order to divide the mixture more completely, and then turning it on the lathe. A third operation is the baking, or firing, which is done in furnaces of a particular construction The firing generally lasts from thirty-six to forty-eight hours; and we judge of the state of the baking by proof-pieces, as they are called, placed in convenient situations, and which we can draw out and examine from time to time. The porcelain in this state

is

is named *biscuit porcelain* by workmen *. A fourth operation is the covering the surface of the biscuit with a varnish or enamel, which must be applied exactly over all the points of the surface, and incorporated with the paste, without cracking or flying. This enamel is composed of pure white quartz, white porcelain, and calcined crystals of gypsum, and sometimes principally of felspar: these substances are ground with the greatest care, then diffused through water, and formed into a paste. When we use it, it must be diluted in water, so as to give it considerable liquidity, and we then plunge into it the biscuit porcelain. The porcelain is now exposed to heat, sufficient to melt the enamel or covering, and then it constitutes white porcelain; and in this state it may be applied to every purpose. If the porcelain is to be painted, it must again be exposed to heat in the furnace. The colours used are all derived from metals; and many of them, though dull when applied, acquire a considerable lustre by the action of the fire. The colours are mixed with a flux, which varies in the different manufactories: in some, a mixture of glass, borax, and nitre, is employed; this mixture is melted in a crucible, and the glass is afterwards ground, and incorporated with the colour. Gum, or oil of lavender, is used as a vehicle, when we wish to lay it on the porcelain. When the painting is finished, the ware is exposed to a heat sufficient to melt the flux containing the colour.

The beautiful purple colours on porcelain, are from oxide of gold, called powder or *precipitate of Cassius*; the violet

* Figures, and generally all porcelain articles which are neither to be painted nor exposed to water, have no occasion for any covering; they are then sold in the state of biscuit.

violet colours from gold precipitated by tin and silver; certain green colours by copper, precipitated from its solutions in the acids by alkalies; red colours from oxides of iron; blue from zaffre; yellow from diaphoretic antimony, mixed with glass of lead; brown and black colours from iron-filings and zaffre; and the finest green tints from oxide of chrome.

It is said that the art of making porcelain was discovered in the year 1703, by Bötticher in Dresden.

Observations.

This mineral is distinguished from the other *Clays*, by the fineness of its particles, its soiling strongly, and its fine but meagre feel.

4. Common Clay.

This species is divided into the following subspecies, viz. Loam, Potters-Clay, Variegated Clay, and Slate-Clay.

First Subspecies.

Loam.

Leim, *Werner*.

Magerer Thon, *Karsten*, Tabel. s. 28.—Leimen, *Hab*. s. 42. *Id. Steffens*, b. i. s. 197. *Id. Lenz*, b. ii. s. 549. *Id. Oken*, b. i. s. 370.

External Characters.

Its colour is yellowish-grey, sometimes inclining to greenish-grey, and is spotted yellow and brown.

It

It occurs massive, and in great masses.

It is dull, and feebly glimmering when small scales of mica are present.

The fracture is intermediate between uneven and coarse earthy.

It soils slightly.

It is very easily frangible.

It is sectile, and the streak is dull.

It is intermediate between friable and soft, but inclining more to the first.

It adheres strongly to the tongue.

It feels somewhat greasy.

It is rather heavy, bordering on light.

Geognostic and Geographic Situations.

It occurs in great beds in alluvial districts, when it sometimes contains remains of elephants, &c.; also in flœtz mountains, along with wacke and basalt, and in fissures, forming veins. It appears in general to be an alluvial deposite, only a comparatively small portion of it occurring in flœtz rocks.

It is so very widely and generally distributed, that it is not necessary to specify any locality.

Uses.

The mud-houses we meet with in different countries, are built of loam. They are generally reared on a foundation of stone and lime, to secure them from damp. It is the practice to build them in spring, and allow them to dry during the summer: they are plastered with lime in autumn, in order to protect them from rain. The loam is mixed with straw or hair, to prevent its cracking. The most advantageous practice is to form the

loam

loam into bricks, to dry these in the shade, and afterwards in the sun. The use of loam-bricks is of high antiquity; for we are told that the ancient city of Damascus, and the walls of Babylon, were built of bricks of this substance.

Second Subspecies.

Potters-Clay.

Töpferthon, *Werner.*

Of this subspecies there are two kinds, viz. Earthy Potters-Clay, and Slaty Potters-Clay.

First Kind.

Earthy Potters-Clay.

Erdiger Töpferthon, *Werner.*

Erdiger Töpferthon, *Steffens,* b. i. s. 198. *Id. Lenz,* b. ii. s. 550.

External Characters.

Its colours are greyish and yellowish white; also light smoke, greenish, and bluish grey.

It occurs massive; and is friable, approaching to solid.

Internally it is generally feebly glimmering.

The fracture in the large is fine-grained uneven; in the small fine earthy.

It is more or less shining in the streak.

It is very soft, passing into friable.

It adheres strongly to the tongue; more strongly than loam.

It

It feels rather greasy.

Specific gravity, 2.085, *K.*

Chemical Characters.

It is infusible.

Constituent Parts.

Silica,	- -	63.00
Alumina,	-	37.00
		100.00 *Kirwan.*

Geognostic Situation.

It is a frequent mineral in alluvial districts, where it sometimes occurs in beds of considerable thickness; it has also been observed in flœtz-trap rocks.

Uses.

It is used in potteries, in the manufacture of the different kinds of earthen-ware: it is also made into bricks, tiles, crucibles, and tobacco-pipes; and is employed in improving sandy and calcareous soils.

Observations.

1. It is distinguished from *Loam* by its colour, fracture, its shining streak, and its stronger adhesion to the tongue.

2. The finer varieties are named *Pipe-Clay.*

Second

Second Kind.

Slaty Potters-Clay.

Schiefriger Töpferthon, *Werner.*

Schiefriger Töpferthon, *Steffens,* b. i. s. 200. *Id. Lenz,* b. ii. s. 554.

External Characters.

Its colour is bluish-grey, which sometimes passes into pearl-grey, also smoke-grey.

It occurs massive.

The lustre of the principal fracture is glistening; the cross fracture dull.

The principal fracture is imperfect slaty; the cross fracture fine earthy.

It agrees in other characters with the preceding kind, being only more greasy to the feel.

Geognostic Situation.

It occurs in considerable beds in alluvial districts, along with Earthy Potters-Clay.

Third Subspecies.

Variegated Clay.

Bunter Thon, *Werner.*

Bunter Thon, *Steffens,* b. i. s. 200. *Id. Lenz,* b. ii. s. 554.

External Characters.

Its colours are yellowish-white, reddish-white, flesh and peach-blossom red, ochre-yellow, and yellowish-brown.

These colours are generally arranged in broad stripes, and often in veined delineations.

It occurs massive.

The fracture is coarse earthy, inclining to slaty.

It is dull, both externally and internally.

It becomes strongly resinous in the streak.

The other characters as in the preceding subspecies.

Geognostic and Geographic Situations.

It occurs in alluvial deposites near Wehrau, in Upper Lusatia.

Observations.

It is closely allied to Lithomarge, and even passes into it.

Fourth Subspecies.

Slate-Clay.

Schiefer Thon, *Werner.*

Slate-clay, Shale, *Kirwan,* vol. i. p. 182.—L'Argile schisteuse, *Broch.* t. i. p. 327. *Id. Hauy,* t. iv. p. 446.—Schiefer Thon, *Reuss,* b. ii. s. 99. *Id. Lud.* b. i. s. 107. *Id. Suck.* 1r th. s. 490. *Id. Bert.* s. 211. *Id. Mohs,* b. i. s. 440. *Id. Hab.* s. 47. *Id. Leonhard,* Tabel. s. 22.—Argile feuilletée, *Brong.* t. i. p. 525. ?—Schiefriger Thon, *Karsten,* Tabel. s. 28.—Schiefer Thon, *Steffens,* b. i. s. 201. *Id. Lenz,* b. ii. s. 555.

External Characters.

Its colours are smoke and ash grey, greyish-black, and sometimes brownish-red.

It

It occasionally contains impressions of unknown ferns and reeds.

It occurs massive.

It is dull, or glimmering, owing to intermixed scales of mica.

The fracture in the large is more or less perfectly slaty; in the small, earthy.

The fragments are tabular.

It is opaque.

It is intermediate between soft and very soft.

It affords a dull streak.

It is easily frangible.

It adheres slightly to the tongue.

It feels somewhat meagre.

Specific gravity, 2.636, *Karsten*.

Geognostic Situation.

It occurs in beds in all the flœtz coal-formations. It passes into claystone, sandstone, and bituminous-shale, and sometimes inclines to clay-slate.

Geographic Situation.

It occurs more or less abundantly in all the coal districts in this island; and in other parts of the world where coal, and its accompanying rocks, have been particularly attended to.

5. Claystone.

Thonstein, *Werner*.

L'Argile endurcie, *Broch.* t. i. p. 325.—Thonstein, *Reuss,* b. ii. s. 96. *Id. Lud.* b. i. s. 106.—Verharteter Thon, *Suck.* 1r th. s. 489.—Thonstein, *Bert.* s. 210. *Id. Mohs,* b. i. s. 442. *Id. Hab.* s. 48. *Id. Leonhard,* Tabel. s. 22. *Id. Karst.* Tabel. s. 36. *Id. Steffens,* b. i. s. 192. *Id. Lenz,* b. ii, s. 559.—Roche argilleuse, *Hauy.*

External Characters.

Its colours are pearl, bluish, smoke, and yellowish grey: from yellowish-grey it passes into yellowish-white; from pearl-grey into lavender-blue, flesh and brownish red. It is sometimes spotted and striped.

It occasionally contains vegetable impressions.

Internally it is dull.

The fracture is fine earthy, sometimes inclining to slaty, and conchoidal.

The fragments are indeterminate angular, and rather blunt-edged.

It is opaque.

It is soft, inclining to semi-hard.

Rather easily frangible; and brittle in a low degree.

Specific gravity, 2.210, *Karsten.*

Geognostic Situation.

It occurs in beds, along with primitive porphyry; also forming the basis of clay-porphyry in primitive mountains; it forms the basis of flœtz clay-porphyries, and appears in beds, along with black coal, and is a constituent of some kinds of tuff.

Geognostic

Geographic Situation.

It occurs along with flœtz-porphyry in the Pentland Hills; in a similar situation in the Island of Arran; on the mountain of Tinto; in the Ochil Hills; and in many other places in Scotland. It occurs frequently on the Continent of Europe; and it has been observed associated with the porphyries of Asia and America.

6. Adhesive-Slate.

Klebschiefer, *Werner*.

Le Schiste à polir, *Broch.* t. i. p. 376.—Klebschiefer, *Reuss*, b. iv. s. 159.—Polierschiefer, *Leonhard*, Tabel. s. 22.—Klebschiefer, *Karsten*, Tabel. s. 26. *Id. Steffens*, b. i. s. 151. *Id. Lenz*, b. ii. s. 560.

External Characters.

Its colour is intermediate between greenish and yellowish, and also smoke-grey.

It occurs massive.

It is dull.

The fracture is straight slaty.

The fragments are tabular.

It exfoliates on exposure to the air, but becomes compact again when immersed in water.

It becomes shining in the streak.

It is soft, passing into very soft.

It is sectile.

It splits very easily.

It adheres strongly to the tongue.

Feels somewhat greasy.

Specific gravity, 2.080, *Klaproth*.

Chemical Characters.

It is infusible before the blowpipe.

Constituent Parts.

Silica, -	62.50	Silica, -	58.0
Alumina,	00.75	Alumina, -	5.0
Magnesia,	8.00	Magnesia,	6.5
Lime, -	00.25	Lime, -	1.5
Carbon, -	00.75	Iron & Manganese,	9.0
Iron, -	4.00	Water, -	19.0
			100.0
	Klaproth.		*Bucholz.*

Geognostic Situation.

It occurs in beds in the third flœtz gypsum, and contains imbedded menilite.

Geographic Situation.

It has hitherto been found only in the gypsum formation around Paris.

Observations.

It has been frequently confounded with Polier Slate, from which it is distinguished by its greater specific gravity.

7. Polier,

7. Polier, or Polishing-Slate.

Polierschiefer, *Werner.*

Polierschiefer, *Mohs,* b. i. s. 451. *Id. Karsten,* Tabel. s. 26. *Id. Steffens,* b. i. s. 149. *Id. Lenz,* b. ii. s. 562. *Id. Oken,* b. i. s. 279.

It is divided by Karsten into three subspecies, viz. Common Polier-Slate, Earthy Polier-Slate, and Friable Polier-Slate.

First Subspecies.

Common Polier-Slate.

Gemeiner Polierschiefer, *Karsten.*

Id. Lenz, b. ii. s. 562.

External Characters.

Its colours are snow and yellowish white; also straw, cream, and ochre yellow.

It occurs massive.

It is dull.

The principal fracture is thin slaty; the cross fracture even.

The fragments are tabular.

It is opaque.

It is semi-hard.

Feels meagre; and soils strongly.

It is very brittle; and easily frangible.

It is so light as to swim on the surface of water.

Constituent Parts.

Silica, - -	79.00
Alumina, - -	1.00
Lime, - -	1.00
Oxide of Iron, -	4.00
Water, - -	14.00
	100.00

Bucholz, in Gehlen's Journal für Chemie und Physique, b. ii. s. 31.

Second Subspecies.

Earthy Polier-Slate.

Erdiger Polierschiefer, *Karsten*.

External Characters.

The colours are the same as in the preceding subspecies.

It occurs massive.

The principal fracture is thick and straight slaty; the cros fracture earthy.

It is opaque.

Rather brittle.

It adheres strongly to the tongue; and absorbs water with an audible noise, and the emission of numerous air-bubbles.

It soils strongly.

It is soft; and feels meagre.

It is easily frangible; and

Is so light as nearly to swim on the surface of water.

Constituent

Constituent Parts.

Silica, - -	83.50
Alumina, - -	4.00
Lime, - -	9.50
Oxide of Iron, -	1.50
Water, - -	9.00

Bucholz, Gehlen's Journal, b. ii. s. 34.

Third Subspecies.

Friable Polier-Slate.

Zerreiblicher Polierschiefer, *Karsten.*

Id. Lenz, b. ii. s. 563.

External Characters.

Same colours as preceding.

It occurs massive.

The principal fracture is thin slaty; the cross fracture fine earthy.

It is very soft.

It is friable, and sectile.

Constituent Parts.

Silica, - -	87.00
Alumina, - -	0.50
Lime, - -	0.50
Oxide of Iron, -	1.50
Water, - -	10.00

Bucholz, Gehlen's Journ. b ii. s. 34.

Geognostic

Geognostic and Geographic Situations.

It is found at Kritchelberg, near Kitschlin, in the vicinity of Bilin in Bohemia, sometimes with impressions of leaves, more rarely inclosing skeletons of fish, or petrified wood, in a bed resting on marl. It is said also to have been found near Zwickau in Saxony; and in Auvergne.

Use.

It is used for polishing glass, marble, and metals.

Observations.

1. It is conjectured to be a pseudo-volcanic production.

2. Werner first established it as a distinct species, and separated it from Adhesive-Slate with which it had been confounded.

8. Tripoli.

Tripel, *Werner.*

Tripela, *Wall.* t. i. p. 94.—Trippel, *Wid.* p. 353.—Tripoli, *Kirw.* vol. i. p. 202.—Tripel, *Estner,* b. ii. s. 631. *Id. Emm.* b. i. s. 307. *Id. Nap.* p. 210. *Id. La Meth.* t. ii. p. 457.—Le Tripoli, *Broch.* t. i. p. 379.—Quartz aluminifere Tripoléen, *Hauy,* t. iv. p. 467.—Tripel, *Reuss,* b. ii. s. 446. *Id. Lud.* b. i. s. 108. *Id. Suck.* 1r th. s. 428. *Id. Bert.* s. 247. *Id. Mohs,* b. i. s. 449. *Id. Hab.* s. 6.—Tripoli, *Lucas,* s. 60.—Tripel, *Leonhard,* Tabel. s. 22.—Tripoli, *Brong.* t. i. p. 329. —Tripel, *Karsten,* Tabel, s. 24.—Tripoli, *Kid,* Appendix, p. 31.—Tripil, *Steffens,* b. i. s. 147. *Id. Lenz,* b. i. s. 564. *Id. Oken,* b. i. s. 278.

External Characters.

Its colours are yellowish, ash, and smoke grey; also cream and ochre yellow. It is spotted and striped.

It occurs massive.
It is dull.
The fracture is coarse earthy, inclining to slaty.
The fragments are angular.
It is very soft, passing into friable.
It feels meagre, and rather rough.
It does not adhere to the tongue.
Specific gravity, 2.202, *Bucholz*.

Chemical Characters.

It is infusible before the blowpipe.

Constituent Parts.

Silica, - - -	81.00
Alumina, - -	1.50
Trace of Lime.	
Black and Red Oxide of Iron,	8.00
Sulphuric Acid, - -	3.45
Water, - - - -	4.55
Loss, - - - -	1.50
	100 *Bucholz*.

The sulphuric acid and water are considered as accidental constituent parts.

Geognostic Situation.

It occurs in beds in coal-fields; also in beds, along with flœtz limestone, and alternating with clay, under basalt.

Geographic Situation.

It is found in Derbyshire, where it is named *Rottenstone;* also in the coal-fields of Dresden and Thuringia; in flœtz-trap districts in Bohemia; in Auvergne, where it it is said to be associated with pseudo-volcanic rocks; in the island of Corfu; at Ronneburg and Kerms in Austria; near Burgos in Spain; and Tripoli in Barbary.

Uses.

It is used for polishing metals and stones: for these purposes, it is mixed with sulphur, in the proportion of two parts of tripoli to one of sulphur; these are well rubbed together on a marble-slab, and applied to the stone or metal by means of a piece of leather. When combined with red ironstone, it is used for polishing optical glasses. It is sometimes used for moulds, in which are cast small metallic or glass figures, and medallions.

It is said that a fine species of tripoli, found near Burgos in Spain, is used as an ingredient in the manufacture of porcelain.

Observations.

1. The yellow varieties of tripoli are held to be the best.

2. Some mineralogists are of opinion, that it is a mixture of fine sand and clay, therefore that it is a mechanical deposite: others are inclined to view it as a chemical formation,—an opinion which appears to be countenanced by its geognostic relations.

9. Floatstone.

9. Floatstone.

Schwimmstein, *Werner*.

Schwimmstein, *Reuss*, b. iv. s. 202. *Id. Mohs*, b. i. s. 443.—Quartz nectique, *Lucas*, p. 35.—Schwimmstein, *Leonhard*, Tabel. s. 22.—Quartz nectique, *Brard*, p. 98.—Schwimmstein, *Karsten*, Tabel. s. 24.—Quartz nectique, *Hauy*, Tabl. p. 27.—Schwimmstein, *Steffens*, b. i. s. 166. *Id. Lenz*, b. ii. s. 566. *Id. Oken*, b. i. s. 280.

External Characters.

Its colours are yellowish-white and yellowish-grey, with yellowish-brown spots.

It occurs massive, and in tuberose masses, which are frequently vesicular, and often contain portions of flint.

It is dull.

The fracture is earthy.

The fragments are blunt-angular.

It is opaque.

It is very soft, passing into friable.

Feels meagre and rough.

It is brittle, and very easily frangible.

It absorbs water with an audible noise, and with the emission of numerous air-bubbles, and becomes translucent.

Specific gravity, 0.448, *Karsten*. 0.512, *Tralles*. 0.797, *Kopp*.

Chemical Characters.

It is infusible before the blowpipe.

Constituent

Constituent Parts.

Silica, -	98.0	94.0	91.00
Water, -		5.0	6.00
Carbonate of Lime,	2.0		2.00
Oxide of Iron, with Alumina, -		0.5	2.25
	100.0	99.5	99.25
	Vaquelin, in Hauy's Min. t. ii. p. 432.	*Bucholz*, in Leonhard's Min. Taschenb. for 1812, s. 5. & 8.	

Geognostic and Geographic Situations.

It has hitherto been found only at St Oien, in the vicinity of Paris, the Spanish localities given by some mineralogists being erroneous. It occurs, along with flint, in flœtz limestone, and sometimes contains petrifactions of the same species as those found in flint. Dr Schneider has shewn that there is an uninterrupted transition from flint into floatstone, in such a manner, that the centre of a mass will be pure and solid flint, but becomes gradually more porous as we approach the surface, when it passes into floatstone. Hence, this mineral may on a general view, be considered as an uncommonly vesicular or porous flint; and it would be an improvement in the arrangement, to follow the method of Karsten and Steffens, to place it beside flint.

XII. CLAY-

XII. CLAY-SLATE FAMILY.

This Family contains the following Species: Alum-Slate, Bituminous-Shale, Drawing-Slate, Whet-Slate, and Clay-Slate.

1. Alum-Slate.

This species is divided into two subspecies, viz. Common Alum-Slate, and Glossy Alum-Slate.

First Subspecies.

Common Alum-Slate.

Gemeiner Alaunschiefer, *Werner.*

Schistus aluminaris? *Wall.* t. ii. p. 32.—Var. of Alaunschiefer, *Wid.* s. 396. *Id. Estner,* b. ii. s. 651.—Gemeiner Alaunschiefer, *Emm.* b. i. s. 296.—Schisto aluminoso, *Nnp.* p. 264.—Varieté de l'Argile schisteuse, *Hauy.*—Le Schiste alumineux commune, *Broch.* t. i. p. 386.—Gemeiner Alaunschiefer, *Reuss,* b. ii. s. 143.—Alaunschiefer, *Lud.* b. i. s. 110. *Id. Suck.* 1r th. s. 529.—Schiefriger Aluminit, *Bert.* s. 219.—Alaunschiefer, *Mohs,* b. i. s. 454.—Gemeiner Alaunschiefer, *Leonhard,* Tabel. s. 22.—Alaunschiefer, *Haus.* s. 86.—Gemeiner Alaunschiefer, *Karsten,* Tabel. s. 36. *Id. Steffens,* b. i. s. 205. *Id. Lenz,* b. ii. s. 571. *Id. Oken,* b. i. s. 362.

External Characters.

Its colour is intermediate between bluish and greyish black.

It

It occurs massive, and sometimes in roundish balls, which are imbedded in the massive varieties.

Its lustre is more or less glimmering.

The fracture is nearly perfect straight slaty.

The fragments are tabular.

It is opaque.

It retains its colour in the streak.

It is soft.

It is easily frangible, and rather brittle.

Specific gravity, 2.384, *Kirwan*. 2.017, *Karsten*.

Second Subspecies.

Glossy Alum-Slate.

Glänzender Alaunschiefer, *Werner*.

Var. Alaunschiefer, *Wid.* s. 395.—Glänzender Alaunschiefer, *Emm.* b. i. s. 297.—Alaunschiefer, *Estner*, b. ii. s. 651.—Varieté de l'Argile schisteuse, *Hauy*.—La Schiste alumineux eclatante, *Broch.* t. i. p. 388.—Glänzender Alaunschiefer, *Reuss*, b. ii. s. 145. *Id. Hab.* s. 49. *Id. Leonhard*, Tabl. s. 22. *Id. Karst.* Tabel. s. 36. *Id. Steffens*, b. i. s. 206. *Id. Lenz*, b. ii. s. 572. *Id. Oken*, b. i. s. 362.

External Characters.

Its colour is intermediate between bluish and iron black, and it sometimes exhibits on the surface of fissures the pavonine or temper-steel tarnish.

It occurs massive.

Its lustre is semi-metallic and shining on the principal fracture, and glimmering or dull on the cross fracture.

The fracture is partly straight, partly undulating curved slaty.

The

The fragments are tabular, and these run into wedge-shaped fragments.

Specific gravity, 2.588.

In all the other characters it agrees with the preceding subspecies.

Geognostic Situation.

Both subspecies agree in geognostic situation: they occur in primitive, and also in transition clay-slate, and more rarely, in veins traversing these rocks. It is said that some varieties of alum-slate have been observed associated with old flœtz rocks.

Geographic Situation.

It occurs along with grey wacke and grey wacke-slate, in the vicinity of Moffat, in Dumfriesshire; in the transition districts of Lanarkshire, particularly in the neighbourhood of Lead Hills; and near the Ferry-town of Cree in Galloway: there are considerable beds of alum-slate on the Continent of Europe, as in Saxony, Bohemia, France, and Hungary. Esmark observed a vein of alum-slate, about two fathoms wide, at Telkobanya in Hungary; and similar veins are to be seen near Freyberg in Saxony.

Uses.

This mineral, when roasted and lixiviated, affords alum.

Observations.

1. Alum-Slate is distinguished from *Clay-Slate*, by its streak always remaining unaltered in the colour.

2. The two subspecies were distinguished by Wallerius and Cronstedt.

 2. Bituminous-

2. Bituminous Shale.

Brandschiefer, *Werner*.

Shistus pinguis? *Wall.* t. i. p. 3.4.; Schistus carbonarius, *Id.* p. 358.—Brandschiefer, *Wid.* s. 394.—Bituminous Shale, *Kirw.* vol. i. p. 183.—Brandschiefer, *Estner,* b ii. s. 658. *Id. Emm.* b. i. s. 289.—Schisto bituminoso, *Nap* p. 263.—Argillite bitumineux, *Lam.* t. ii. p. 116.—Varieté de l'Argile schisteuse, *Hauy.*—Le Schiste bitumineux, *Broch.* t. i. p. 389.—Brandschiefer, *Reuss,* b. ii. s. 120. *Id. Lud.* b. i. s. 111. *Id. Suck.* 1r th. s. 504. *Id. Bert.* s. 218. *Id. Mohs,* b. i. s. 456. *Id. Leonhard,* Tabel. s. 23. *Id. Karsten,* Tabel. s. 36.—Bituminous Shale, *Kid,* vol. i. p. 189.—Brandschiefer, *Steffens,* b. i. s. 204. *Id. Lenz,* b. i. s. 573. *Id. Oken,* b. i. s. 361.

External Characters.

Its colour is light brownish-black.

It occurs only massive.

Internally its lustre is glimmering.

The fracture is rather thin and straight slaty.

The fragments are tabular.

It becomes resinous in the streak, but the colour is not changed.

It is opake.

It is very soft, approaching to soft.

It is rather sectile, and easily frangible.

Specific gravity, 1.991, 2.049, *Kirwan.* 2.060, *Karsten.*

Constituent

Constituent Parts.

Two hundred grains afforded the following parts, partly as educts, partly as products:

Carbonated Hydrogen Gas,	80 cubic inches.
Empyreumatic Oil, -	30 grains.
Thick Pitchy Oil, - -	5 do.
Ammoniacal Water, -	4
Carbon, - - -	20
Silica, - - -	$87\frac{1}{2}$
Alumina, - - -	$6\frac{1}{2}$
Lime, - - -	$10\frac{1}{2}$
Magnesia, - - -	1
Oxide of Iron, - -	3

Klaproth, Beit. B. v. s. 184.

Geognostic Situation.

It occurs principally in rocks of the coal-formation, where it frequently alternates with, and passes into, slate-clay, and also into coal. It sometimes contains vegetable impressions, and also animal remains, particularly of shells. It occurs in beds of considerable magnitude in hills of iron-clay.

Geographic Situation.

It occurs in all the coal districts in this island, and also in those in Bohemia, Poland, Silesia, and other countries.

Observations.

1. It is distinguished from *Slate-Clay*, with which it has been confounded, by the streak: in Slate-Clay, the

 streak

streak is always dull; whereas it is invariably shining and resinous in Bituminous-Shale.

2. In this species, the clay is combined with bitumen, but in alum-slate with Carbon.

3. Some mineralogists consider the *Ampelitis* of the ancients as drawing-slate, others as bituminous shale.

3. Drawing-Slate, or Black Chalk.

Zeichenschiefer, *Werner*.

Schistus pictorius nigrica, *Wall.* t. i. p. 358.—Zeichenschiefer, *Wern.* Cronst. s. 208.—Black Chalk, *Kirw.* vol i p 195.—Schwarze Kreide, *Estner*, b. ii. s. 661.—Zeichenschiefer, *Emm.* b. i. s. 303.—Schisto pittorio, *Nap.* p. 269.—Melantirite, ou Crayon noire, *Lam.* t. ii. p. 112.—Argile schisteuse graphique, *Hauy*, t. iv. p. 447.—Le Schiste a dessiner, *Broch.* t. i. p. 391.—Zeichenschiefer, *Reuss*, b. ii. s. 146. *Id. Lud.* b. i. s. 112. *Id. Suck.* 1r th. s. 505. *Id. Bert.* s. 217. *Id. Mohs*, b. i. s. 458. *Id. Hab.* s. 43. *Id. Leonhard*, Tabel. s. 23.—Ampelite graphique, *Brong.* t. i. p. 563.—Zeichenschiefer, *Haus.* s. 85. *Id. Karst.* Tabel. s. 36.—Black Crayon, *Kid*, vol. i. p. 190.—Zeichenschiefer, *Steffens*, b. i. s. 208. *Id. Lenz*, b. ii. s. 575. *Id. Oken*, b. i. s. 361.

External Characters.

Its colour is intermediate between bluish and greyish black, but rather more inclining to the latter colour.

It is massive.

The lustre of the principal fracture is glimmering, of the cross fracture dull.

The principal fracture is imperfect and curved slaty; the cross fracture fine earthy

The fragments are partly tabular, partly splintery.

It is opaque.

It soils lightly, and writes.

It retains its colour in the streak, and becomes glistening.

It is very soft.

It is sectile.

Does not adhere to the tongue.

Feels fine, but meagre

Specific gravity, 2.110, *Kirwan*. 2.111, *Karsten*.

Chemical Character.

It is infusible.

Constituent Parts.

Silica, - -	64.06
Alumina, - -	11.00
Carbon, - -	11.00
Water, - -	7.20
Iron, - -	2.75

According to *Wiegleb*, Crell's Ann. 1797, s. 485.

Geognostic Situation.

It occurs in beds in primitive and transition clay-slate.

Geographic Situation.

It is found at Marilla in Spain, Brittany in France, and in Italy; also in Germany, as in the mountains of Bareuth.

Uses.

It is used for drawing, and also as a black colour in painting When used for drawing, it is cut into square pencils, which are sometimes inclosed in wooden cases, like pencils of graphite or black-lead. We must select for this purpose those varieties having the darkest colour, the finest earthy fracture, and which are free of quartzy particles and veins. It has been found, that these pencils become dry and hard, and unfit for drawing by long keeping. To prevent this evil, the pencils should be kept in a moist place; or, what is better, the slate should be ground, and mixed with gum-water, and run into moulds; and pencils of this kind, if well prepared, will remain long fit for use. We must be careful that too much gum-water is not added, otherwise the particles will be so closely aggregated that the pencils will not leave a trace on the paper; and on the other hand, we must see that too little gum is not added; for if this be the case, the pencil will soil the paper, and no regular or well formed trace will be left on it.

When black chalk is used for painting, it is first pounded and ground, and then mixed with oil or size, and is used as a black paint. It is, however, not much valued, as it is at best but a coarse colour. Certain varieties burn red, or reddish-brown, and these are sometimes used for red or brown colours.

Observations.

1. Some varieties of Bituminous Shale have been confounded with Black Chalk; but a comparison of their trace on paper, enables us at once to distinguish them: the trace of Bituminous Shale being brownish and irregular, whereas that of Black Chalk is regular and black.

2. The most highly prized varieties of this mineral, are those found in Spain, Italy, and France.

4. Whet-

4. Whet-Slate.

Wetzschiefer, *Werner.*

Schistus coticula, *Wall.* t. ii. p. 353.—Wetzschiefer, *Wid.* s. 402. —Novaculite, *Kirw.* vol. i. p. 238.—Wetzschiefer, *Estner,* b. ii. s. 664. *Id. Emm.* b. i. s. 305.—Pietra cote, *Nap.* p. 270. Cos, *Lam.* t. ii. p. 105.—Le Schiste a aiguiser, *Broch.* t. i. p. 393.—Argile schisteuse novaculaire, *Hauy,* t. iv. p. 448. Wetzschiefer, *Reuss,* b. ii s. 149. *Id. Lud* b. i. s. 112. *Id. Suck.* 1r th. s. 506. *Id. Bert.* s 216. *Id. Mohs,* b. i. s. 460. *Id. Hab.* s. 42. *Id. Leonhard,* Tabl. s. 23.—Schiste coticule, *Brong.* t. i. p. 558.—Wetzschiefer, *Haus.* s. 87. *Id. Karst.* Tabel. s. 38.—Novaculite, or Honestone, *Kid,* vol. i. p. 216.—Wetzschiefer, *Steffens,* b. i. s. 211. *Id. Lenz,* b. ii. s. 576. *Id. Oken,* b. i. s. 359.

External Characters.

Its most common colour is greenish-grey; it is found also mountain, asparagus, olive, and oil green.

It occurs massive.

Internally it is feebly glimmering.

The fracture in the large is straight slaty; in the small, splintery.

The fragments are tabular.

It is translucent on the edges.

It is soft, inclining to semi-hard.

It feels rather greasy.

Specific gravity, 2.677, *Karsten.*

Geognostic Situation.

It occurs in beds in primitive and transition clay-slate

Geographic

Geographic Situation.

It is found at Seifersdorf, near Freyberg; at Lauenstein and Sonnenberg, in the district of Meinengen; and also in the Hartz. Very fine varieties are brought from Turkey.

Uses.

When cut and polished, it is used for sharping iron and steel instruments. For these purposes, it is necessary that it contain no intermixed hard minerals, such as quartz. The light-green coloured varieties from the Levant are the most highly prized: those from Bohemia are also much esteemed in commerce. The Levant whet-slate is brought in masses to Marseilles, and is there cut into pieces of various sizes. It is ground by means of sand or sandstone, and polished with pumice and tripoli. These whet-stones, or *hones*, as they are called, ought to be kept in damp and cool places; for when much exposed to the sun, they become too hard and dry for many purposes.

The powder of whet-slate is used for cutting and polishing metals, and is by artists considered as a variety of emery.

Observations.

1. It is distinguished from other minerals by colour, fracture, transparency, and hardness.

2. This species does not include every kind of mineral used as whet-stone; for some varieties of clay-slate, of sandstone, and of slate-clay, are used for that purpose.

5. Clay-

5. Clay-Slate.

Thonschiefer, *Werner*.

Schistus ardesia tegularis, *Wall.* t. i. p. 351.—Thonschiefer, *Wid.* s. 391.—Argillite, *Kirw.* vol. i. p. 234.—Killas, *Id.* p. 237. *Id. Emm.* b. i. s. 284. *Id. Estner,* b. ii. s. 667.—Ardoise, *Lam.* t. ii. p. 110.—Le Schiste argileux, *Broch.* t. i. p. 395.—Argile schisteuse tegulaire tabulaire, *Hauy,* t. iv. p. 447.—Thonschiefer, *Reuss,* b. ii. s. 151. *Id. Lud.* b. i. s. 113. *Id. Suck.* 1r th. s. 508. *Id. Bert.* s. 215. *Id. Mohs,* b. i. s. 462. *Id. Hab.* s. 42. *Id. Leonhard,* Tabel. s. 23.—Schiste argileux, *Brong.* t. i. p. 557. *Id. Haus.* s. 87. *Id. Karst.* Tabel. s. 38.—Schistus, or Slate, *Kid,* vol. i. p. 186.—Thonschiefer, *Steffens,* b. i. s. 210. *Id. Lenz,* b. ii. s. 578. *Id. Oken,* b. i. s. 359.

External Characters.

Its colours are yellowish, ash, smoke, bluish, pearl, and greenish-grey; from greenish-grey it passes into blackish-green, and greenish-black; from dark smoke-grey into greyish-black; and from pearl-grey into brownish-red *, It is sometimes spotted.

It occurs massive.

Its lustre is pearly, inclining to resinous, and is glistening, or glimmering.

The fracture is more or less perfect slaty; and some varieties approach to foliated, and others to compact. The slaty is either straight, or undulating curved, and the latter has a twofold obliquely insersecting cleavage.

The

* Houses roofed with the red variety of clay-slate, appear as if covered with copper.

The fragments are generally tabular, seldom splintery or trapezoidal.

It is opaque.

It affords a greyish-white dull streak.

It is soft.

It is sectile, and easily split.

It feels rather greasy.

Specific gravity, 2.661, *Kirwan.* 2.786, *Karsten.*

Chemical Characters.

It is fusible into a slag before the blowpipe.

Constituent Parts.

Silica, -	48.6	Silica, -	38.0
Alumina, -	23.5	Alumina, -	26.0
Magnesia, -	1.6	Magnesia, -	8.0
Peroxide of iron,	11.3	Lime, -	4.0
Oxide of manganese,	0.5	Peroxide of iron,	14.0
Potash, -	4.7		
Carbon, -	0.3		*Kirwan.*
Sulphur, -	0.1		
Water, and Volatile Matter, -	7.6		
Loss, -	1.8		
	100 *Daubuisson.*		

Geognostic Situation.

It occurs in primitive and transition mountains: in primitive mountains it generally rests on mica-slate, and the older strata of clay-slate alternate with the newer ones of mica-slate; when the mica-slate is awanting, it rests on gneiss, and alternates with it in the same manner as it does with mica-slate; when the gneiss is awanting,

ing, it rests on granite, and also alternates with it. These facts shew, that clay-slate is sometimes of cotemporaneous formation with mica-slate, sometimes with gneiss, and even with granite: In transition mountains, it rests on and alternates with grey wacke, grey wacke-slate, transition trap, transition limestone, and other rocks of the transition class.

Transition clay-slate is sometimes scarcely to be distinguished from the primitive varieties of this rock, otherwise than by its geognostic characters: Transition clay-slate alternates with, and passes into, grey wacke-slate, and contains petrifactions: Primitive clay-slate alternates with, and passes into, mica-slate, and never contains petrifactions; and these are some of the geognostic characters by which we are enabled to distinguish the one from the other.

Geographic Situation.

It is a very generally distributed rock throughout the mountainous regions in the different quarters of the globe. It abounds in many of the highland districts in Great Britain and Ireland, and in several of the smaller islands that lie near their coasts. On the Continent of Europe, it forms a considerable portion of the Hartz, the Erzgebirge, the Fichtelgebirge, the Thuringerwaldgebirge, and of many other great groups of mountains.

Uses.

It is principally used for roofing of houses. Those varieties of clay-slate used for roofing houses, are named *Roofing-Slate*, and should possess the following properties.

1. They must split easily and regularly into thin and straight plates of the requisite magnitude. This is only

the

the case, however, with such varieties of clay-slate as possess a regular and perfect slaty fracture, without rents, or intermixed foreign parts. A clay-slate which contains grains, crystals, or veins of quartz, garnet, schorl, hornblende, or iron-pyrites, will not split into regular plates or *slates*, because these hard bodies do not yield on splitting the mass, and hence the slate generally breaks at such places. If the clay-slate is very thick-slaty, it cannot be split into slates of sufficient thinness, and hence is of but little use, because when the slates are beyond a certain thickness, they are too heavy for roofs. When the clay-slate is curved slaty, it does not split into useful slates. It may be noticed, that care must be taken to keep the slate in a damp place, previous to splitting, otherwise, if it becomes dry, it will not split without difficulty. It is therefore advisable to split the masses as soon as possible after separating them from the rock.

2. A good roof slate must be sufficiently compact, and not porous, so that the rain and snow water may not percolate through and destroy the wooden work of the roof. Some varieties of clay slate are so porous that they imbibe much water, do not dry easily, and hence afford opportunity for the growth of mosses and lichens, which in time cover the surface of the slate. These plants retain moisture long, and keep the surface, and even the interior of the slate moist, so that during the winter season, by the freezing of the moisture, the slate splits and falls into pieces In order to ascertain whether or not the slate has the requisite compactness, we have only to dry it completely, then weigh it, afterwards plunge it into water, and allow it to remain for some time. If after wiping it with a cloth, it has not acquired any considerable increase of weight, it is a proof of its being sufficiently compact; on the contrary, if it absorbs much

much water, and becomes considerably heavier by immersion, it shews that it is of a porous and loose texture.

It is remarked, that the slates in the upper strata in quarries are generally porous and loose in their texture, and hence these are generally thrown away as useless.

3. A good slate must be sufficiently solid, and not brittle and shattery; for such slates break in pieces on the application of but a weak force, and do not form a firm roof. When the slate is too brittle, it flies into pieces during the dressing and boring: if it emits a pretty clear sound when struck with a hammer, it is a proof that it is not over brittle; but if it emits a dull sound, it shews that it is soft and shattery. Lastly, if a slate of inconsiderable thickness breaks easily with the hands, it is a proof of its being too soft.

4. No slate can be used with advantage for roofing houses, which readily decomposes by the action of the weather. The decomposition observed to take place in roof-slates is of two kinds: the one is mechanical, the other is chemical. The mechanical decomposition is principally caused by the freezing of water in the porous and softer varieties, by which they are split in pieces: the chemical decomposition is caused by the decay of disseminated iron-pyrites, or the increased oxidation of intermixed iron.

5. Lastly, a good slate ought to resist the action of a considerable degree of heat.

The best roof-slates found in Scotland, are those of Easdale, and some neighbouring islands off the coast of Lorn in Argyle, and of Ballihulish in Appin, also in Argyle. The quantity manufactured annually at Easdale and its vicinity, is about five millions, which gives employment to 300 men; and at Ballihulish, it is estimated that about half a million of slates are prepared

every

every year. There are also considerable slate-quarries in the parish of Luss in Dunbartonshire, in Monteath, Strathearn, Strathmore, the Garioch, and other places. The slate principally in use in London, is brought from Wales, from quarries which are worked at Bangor in Caernarvonshire. There are also extensive slate-quarries near Kendal in Westmoreland, and the slates from that quarter, which are of a bluish-green colour, are more highly esteemed by the London builders than those from Wales. They are not of a large size, but they possess great durability, and are well calculated to give a neat appearance to the roof on which they may be placed. French slates were very much in use, in London, about seventy years ago; they are of small size, very thin, and consequently light, and therefore much less calculated for the climate of this island than the heavier and more durable slates of England and Scotland.

We shall next mention some other uses of clay-slate.

The dark-coloured, most compact, and solid varieties, named *Table-slate*, are used for writing on, but are previously prepared in the following manner. The plate or slate is first smoothed, by means of an iron instrument: it is afterwards ground with sandstone, and slightly polished with tripoli, and lastly rubbed with charcoal powder. It is cut into the required shape, set in a wooden frame, and is then ready for use. When these table-slates are first taken from the quarry, they are rather soft, hence are easily worked; but they become hard by drying.

The small pieces of slate used for writing with, are obtained from a particular variety of clay-slate, named *Writing-slate*, which, on splitting, falls into prismatic or splintery fragments. In order to form a good writing material,

material, it must be more sectile, and softer than table-slate, so that it may leave a coloured streak on its surface, without scratching it. This variety of slate does not occur either frequently or abundantly; and it is remarked, that the strata in which it is contained, are generally traversed by vertical rents, and that the best kinds are found between them. When the slate is separated from the stratum in which it is contained, and laid in heaps, it soon falls into long splintery pieces, which are from a quarter to half an inch thick, and from a few inches to upwards of a foot in length. It is said, that if these pieces are exposed for some time to the action of the sun or frost, they are rendered useless: hence workmen are careful to cover them up, and sprinkle them with water as soon as extracted from the quarry, and preserve them in damp cellars. The pieces are afterwards split, by means of a particular instrument, and then made into the required shape.

In some places in Wales, and also in Germany, clay-slate is used for grave-stones; and is sometimes turned into vases, and other similar articles. The masses used for grave-stones, are cut smooth with sandstone, polished with tripoli, and lastly, rubbed with charcoal-powder, or lamp-black, or graphite, in order to deepen the black colour. On account of its softness, it receives but an indifferent polish: hence, in order to give it a higher degree of lustre, it is a practice to dip it into oil, after polishing, by which process its lustre is improved, and it is also rendered more durable. It is remarked, that if a window or door is opened in the apartment where the workmen are turning the clay-slate into any particular form, it very frequently flies in pieces, although, after the work is finished, it may be exposed to the usual alternations of temperature without risk of injury.

Pounded

Pounded or ground clay-slate is used for cleaning the surface of iron, and other kinds of metallic ware. It scarcely acts on the metal, but unites with the adventitious soiling-matter on its surface. Clay-slate, when well ground, and mixed in certain proportions with loam, forms a compound excellently fitted for moulds, as it receives the most delicate impressions, and with the greatest accuracy: hence it is very advantageously employed in cast-iron works. When it is burnt, and afterwards coarsely ground, it may be used in place of sand, in the making of mortar: mortar of this kind is said to become very solid and impermeable under water.

In smelting houses, it is sometimes employed as a flux, with ores that contain much calcareous earth.

Observations.

It passes into Mica-slate, Chlorite-slate, Talc-slate, Whet-slate, Alum-slate, Drawing slate, and probably into Compact Felspar.

XIII. MICA

XIII. MICA FAMILY.

THIS Family contains the following species: Lepidolite, Mica, Pinite, and Chlorite.

1. Lepidolite.

Lepidolith, *Werner.*

Lepidolith, *Wid.* s. 378. *Id. Kirw.* vol. i. p. 208. *Id. Emm.* b. iii. s. 324. *Id. Estner,* b. ii. s. 228. *Id. Nap.* p 167. *Id. Lam.* t ii. p. 315. *Id. Broch.* t. i. p. 399. *Id. Hauy,* t iv. p. 375. *Id. Reuss,* b. ii. s. 402. *Id. Lud.* b. i. s. 114. *Id. Suck.* 1r th. s. 397. *Id. Bert.* s. 17. *Id. Mohs,* b. i. 465. *Id. Hab.* s. 40. *Id Lucas,* p. 199. *Id. Leonhard,* Tabel. s. 23. *Id. Brong.* t. i. p. 506. *Id. Brard,* p. 411. *Id. Haus.* s. 91. *Id. Karst.* Tabel. s. 30 *Id. Kid,* vol. ii. p. 246. *Id. Hauy.* Tab. p. 64. *Id. Steffens,* b. i. s. 213. *Id. Lenz,* b. ii. s. 582. *Id. Oken,* b. i. s. 390.

External Characters.

Its colour is peach blossom-red, inclining sometimes to flesh-red, sometimes to lilac-blue; it also passes into pearl-grey and yellowish-grey.

It occurs massive, and crystallised in equiangular six-sided prisms.

Internally its lustre is shining, and semi-metallic, inclining to pearly.

The fracture in the large is coarse splintery; in the small, fine foliated.

The fragments are blunt-angular.

It occurs in small and fine granular distinct concretions.

It is translucent.

It is soft; the crystals, however, will scratch glass.

It is rather sectile.

It is rather easily frangible.

Specific gravity, 2.816, *Klaproth*. 2.58, *Karsten*.

Chemical Characters.

Before the blowpipe it intumesces, and melts very easily into a milk-white nearly translucent globule.

Constituent Parts.

Silica, - -	54.50	Silica, -	54.00
Alumina, -	38.25	Alumina,	20.00
Potash, - -	4.00	Potash,	18.00
Manganese & Iron,	0.75	Fluat of Lime,	4.00
Loss, partly Water,	2.50	Manganese,	3.00
		Iron, -	1.00
	100		100
	Klaproth, Beit. b. ii. s. 195.		*Vauquelin*, Jour. de Min. t. ix. p. 235.

Geognostic and Geographic Situations.

It occurs imbedded in granite, in the mountain of Hradisko, near to Rosena in Moravia; in quartz which is contained in granite, in the Riesengebirge in Silesia; in granite at Penig in Saxony; also in primitive rocks in Uton in Sweden; in Norway; in the vicinity of Limoges in France; and in the Isle of Elba*.

Uses.

* Hauy and Tondi received, through Mr Schultz, a mineral from Bavaria, which they consider as a variety of lepidolite. Von Moll's Neue Jahrbuch d. Berg & Hüttenkunde, 3. B. 1. Lif. s. 111.

Uses.

It is sometimes cut into snuff-boxes, which are admired for their colour; but owing to the softness of the mineral, they have rather a dull greasy-like surface.

Observations.

1. It is nearly allied to Mica, from which, however, it is distinguished by colour, fracture, and also by its chemical properties: thus mica contains at least 7 *per cent.* of iron, while lepidolite does not contain more than 1 *per cent.*; and mica is difficultly fusible, whereas lepidolite fuses with great ease.

2. The first account pub'ished of this mineral, was by M. Von Born, in the Chem. Annalen for 1791.

3. The grey variety from Uton has been described as a distinct species, under the name *Petalite.*

2. Mica.

Glimmer, *Werner.*

Mica, *Wall.* t. i. p. 383.—Glimmer, *Wid.* s. 403.—Mica, *Kirw.* vol. i. p. 210.—Glimmer, *Estner,* b ii s. 673. *Id. Emm.* b. i. s. 31.—Mica, *Lam.* t. ii. p. 337. *Id. Nap.* p. 272. *Id. Broch.* t i p. 402. *Id. Hauy,* t. iii. p. 208.—Glimmer, *Reuss,* b. ii. s. 72. *Id. Lud.* b i s 114. *Id. Suck.* Ir th. s. 474. *Id. Bert.* s. 202. *Id. Hab* s 41 — ica, *Lucas,* p. 75. Glimmer, *Leonhard,* Tabel. s. 23.—Mica, *Brong.* t. i. p. 508, *Id. Brard,* p. 182.—Glimmer, *Haus* s. 89. *Id. Karsten,* Tabel, s 30.—Mica, *Kid,* vol i. p. 183. *Id Hauy,* Tabl. p. 53 —Glimmer, *Steffens,* b. i. s. 215. *Id. Lenz,* b. ii. s. 585. *Id. Oken,* b. i. s. 387.

External Characters.

Its most common colour is grey, from which it passes on

on the one side into brown, on the other into black. It exhibits the following varieties of grey, viz. yellowish, ash, and greenish grey; which latter passes into blackish-green; the yellowish-grey passes into silver-white, and into pinchbeck-brown, and brownish-black.

It occurs massive, disseminated; and crystallised,

1. Rhomboidal four-sided prism: this prism is in general so very low, that it may be described as a four-sided table *, fig. 103.
2. Regular six-sided prism: this prism is in general very short, and then it appears as a six-sided table. Fig. 104. the prism; fig. 105. the table.
3. Long four-sided table, fig. 106.
4. Six-sided table, in which the terminal edges are truncated, fig. 107.
4. Oblique six-sided pyramid, with alternate broader and narrower lateral planes, fig. 108.

The crystals are middle-sized and small, generally adhere by their terminal planes, and are sometimes aggregated in a scopiform, but rarely in a rose-like manner. The terminal planes of the prisms, and the lateral planes of the tables, are smooth and splendent; but the lateral planes of the prism are generally streaked and shining.

Internally it is splendent, and the lustre is pearly, semi-metallic, and metallic.

The fracture is foliated, either common or scaly foliated, sometimes undulating or floriform, which latter passes into broad or narrow radiated. One of the cleavages parallel with the lateral planes of the table is very distinct; several others, which are concealed, are parallel with the terminal planes of the table, and meet the distinct cleavage at right angles.

The

* According to Hauy, the primitive form is a rhomboidal prism, in which the angles of the bases are 60° and 120°.

The radiated variety is plumosely streaked.

The fragments are tabular.

It occurs in large, coarse, and small granular concretions; also in columnar concretions.

It is translucent or transparent in thin plates, but rarely in crystals of considerable thickness or length *.

It splits easily in one direction.

It is sectile.

It affords a grey-coloured dull streak.

It is intermediate between soft and semi-hard †.

It feels smooth, but not greasy.

It is elastic-flexible.

Specific gravity, 2.654, 2.034, *Hauy*. 2.726, *Karsten*.

Chemical Characters.

Before the blowpipe, it melts into a greyish-white enamel.

* Count de Bournon mentions crystals of mica in his valuable collection, of considerable thickness, which are transparent in the direction of their axes. He also notices particularly the difference of colour observed as we look in the direction of the axis, or across the crystal: thus, he observed in a transparent crystal from Pegu, that the colour in the direction of the axis was yellowish-green; but at right angles to the axis, was beautiful *vert d'herbe*. In other crystals, the colour in the line of the axis was of a beautiful green, whilst in the opposite direction it was orange; and in some other crystals, the colour parallel with the axis was white, but perpendicular to it flesh-red.

† Count de Bournon, at page 120. of his *Catalogue Mineralogique*, remarks, that a thick piece or crystal of mica, will scratch glass, and even quartz.

Constituent Parts.

Common Mica of Zinnwald.	
Silica, -	47.00
Alumina,	22.00
Oxide of Iron,	15.50
Oxide of Manganese,	1.75
Potash, -	14.50
	98.75

Klaproth, Beit. b. v. s. 69.

Large foliated Mica from Siberia.	
Silica, -	48.00
Alumina,	34.25
Oxide of Iron,	4.50
Oxide of Manganese,	0.50
Potash, -	8.75
Loss by heating,	1.25
	97.25

Klaproth, Id. s. 73.

Black Mica from Siberia,	
Silica,	42.50
Alumina,	11.50
Oxide of Iron,	22.00
Oxide of Manganese,	2.00
Potash, -	10.00
Magnesia,	9.00
Loss by heating,	1.00
	98.00

Klaproth, Id. s. 78.

Geognostic Situation.

This mineral occurs as an essential constituent part of several primitive rocks, and accidentally intermixed with others, both of the primitive, transition, flœtz, and alluvial classes. Thus, along with felspar and quartz, it forms granite and gneiss, and with quartz mica-slate: it is occasionally intermixed with clay-slate, primitive limestone, sienite, porphyry, greenstone, hornblende-slate and hornblende rock, whitestone, grey wacke, grey wacke-slate, sandstone, wacke, amygdaloid, basalt, and various alluvial deposites. It sometimes forms short beds in granite, and other primitive rocks; or it appears in globular, oval, tuberose, or irregular-shaped cotemporaneous masses, in granite or gneiss. It also occurs in veins, as in those formed of granite or quartz, or in such as contain ores of different kinds, as tinstone and copper-pyrites.

Geographic Situation.

The rocks in which mica occurs, are so universally distributed, that it is not necessary to enter into any detail

tail of localities: we may merely mention, that most of the mica of commerce is brought from Siberia, and the borders of the Caspian Sea, where it occurs in large plates or crystals, in granite.

Uses.

In some countries, as in Siberia, mica is an article of commerce, and is regularly mined. In Siberia, the principal mica mines are those on the banks of the Wettin, the Aldan, and other rivers that fall into the Lena. It occurs in nests, often of considerable magnitude, imbedded in granite. The mica is extracted by means of hammers and chisels, is then washed of the adhering earth, and assorted into different kinds, according to goodness, purity, and size. The plates or tables intended for sale, must be clear, well coloured, and as free as possible from spots. The greenish-coloured and imperfectly transparent, or the spotted varieties, are laid aside, and sold at a low rate. It is exported in considerable quantity from Russia. In 1781, 200 puds were sent from St Petersburgh to Lübec, and a very considerable quantity to England and Ireland.

In Siberia, where window-glass is scarce, it is used for windows; also for a similar purpose in Peru, and, I believe, also in New Spain, as it appears that the mineral named *Teculi* by Ulloa, and which is used for that purpose, is a variety of mica. It is also used in lanterns, in place of glass, as it resists the alternations of heat and cold better than that substance. In Russia, it is employed in different kinds of inlaid work. It is sometimes intermixed with the glaze in particular kinds of earthen-ware: the heat which melts the glaze has no effect on the mica; hence it appears dispersed throughout the glaze, like plates or scales of silver or gold, and

 thus

thus gives to the surface of the ware a very agreeable appearance. Some artists use it in the making of artificial avanturines.

3. Pinite.

Pinit, *Werner*.

Pinit, *Reuss*, b. ii. s. 69. *Id. Lud.* b. ii. s. 149. *Id. Suck.* 1r th s. 469. *Id. Bert.* s. 298. *Id. Mohs*, b. i. s. 480. *Id. Lucas*, p. 280. *Id. Leonhard*, Tabel. s. 24. *Id. Brong.* t. i. p. 507. *Id. Brard*, p. 185. *Id Karsten*, Tabel. s. 48. *Id. Hauy*, Tabl. p. 53. *Id. Steffens*, b. i. s. 219. *Id. Lenz*, b. ii. s. 592. *Id. Oken*, b. i. s. 389.

External Characters.

Its colour is blackish-green, altered on the surface by brown or red iron-ochre into brownish-red.

It occurs massive; and crystallised in the following figures:

1. Equiangular six-sided prism *.
2. The preceding figure, truncated on all the lateral edges.
3. N° 1. truncated on the angles.

The crystals are middle sized and small, and generally imbedded.

The longitudinal fracture is small-grained uneven, and glistening; the cross fracture is imperfect foliated, and shining, inclining to glistening, and resinous.

The fragments are blunt-angular, seldom tabular.

It is opaque.

The

* The primitive form, according to Hauy, is a regular six-sided prism.

The massive varieties sometimes occur in thick and straight lamellar concretions.

It is soft, passing into very soft.

It is sectile.

It is very easily frangible.

It feels somewhat greasy.

Specific gravity, 2.914, *Hauy*. 2.974, *Kirwan*.

Chemical Characters.

It is infusible before the blowpipe.

Constituent Parts.

Silica, -	29.50	Silica, -	46.0
Alumina,	63.75	Alumina, -	42.0
Oxide of Iron,	6.75	Oxide of Iron,	2.5
		Loss by calcination,	7.0
	100.00	Loss, - -	2.5
			100.0
Klaproth, Jour. des Mines, N. 100. p. 311.		*Drappier*, Jour. des Mines, N. 100. p. 311.	

Geognostic and Geographic Situations.

It is found imbedded in the granite of St Michael's Mount in Cornwall: in granite at Schneeberg in Saxony, and in the porcelain-earth of Aue, also in Saxony; in a greyish porous felspar-porphyry in the Puy de Dome, in Auvergne; in Dauphiny, along with epidote, axinite, rock-crystal, chlorite, and iron-ochre; and in the Bavarian Waldgebirge.

Observations.

1. It is distinguished from *Mica*, with which it has been

been confounded, by its circumscribed series of colour, its peculiar truncations, its never inclining to the tabular form, and its fracture.

2. It was first established as a distinct species by Werner, and named *Pinite*, from the Pini Gallery in the mines of Schneeberg, where it was first found.

3. According to Bernhardi, in Von Moll's Ephemerid. B. iii. st. 1. pinite is nearly allied to Schorl.

4. Chlorite.

This species is subdivided into four subspecies, viz. Earthy Chlorite, Common Chlorite, Slaty Chlorite, and Foliated Chlorite.

First Subspecies.

Earthy Chlorite.

Erdiger Chlorit, *Karsten.*

Chlorite in a loose form. Peach of the Cornish Miners, *Kirw.* vol. i. p. 147.—Erdiger Chlorit, *Reuss,* b. ii. s. 81. *Id. Lud.* b. i. s. 116. *Id. Suck.* 1r th. s. 479. *Id. Bert.* s. 426. *Id. Mohs,* b. i. s. 484. *Id. Leonhard,* Tabel. s. 24. *Id. Haus,* s. 90. *Id. Karst.* Tabel. s. 12.—Talc Chlorite terreux, *Hauy,* Tabl. p. 56.—Erdiger Chlorit, *Steffens,* b. i. s. 221. *Id. Lenz,* b. ii. s. 600. *Id. Oken,* b. i. s. 382.

External Characters.

Its colours are dark mountain and leek green, and sometimes olive-green.

It occurs massive, disseminated, in crusts, moss-like, inclosed in adularia and rock-crystal.

It

It is glimmering or glistening, and the lustre is pearly.

It consists of fine scaly particles, which are seldom loose, and feel rather greasy.

It becomes lighter in the streak.

Specific gravity, 2.612, 2.699.

Chemical Characters.

It melts before the blowpipe into a blackish slag.

Constituent Parts.

Silica, - -	26.00
Alumina, - -	18.50
Magnesia, - -	8.00
Muriate of Soda, or Potash,	2.00
Oxide of Iron, -	43.00
Loss, - - -	2.50
	99.00

Vauquelin, Journ. des Mines, N. 39. p. 167.

Geognostic and Geographic Situations.

It occurs in felspar and adularia veins in St Gothard; also in Dauphiny, where it encrusts rock-crystal, axinite, and sphene. It is found also in Salzburg, Norway, Harzebirge Forest, Dognazka in the Bannat, Tyrol, and many other places.

Observations.

1. It is characterised by its green colour, scaly glimmering particles, slightly greasy feel, and its not soiling.

2. The great quantity of iron it contains, is by Karsten considered more as an accidental than regular constituent part.

3. The scaly parts, according to Hauy, when viewed by the microscope, appear to be regular six-sided prisms.

Second

Second Subspecies.

Common Chlorite.

Gemeiner Chlorit, *Werner*.

Indurated Chlorite, *Kirwan*, vol. i. p. 148.—Gemeiner Chlorit, *Reuss*, b. ii. s. 84. *Id. Lud.* b. i. s. 117. *Id. Suck.* 1r th. s. 483. *Id. Bert.* s. 426. *Id. Mohs*, b. i. s. 485. *Id. Hab.* s. 59. *Id. Leonhard*, Tabel. s. 24.—Chlorite commune, *Brong.* t. i. p. 500.—Blättricher Chlorit, *Haus.* s. 90. *Id. Karsten*, Tabel. s. 42. *Id. Steffens*, b. i. s. 222. *Id. Lenz*, b. ii. s. 60. *Id. Oken*, b. i. s. 382.

External Characters.

Its colour is intermediate between dark mountain and leek green, with much intermixed black.

It occurs massive.

Its lustre is glimmering.

The fracture is fine earthy, also splintery, and fine granular foliated, or scaly.

The fragments are blunt cornered.

It becomes light mountain-green in the streak.

It is soft, passing into very soft.

It is opaque.

It feels somewhat greasy.

Specific gravity, 2.832, *Wid.*

Geognostic and Geographic Situations.

It occurs not only disseminated through rocks of different kinds, as granite, and mica-slate, but also in beds and veins. The granite of Mont Blanc contains common chlorite in veins, or disseminated through it: in Saxony,

Saxony, Salzburg, and other countries, it occurs in beds, which contain magnetic ironstone, copper-pyrites, iron-pyrites, arsenical-pyrites, hornblende, actynolite, and calcareous-spar. In the island of Arran, it occurs in quartz veins that traverse clay-slate; in similar repositories in the island of Bute, and in several other districts in Scotland. In England, it occurs in the Wherry Mine, Penzance, and other places in Cornwall *.

Third Subspecies.

Slaty Chlorite, or Chlorite-Slate.

Chlorit-Schiefer, *Werner.*

Schiefriger Chlorit, *Karsten.*

Chlorit Schiefer, *Reuss,* b. ii. s. 88. *Id. Lud.* b. i. s. 117. *Id. Suck.* 1r th. s. 484. *Id. Bert.* s. 427. *Id. Mohs,* b. i. s. 487. *Id. Hab.* s. 59. *Id. Leonhard,* Tabel. s. 24.—Chlorit schisteuse, *Brong.* t. i. p. 501.—Schiefriger Chlorit, *Haus.* s. 90. *Id. Karst.* Tabel. s. 42.—Talc Chlorite fissile, *Hauy,* Tabl. p. 56.—Schiefriger Chlorit, *Steffens,* b. i. s. 223.—Chlorit Schiefer, *Lenz,* b. ii. s. 605. *Id. Oken,* b. i. s. 383.

External Characters.

Its colour is intermediate between mountain and leek green, and sometimes passes into blackish-green.

It is massive.

The lustre is glistening and resinous.

The fracture is more or less perfectly slaty, inclining to scaly foliated.

The fragments are tabular.

Specific gravity, 2.905, *Saussure.* 2.822, *Karsten.* 2.794, *Grüner.*

Chemical

* Greenough.

Chemical Characters.

It is said to be infusible before the blowpipe.

Constituent Parts.

Silica, - -	29.50
Alumina, - -	15.62
Magnesia, - -	21.39
Lime, - -	1.50
Iron, - -	23.39
Water, - -	7.38

Grüner.

Geognostic Situation.

It occurs principally in beds, subordinate to clay-slate, and is occasionally associated with potstone and talc-slate. It frequently contains octahedral crystals of magnetic ironstone; also garnets, schorl, and cube-spar.

Geographic Situation.

It occurs in beds, in the clay slate districts of the Grampians, and other parts of Scotland. On the Continent, it is found in Norway, Sweden, Saxony, Switzerland, Corsica, and other countries.

Observations.

It passes into Common Chlorite; and in mountains, it is to be observed passing into Hornblende-slate and Clay-slate, and appears allied to Potstone and Talc.

Fourth

Fourth Subspecies.

Foliated Chlorite.

Blättriger Chlorit, *Werner*.

Blättriger Chlorit, *Reuss*, b. ii. s. 86. *Id. Lud.* b. i. s. 118. *Id. Suck.* 1r th. s. 481. *Id. Mohs,* b. i. s. 486. *Id. Leonhard,* Tabel. s. 24. *Id. Haus.* s. 90. *Id. Karsten,* Tabel. s. 62.—Talc Chlorit, *Hauy,* Tabl. p. 56.—Blättriger Chlorit, *Steffens,* b. i. s. 224. *Id. Lenz,* b. ii. s. 603. *Id. Oken,* b. i. s. 383.

External Characters.

Its colour is dark leek-green.

It occurs massive, and crystallised in six-sided tables. These tables are aggregated together, in such a manner as to form the two following figures:

A. Cylinder terminated by two cones.

B. Two truncated cones, joined base to base.

If we suppose the six-sided table N° 1. to revolve around an axis which passes through its two opposite angles, the figure A will be formed; but if it revolves around an axis which passes through two opposite sides, the figure B will be formed. The streaking on the surfaces shews the mode of aggregation of the tables.

The crystals are generally longitudinally streaked, and are small or middle-sized.

Externally it is glistening, approaching to shining and resinous; internally it is shining and resinous.

The fracture is foliated, generally curved, and with a single cleavage.

The fragments are indeterminate angular, or tabular.

The

The massive varieties occur in small and fine granular concretions.

It is opaque, or translucent on the edges.

It is soft.

It is sectile.

The folia are common flexible.

It feeels rather greasy.

It is rather difficultly frangible.

Its colour is lighter in the streak.

Specific gravity, 2.823, *Karsten.*

Constituent Parts.

Silica, - -	35.00
Alumina, - -	18.00
Magnesia, - -	29.90
Iron, - -	9.70
Water, - -	2.70

Lampadius, Handbuch zur Chem. Analyse der Mineral Körper, s. 229.

Geognostic and Geographic Situations.

Europe.—It occurs in the island of Jura, one of the Hebrides, in a micaceous quartzose rock. On the continent of Europe, it is found in St Gothard, where it is associated with adularia, rock-crystal, and rutile; also in the valley of Fusch in Salzburg, where it occurs along with amethyst and adularia, and seldom with prehnite; and in Sweden, Saxony, and Corsica.

Asia.—It occurs in Siberia, along with slaty chlorite.

OBSERVATIONS.

OBSERVATIONS ON THE SPECIES.

1. It was Saussure the Father, who first directed the attention of mineralogists to this species; and Werner was the first who ascertained its oryctognostic relations.

2. It is nearly allied to Talc and Mica, and also to Potstone. The foliated subspecies approaches the nearest to mica, the common and slaty to Potstone.

3. Hausmann, in his "Entwurf eines Systems der Unorganisirten Naturkörper," describes a substance under the name *Conchoidal Chlorite*, which deserves to be more particularly examined. The following is his account of it:

"*Conchoidal Chlorite.* Colour leek-green; internally dull; but shining and resinous on the surface of the fissures. Fracture flat conchoidal, inclining to splintery and earthy, even sometimes approaching to slaty. Becomes resinous and shining in the streak. Translucent on the edges. Soft. It occurs in the Hartz, disseminated through transition amygdaloid and greenstone."

 XIV. LITHO-

XIV. LITHOMARGE FAMILY.

This Family contains the following species: Green Earth, Pimelite, Lithomarge, Mountain-Soap, Yellow Earth, Cimolite, and Kollyrite.

1. Green Earth.

Grünerde, *Werner*.

Green Earth, *Kirwan*, vol. i. p. 196.—Grünerde, *Emm.* b. i. s. 353.—La Terre verte, *Broch.* t. i. p. 445.—Grünerde, *Reuss*, b. ii. s. 157. *Id. Lud.* b. i. s. 126. *Id. Suck.* 1r th. s. 522. *Id. Bert.* s. 214. *Id. Mohs*, b. i. s. 515. *Id. Hab.* s. 39.—Talc Chlorite zographique, *Lucas*, p. 84.—Grünerde, *Leonhard*, Tabel. s. 26.—Chlorite Baldogée, *Brong.* t. i. p. 501.—Talc Chlorite zographique, *Brard*, p. 198.—Erdiger Chlorit, *Haus.* s. 90.—Grünerde, *Karsten*, Tabel. s. 26.—Talc Chlorite zographique, *Hauy*, Tabl. p. 56.—Grünerde, *Steffens*, b. i. s. 257. *Id. Lenz*, b. ii. s. 621. *Id. Oken*, b. i. s. 277.

External Characters.

Its colour is celandine-green, of various degrees of intensity, which passes into mountain-green, blackish-green, and olive-green.

It occurs massive, seldom disseminated, more frequently in globular and amygdaloidal shaped pieces, which are sometimes hollow, in crusts lining the vesicular cavities in amygdaloid, or on the surface of agate balls.

Internally it is dull.

The

The fracture is fine earthy, sometimes flat conchoidal.
It adheres slightly to the tongue.
It is glistening in the streak.
It feels rather greasy.
It is very soft, and rather sectile.
Specific gravity, 2.598, *Karsten.* 2.632, *Kirwan.*

Chemical Characters.

Before the blowpipe, it is converted into a black vesicular slag.

Constituent Parts.

	From Cyprus.	From the Veronese.
Silica, - -	51.50	53.0
Oxide of Iron,	20.50	28.0
Magnesia, -	1.50	2.0
Potash, -	18.00	10.0
Water, -	8.00	6.0
Loss, - -	0.50	
		99.0
	100.00	
	Klaproth, Beit. B. iv. p. 244.	*Klaproth*, Id. s. 241.

Geognostic Situation.

It occurs principally in the amygdaloidal cavities of amygdaloid, and incrusting the agates often found in that rock. It appears sometimes to colour sandstone.

Geographic Situation.

It is a frequent mineral in the amygdaloid of Scotland; it occurs also in that of England and Ireland. It is found in the amygdaloid of Iceland and the Feroe Islands; and on the Continent of Europe, it occurs in Saxony, Bohemia, near Verona, the Tyrol, and Hungary.

Uses.

It is used as a green colour in water-painting. When slightly burnt, it affords a beautiful and durable brown water-colour: when burnt with oil, it affords a black colour; and the green colour is improved if the mineral is dissolved in muriatic acid. It may be imitated, by mixing together, in proper portions, well ground yellow earth, chalk, and indigo.

Observations.

This mineral was first established as a distinct species by Werner. It is distinguished by its colour, shape, fracture, streak, hardness, and geognostic situation.

2. Pimelite.

Pimelit, *Werner.*

Pimelit, *Broch.* t. ii. p. 412. *Id. Reuss,* b. ii. s. 452. *Id. Lud.* b. ii. s. 145. *Id. Suck.* 1r th. s. 430. *Id. Bert.* s. 273. *Id. Mohs,* b. i. s. 304. *Id. Leonhard,* Tabel. s. 11.—Nickel Oxide, *Brong.* t. ii. p. 210. *Id. Karst.* Tabel. s. 26. *Id. Haus.* 84. *Id. Steffens,* b. i. s. 152. *Id. Lenz,* b. ii. s. 623. *Id. Oken,* b. i. s. 277. 381.

It is divided into two subspecies, viz. Friable Pimelite, and Indurated Pimelite.

First Subspecies.

Friable Pimelite.

Zerreiblicher Pimelit, *Karsten.*

External Characters.

Its colour is siskin-green.

It

It occurs massive.
It is dull.
The fracture is earthy.
The fragments are blunt-edged.
It is very soft, passing into friable.
It feels somewhat greasy.

Constituent Parts.

Silica, - - -	35.00
Water, - - -	37.91
Alumina, - -	5.10
Magnesia, - -	1.25
Lime, - - -	0.40
Oxide of Nickel, -	15.62

Klaproth, Beit. b. ii. s. 139 *.

Second Subspecies.

Indurated Pimelite.

Verhärteter Pimelit, *Karsten.*

External Characters.

Its colour is apple-green.
It occurs partly earthy, partly in crusts.
Internally it is feebly glimmering.
The fracture is even.
The fragments are rather sharp-edged.
It is soft.
It feels very greasy.

Chemical Characters.

It is infusible before the blowpipe, but loses part of its weight.

 Geognostic

* It is the Chrysopras Erde of Klaproth.

Geognostic and Geographic Situations.

Pimelite occurs at Kosemütz, Grachau, and Glässendorf, in Silesia. It is associated with crysoprase; traverses serpentine in veins, lines its fissures, and invests it in the form of crusts *.

Observations.

1. It was first established as a distinct species by Karsten.

2. It is named *Pimelite*, from the greasy feel of the second subspecies.

3. Lithomarge.

Steinmark, *Werner.*

This species is divided into two subspecies, viz. Friable Lithomarge, and Indurated Lithomarge.

First Subspecies.

Friable Lithomarge.

Zerreiblicher Steinmark, *Werner.*

Friable Lithomarge, *Kirwan,* vol. i. p. 187.—Zerreiblicher Steinmark, *Reuss,* b ii. s. 49. *Id. Leonhard,* Tabel. s. 26. *Id. Karsten,* Tabel. s. 28. *Id. Steffens,* b. i. s. 246. *Id. Lenz,* b. ii. s. 618.

External Characters.

Its colours are snow-white, yellowish-white, and rarely reddish-white.

It

* Dr Macknight suspects that it occurs in the flœtz-trap rocks of Tinto in Lanarkshire. Vid. Wernerian Memoirs, vol. ii. p. 131.

It occurs massive, and disseminated.

It consists of very fine scaly, feebly glimmering, particles.

It becomes shining in the streak.

It is generally slightly cohering, seldom loose.

It soils slightly.

It feels rather greasy.

It is light.

It phosphoresces in the dark.

Constituent Parts.

Silica, - - -	32.00
Alumina, - -	26.50
Iron, - - -	21.00
Muriat of Soda, - -	1.50
Water, - - - -	17.00
	98.00

Klaproth, Beit, b. iv. s. 349.

Geognostic and Geographic Situations.

Undoubted varieties of this mineral occur in tinstone veins at Ehrenfriedersdorf, also at Penig; in fissures in grey-wacke, in the Hartz; in manganese veins with red ironstone at Walkenried; and it is said also in Nassau, Bavaria, and Transylvania.

Observations.

Klaproth describes a substance under the name *Earth of Sinopis*, which is found in Pontus. It is of a dark red colour, and, according to Karsten, is but a variety of the friable lithomarge. The analysis here given is of the Earth of Sinopis.

Second Subspecies.

Indurated Lithomarge.

Verhärtetes Steinmark, *Werner*.

Terra miraculosa Saxoniæ, *Schütz*, in Nov. Act. Cæs. Nat. Curios. 3. App. p. 93.—Steinmark, *Hoffmann*, Bergm. Journ. 1788, 1. 2. s. 520.—Indurated Lithomarge, *Kirwan*, vol. i. p. 188.—La Moelle de Pierre, ou Lithomarge, *Broch*. t. i. p. 447.—Argil Lithomarge, *Hauy*, t. iv. p. 444.—Steinmark, *Reuss*, b. ii. s. 164. *Id. Leonhard*, Tabel. s. 26.—Argile Lithomarge, *Brong*, t. i. p. 521.—Verhärtetes Steinmark, *Haus*. s. 86. *Id. Karsten*, Tabel. s. 28. *Id. Steffens*, b. i. s. 248. *Id. Lenz*, b. ii. s. 619.

External Characters.

Its colours are yellowish and reddish white, which latter passes from pearl-grey, through lavender-blue, plum-blue, into flesh-red, sometimes ochre-yellow. The white and red varieties are generally uniform, but the others are disposed in clouded and spotted delineations.

It occurs massive.

It is dull.

The fracture is fine earthy in the small, and large conchoidal in the great.

The fragments are rather blunt-angular.

It is opaque.

It soils lightly.

It becomes shining in the streak.

It is very soft, and sectile.

It adheres strongly to the tongue.

It feels greasy.

Specific gravity, 2.209, *Karsten*.

Chemical

Chemical Characters.

It is infusible before the blowpipe. Several of the varieties phosphoresce when heated.

Geognostic Situation.

It occurs in veins in porphyry, gneiss, grey-wacke, and serpentine: in drusy cavities in topaz-rock; or nidular, in basalt, amygdaloid, and serpentine; and it is said also in beds, in a coal formation.

Geographic Situation.

At Rochlitz in Saxony, it occurs in cotemporaneous veins, traversing clay-porphyry; at Ehrenfriedersdorf and Altenberg, also in Saxony, the white and red varieties occur in veins in gneiss; at Zöblitz, it traverses serpentine, in the form of veins; the yellow variety lines the drusy cavities of the topaz-rock; and the Saxon Terra miraculosa, a variety of this mineral, appears to occur in small beds, in a coal formation near Planitz. In the Hartz, it occurs in veins that traverse grey-wacke; and is described as a production of the rocks of Bavaria, Bohemia, and Norway.

Uses.

The Chinese are said to use it, when mixed with the the root of Veratrum album, in place of snuff; and in Germany it is employed for polishing serpentine.

Observations.

1. The friable subspecies is characterised by its scaly particles, soiling, and low degree of coherence; the indurated lithomarge by fracture, streak, softness, and sectility.

2 It is intermediate between Steatite and Variegated Clay, and appears sometimes to pass into Meerschaum.

4. Mountain-

4. Mountain-Soap.

Bergseife, *Werner.*

Bergseife, *Wid.* s. 436. *Id. Emm.* b. i. s. 360.—Le Savon de Montagne, *Broch.* t. i. p. 453. *Id. Reuss,* b. ii. s. 171. *Id. Lud.* b. i. s. 127. *Id. Suck.* 1r th. s. 502. *Id. Bert.* s. 208. *Id. Mohs,* b. i. s. 522. *Id. Leonhard,* Tabel. s. 26. *Id. Karst.* Tabel. s. 28. *Id. Haus.* s. 86. *Id. Steffens,* b. i. s. 256. *Id. Lenz,* b. ii. s. 625. *Id. Oken,* b. i. s. 385.

External Characters.

Its colours are pitch, or dark brownish black, but hair and clove brown on the surface of the fissures.

It occurs massive.

It is dull.

The fracture is fine earthy, passing into even, sometimes into imperfect conchoidal.

The fragments are rather blunt-angular.

It is opaque.

It becomes shining in the streak, or even by slight friction.

It writes, but does not soil.

It is very soft, and sectile.

It is easily frangible.

It adheres strongly to the tongue.

It feels greasy.

It is light.

Geognostic and Geographic Situations.

It occurs in trap rocks in the Island of Skye. It was formerly found at Olkutzk in Gallicia, but is now no longer to be met with in that quarter. It is said to occur

cur in a bed in basalt in the district of Nassau; and in a bed, immediately under the soil, along with potters-clay and loam, near Waltershaus, at the foot of the mountains of the Forest of Thuringia.

Use.

It is valued by painters as a crayon.

Observations.

1. This mineral is characterised by its colour, fracture, streak, adherence to the tongue, and its writing without soiling.

2. It is allied to Bole and Lithomarge.

3. It was first established as a distinct species by Werner, and particularly described by Stift, in Moll's Ephemer. 4. 1. s 31.; and by Schlottheim, in the Magaz. Naturf. Fr. in Berlin, 1.4. s. 406.

5. Yellow Earth.

Gelberde, *Werner.*

Gelberde, *Wid.* p. 427.—Yellow Earth, *Kirw.* vol. i. p. 194.—Gelberde, *Estner,* b. i. s. 362.—La Terre jaune, *Broch.* t. i. p. 455.—Gelberde, *Reuss,* b. ii. s. 101. *Id. Lud.* b. i. s. 128. *Id. Suck.* 1r th. s. 524. *Id. Bert.* s. 302. *Id. Mohs,* b. i. s. 524. *Id. Hab.* s. 48. *Id. Leonhard,* Tabel. s. 26. *Id. Karst.* Tabel. s. 48. *Id. Steffens,* b. i. s. 261. *Id. Lenz,* b. i. s. 626. *Id. Oken,* b. i. s. 372.

External Characters.

Its colour is ochre-yellow, of different degrees of intensity.

It

It occurs massive.

It is dull on the cross fracture, but glimmering on the principal fracture.

The fracture in the large inclines to slaty; in the small, it is earthy.

The fragments are tabular, or indeterminate angular.

It become somewhat shining in the streak.

It is opaque.

It soils slightly.

It is very soft, passing into friable.

It adheres to the tongue.

It feels rather greasy.

It is rather light?

Chemical Characters.

Before the blowpipe, it is converted into a black and shining enamel.

Geognostic and Geographic Situations.

It has hitherto been found only at Wehraw, in Upper Lusatia, where it is associated with clay, and clay ironstone. It is mentioned by mineralogists as occurring in Denmark, Norway, Stiria, Austria, and other countries; but most, if not all of these localities, may be omitted, as they appear to refer, not to the true yellow earth, but to varieties of clay, coloured with iron-ochre.

Uses.

It may be employed as a yellow pigment.

6. Cimolite.

6. Cimolite.

Cimolith, *Klaproth.*

Cimolith, *Klaproth,* Beit. b. i. s. 291.—La Cimolite, *Hauy,* t. iv. p. 446. *Id. Reuss,* b. ii. s. 169. *Id. Lud.* b. i. s. 150. *Id. Suck.* 1r th. s. 500. *Id. Bert.* s. 212. *Id. Leonhard,* Tabel. s. 21. *Id. Haus.* s. 86. *Id. Karsten,* Tabel. s. 28. *Id. Steffens,* b. i. s. 260.—Kimolit, *Lenz,* b. ii. s. 544.—Cimolith, *Oken,* b. i. s. 372.

External Characters.

Its colours are greyish-white, and pearl-grey, which become reddish by the action of the weather.

It occurs massive.

It is dull.

The fracture is earthy, sometimes inclining to slaty.

It is opaque.

It becomes shining in the streak.

It soils very slightly.

It is very soft.

It is rather easily frangible.

It adheres pretty strongly to the tongue.

Specific gravity, 2.00, *Klaproth.* 2.187, *Karsten.*

Chemical Character.

It is infusible.

Constituent Parts.

Silica,	63.00
Alumina,	23.00
Iron,	1.25
Water,	12.00
	99.25

Klaproth, Beit. b. i. s. 299.

Geognostic

Geognostic and Geographic Situations.

It appears to occur in beds, in the islands of Argentiera or Cimolia, and Milo, in the Mediterranean Sea.

Uses.

It was highly prized as a medicine by the ancients; they also used it for cleansing woollen and other stuffs, for which purpose it is excellently suited.

Observations.

1. This mineral is mentioned by several ancient writers, as Theophrastus, Dioscorides, Strabo, Pliny, and Ovid; and in modern times, first by Tournefort, and next by Klaproth, who in the year 1794 received specimens of it from Mr Hawkins, who had collected it in the island of Cimolia.

2. It appears to be nearly allied to Fullers Earth.

7. Kollyrite.

Kollyrit, *Karsten.*

Natürlicher Alaunerde, *Klaproth,* Beit. b. i. s. 257.—Kollyrit, *Leonhard,* Tabel, s. 21. *Id. Karst.* Tabel. s. 48 *Id Haus.* s. 85. *Id. Steffens,* b. i. s. 259. *Id. Lenz,* b. ii. s. 543. *Id. Oken,* b. i. s. 370.

External Characters.

Its colours are snow, greyish, reddish, and yellowish white.

It occurs massive.

Internally it is dull; but the reddish-white variety is feebly glimmering.

The

The fracture is fine earthy and even.

The fragments are indeterminate angular and sharp edged.

The snow-white is feebly, the reddish-white is strongly translucent on the edges.

It becomes shining and resinous in the streak.

It soils slightly.

It is very soft; the snow-white variety friable, the reddish-white approaching to soft.

It is brittle, and very easily frangible.

It adheres strongly to the tongue.

It is light.

Chemical Character.

It is infusible.

Constituent Parts.

Silica, - - -	14
Alumina, - -	45
Water, - -	42

Klaproth, Beit. b. i. s. 257.

Geognostic and Geographic Situations.

It is found in the Stephen's pit at Schemnitz in Hungary, where it forms a vein from four to five inches wide in sandstone.

Observations.

Friesleben, in Lempe's Magaz. für Bergbaukunde, 10. s. 99. describes a substance found near Weissenfels, which nearly resembles Kollyrite.

XV. SOAP-

XV. SOAPSTONE FAMILY.

This Family contains the following species: Native Magnesia, Magnesite, Meerschaum, Bole, Sphragide, Fullers Earth, Steatite, and Figure stone.

1. Native Magnesia.

Native Magnesia, *Bruce.*

Bruce on Native Magnesia from New Jersey, American Mineralogical Journal, vol. i. p. 26.–30.

External Characters.

Its colour is snow-white, passing into greenish-white.

It occurs massive.

Its lustre is pearly.

The fracture is foliated, or radiated

It is semi-transparent in the mass, transparent in single folia.

It is soft, and somewhat elastic.

It adheres slightly to the tongue.

Specific gravity, 2.13.

Chemical Characters.

Before the blowpipe, it becomes opaque and friable, and loses weight. It is soluble in the sulphuric, nitric, and muriatic acids.

Constituent Parts.

Magnesia, - -	70
Water of crystallization,	30
	100

Bruce, American Min. Journal, vol. i. p. 30.

Geognostic and Geographic Situations.

It occurs in small veins in serpentine, at Hoboken in New Jersey.

Observations.

It was discovered by Dr Bruce, Professor of Mineralogy in New York.

2. Magnesite.

Reine oder Natürliche Talkerde, *Werner*.

Magnesie native, *Broch.* t. ii. p. 499.—Reine Talkerde, *Reuss*, b. ii. s. 223. *Id. Lud.* b. i. s. 154. *Id. Suck.* 1r th. s. 539. Luftsaure Bittererde, *Bert.* s. 136.—Reine Talkerde, *Mohs*, b. i. s. 528. *Id. Hab.* s. 68. *Id. Leonhard*, Tabel. s. 27.—Magnesite de Mitchell, *Brong.* t. i. p. 490.—Magnesit, *Karsten*, Tabel. s. 48.—Magnesie carbonatée, *Hauy*, Tabl p. 16.—Magnesit, *Steffens*, b. i. s. 243. *Id. Lenz*, b. ii. s. 631. *Id. Oken*, b. i. s. 386.

External Characters.

Its colour is yellowish-grey or yellowish-white, passing into cream-yellow. It is marked with yellowish and ash-grey spots, and also with bluish-grey dots, and dendritic delineations.

It occurs massive, tuberose, and a shape which is intermediate between vesicular and perforated; and the walls of the vesicles are rough and uneven.

Internally it is dull.

The fracture is large and flat conchoidal, which passes into even and coarse splintery.

The fragments are rather sharp-edged.

It is nearly opaque.

It is soft: it is scratched by fluor-spar, but it scratches calcareous-spar.

It feels rather meagre.

It is rather easily frangible.

Specific gravity, 2.881.

Chemical Characters.

It is infusible; but before the blowpipe it becomes so hard as to scratch glass.

Constituent Parts.

Magnesia, -	48.00	46.00	45.42
Carbonic Acid, -	52.00	51.00	47.00
Silica, - - -			4.50
Alumina, - -	Trace.	1.00	0.50
Ferruginous Manganese,	Trace.	0.25	0.50
Lime, - -	Trace.	0.16	0.08
Water, - -		1.00	2.00
	Bucholz.	*Bucholz.*	*Bucholz.*

Geognostic and Geographic Situations.

It is found at Hrubschitz in Moravia, in serpentine rocks, along with meerschaum, common and earthy talc, mountain-

mountain-cork, and rhomb-spar; also at Gulfen, near Kraubat in Upper Stiria, where it occurs in serpentine, along with bronzite; and in serpentine, at Baudissero and Castellamonte in Italy.

Observations.

1. It is distinguished from *Meerschaum*, with which it has been confounded, by its colour, external shape, fracture, meagre feel, and weight.

2. It was first discovered by that excellent mineralogist, the late Dr Mitchell of Belfast.

3. Meerschaum.

Meerschaum, *Werner*.

Meerschaum, *Wid.* s. 456.—Keffekill, *Kirw.* vol. i. p. 144.—Meerschaum, *Emm.* b. i. s. 378.—Schiuma di Mare, *Nap.* p. 307.—Varieté de Talc, *Lam.* t. i. p. 342.—L'Ecume de Mer, *Broch.* t. i. p. 462.—Meerschaum, *Reuss,* b. ii. s. 219. *Id. Lud.* b. i. s. 129. *Id. Suck.* 1r th. s. 566. *Id. Bert.* s. 139. *Id. Mohs,* b. i. s. 529. *Id. Hab.* s. 69. *Id. Leonhard,* Tabel. s. 27. *Id. Kid,* vol. i. p. 99. *Id. Karst.* Tabel. s. 42. *Id. Steffens,* b. i. s. 241. *Id. Lenz,* b. ii. s. 636. *Id. Oken,* b. i. s. 386.

External Characters.

Its colours are yellowish and greyish white, seldom snow-white.

It occurs massive, and in tuberose-shaped pieces.

Internally it is dull.

The fracture is fine earthy, passing on the one side into flat conchoidal, on the other into even.

 The

The fragments are indeterminate angular.

It is opaque; rarely translucent on the edges.

It is very soft.

It is very sectile.

It is rather difficultly frangible.

It adheres strongly to the tongue.

It feels rather greasy.

Specific gravity, 1.209, *Karsten*. 1.600, *Klaproth*.

Chemical Characters.

Before the blowpipe, it melts on the edges into a white enamel.

Constituent Parts.

Silica, - -	41.50
Magnesia, - -	18.25
Lime, - -	0.50
Water and Carbonic Acid,	39.00
	98.25

Klaproth, Beit. B. ii. s. 172.

Geognostic and Geographic Situations.

Europe.—It occurs in veins, in the serpentine of Cornwall; in serpentine, at Hrubschitz in Moravia; at Vallecas, near Madrid in Spain, also in serpentine. It is dug at Sebastopol and Kaffa, in the Crimea *; and near Thebes in Greece.

Asia.—It occurs in beds immediately under the soil, at Kittisch and Bursa in Natolia; and in the mountains of Esekischehir, also in Natolia, from 600 to 700 men are employed in digging meerschaum.

Uses.

* Gallitzin, Descript. Physic de la Contrée de Tauride, p. 85.

Uses.

When first dug from the earth, it is soft and greasy. It lathers with water like soap: hence it is used by some nations, as by the Tartars, for washing. In Turkey, it is made into tobacco-pipes. These pipes are manufactured of the meerschaum of Natolia, and that dug near Thebes. It is prepared for that purpose in the following manner: It is first agitated with water in great reservoirs, and is then allowed to remain at rest for some time. The mixture soon passes into a kind of fermentation, resembling that which porcelain-earth experiences when placed in similar circumstances, and a disagreeable odour, resembling that of rotten eggs, is exhaled. As soon as the smell ceases, the mass is farther diluted with water, which is after a time poured off, and fresh water added repeatedly, until the mass is sufficiently washed and purified: what remains is the mass in a pure state. The pure meerschaum is now dried to a considerable degree, is then pressed into a brass mould, and some days afterwards it is hollowed out. The heads formed in this way are then dried in the shade, and lastly, baked in a furnace constructed for the purpose. The heads in this state are brought to Constantinople, where they are subjected to farther processes: they are first boiled in milk, and next in linseed-oil and wax; when perfectly cool, they are polished with rushes and leather. The boiling in oil and wax makes them denser, and more capable of receiving a higher polish; and further, when thus impregnated, they acquire, by use, various shades of red and brown on their surface, which is thought to add very considerably to their beauty. In Turkey, and even in Germany, pipes which have been much used, are more valued than those newly made, on account of the colouring they possess. Indeed, in those countries, there are people whose sole employment is smoking tobacco-pipes, until they acquire the favourite tints of colour. By long use, the heads become black; but by boiling in milk and soap, they become white again.

When meerschaum is exposed to a very high degree of heat, it becomes so hard as to give sparks with steel. It is alleged that the porcelain of Samos was made of the meerschaum found in that island; and it is supposed that the porcelain knives mentioned by Pliny, as being used by surgeons, were made from this mineral.

Observations.

1. It is distinguished from *Magnesite* by its colour, difficult frangibility, strong adhesion to the tongue, inferior hardness and specific gravity. Its fracture at once distinguishes it from *Native Magnesia.*

2. It is nearly allied to Magnesite, into which it sometimes passes.

4. Bole.

Bol, *Werner.*

Bolus, *Waller.* t. i. p. 51.—Bole, *Kirw.* vol. i. p. 191.—Bol, *Estner,* b. ii. s. 784. *Id. Emm.* b. i. s. 381.—Bolo, *Nap.* p. 256.—Le Bol, *Broch.* t. i. p. 459.—Bol, *Reuss,* b. ii. s. 115. *Id. Lud.* b. i. s. 129. *Id. Suck.* 1r th. s. 495. *Id. Bert.* s. 207. *Id. Mohs,* b. i. s. 525. *Id. Hab.* s. 39. *Id. Leonhard,* Tabel. s. 26.—Le Bole Armenie, *Brong.* t. i. p. 543.—Bol, *Haus.* s. 86. *Id Karsten,* Tabel. s. 28.—Bole, *Kid,* vol. i. p. 179.—Bol, *Steffens,* b. i. s. 253. *Id. Lenz,* b. ii. s. 634.

External Characters.

Its colour is cream-yellow, which passes on the one side

side into flesh-red, on the other into light yellowish-brown, and into a colour which is intermediate between chesnut-brown and brownish-black. Sometimes it is spotted brown and black.

It is massive, and disseminated.

Internally its lustre is glimmering.

The fracture is perfect conchoidal.

The fragments are indeterminate angular, and rather sharp-edged.

The red variety is nearly translucent, the yellow translucent on the edges, and the brown and black opaque.

It is very soft.

It is easily frangible.

It feeels greasy.

It becomes shining and resinous in the streak.

It adheres strongly to the tongue.

Specific gravity, 1.922, *Karsten*. From 1.4 to 2.00, *Kirwan*.

Chemical Characters.

When immersed in water, it breaks in pieces with an audible noise, and the evolution of air-bubbles.

Before the blowpipe, it melts into a greenish-grey coloured slag.

Constituent Parts.

Silica, - -	47.00
Alumina, - -	19.00
Magnesia, - -	6.20
Lime, - - -	5.40
Iron, - -	5.40
Water, - -	7.50

Bergmann, Opusc. t. iv. p. 152.

 It

It is still uncertain whether this analysis of Bergman is of true bole.

Geognostic Situation.

The geognostic situation of this species is rather circumscribed, it having been hitherto observed only in flœtz-trap rocks, principally in trap-tuff, wacke, and basalt, in which it occurs in angular pieces, and disseminated *

Geographic Situation.

Europe.—It is found at Strigau in Silesia; in the Habichtwalde in Hessia; Sienna in Italy.

Asia.—In Armenia.

Uses.

It was formerly an article of the materia medica, and was used as an astringent, and in some places is still employed in veterinary practice. It is said that tobacco-pipes are sometimes made of bole, and that it is an ingredient in the glaze of some kinds of earthen-ware.

Observations.

1. Formerly a number of clayey brick-red and brownish coloured clays, were preserved in collections under the name *bole*. The bole of modern mineralogists, of which we have given a description, was first established as a distinct species by Werner.

2. It inclines sometimes to Lithomarge, sometimes to clay.

5. Sphragide,

* It is said to occur in shell limestone, in the neighbourhood of Waltersdorf in Gotha. Von Schlottheim, in the Magazin Naturf Freunde zu Berlin, 1. 4. s. 305.

5. Sphragide, or Lemnian Earth.

Sphragid, *Werner*.

Sphragid, *Karsten,* Tabel. s. 28.—Lemnische Erde, *Steffens*, b. i. s. 255.—Sphragid, *Lenz,* b. ii. s. 643. *Id. Oken,* b. i. s. 384.

External Characters.

Its colours are yellowish-grey, and yellowish-white. On the surface, it appears frequently marbled with rust-like spots.

It is dull.

The fracture is fine earthy.

It is perfectly meagre to the feel.

It adheres slightly to the tongue.

When immersed in water, it falls into pieces, and numerous air-bubbles are evolved.

Constituent Parts.

Silica, - - -	66.00
Alumina, - -	14.50
Magnesia, - -	0.25
Lime, - -	0.25
Natron, - -	3.50
Oxide of Iron, - -	6.00
Water, - - -	8.50
	99.00

Klaproth, Beit. b iv. s. 333.

Geographic.

Geographic Situation.

Its geognostic situation is unknown, and it has hitherto been found only in the island of Lemnos, in the Mediterranean.

Uses.

In Lemnos it is dug but once a-year, on the 15th of August, in the presence of the clergy and magistrates of the island, after the reading of prayers. The clay is cut into spindle-shaped pieces, of an ounce weight, and each of them is afterwards stamped with a seal, having on it the Turkish name of the mineral. Even so early as the time of Homer, this substance was used as a medicine against poison and the plague, and was then in great repute, as it is at present, in eastern countries. In early times, it was also sold, bearing on it the impression of a seal: hence it was called σφρᾶγις, *sigillum;* and it was in in such estimation, that none but priests durst handle it, and severe punishments were inflicted on those who presumed to dig for it at any other but the stated period. It is mentioned, that Scultetus Montanus, physician to the Emperor Rodolph, in the year 1568, ordered this earth to be kept in apothecaries shops.

Observations.

It was first brought from Lemnos by Mr Hawkins, who presented specimens of it to Klaproth.

6. Fullers

6. Fullers Earth.

Walkerde, *Werner*.

Walkerde, *Wid.* s. 429.—Fullers Earth, *Kirw.* vol. i. p. 184.—Walkerde, *Estner,* b. ii. s. 777. *Id. Emm.* b. ii. s. 375.—Terra da Follone, *Nap.* p. 258.—La Terre à foulon, *Broch.* t. i. p. 464.—Argile smectique, *Hauy,* t. iv. p. 443.—Walkerde, *Reuss,* b. ii. s. 111. *Id. Lud.* b. i. s. 130. *Id. Mohs.* b. i. s. 532. *Id. Hab.* s. 39. *Id. Leonhard,* Tabel. s. 27.—Argile smectique, *Brong.* t. i. p. 522.—Walkthon, *Haus.* s. 86.—Walkerde, *Karst.* Tabel. s. 28.—Fullers Earth, *Kid,* vol. i. p. 175.—Walkerde, *Steffens,* b. i. s. 250. *Id. Lenz,* b. ii. s. 640. *Id. Oken,* b. i. s. 385.

External Characters.

Its colours are greenish-white, greenish-grey, olive-green, and oil-green. Some varieties exhibit clouded and striped colour-delineations.

It occurs massive.

It is dull.

The fracture is coarse and fine grained uneven: some varieties are large conchoidal; and others incline to slaty.

The fragments are blunt-edged, and occasionally incline to slaty.

It is opaque; but when it inclines to sectile, translucent on the edges.

It becomes shining in the streak.

It is very soft.

It is sectile.

It scarcely adheres to the tongue.

It feels greasy.

Specific gravity, 1.72, *Karsten.*

Chemical

Chemical Characters.

It falls into powder in water, without the crackling noise which accompanies bole.

It melts into a brown spongy scoria before the blow-pipe.

Constituent Parts.

	Fullers Earth of Rygate.	
Silica, - -	53.00	51.8
Alumina, -	10.00	25.0
Magnesia, -	1.25	0.7
Lime, - -	0.50	3.3
Muriate of Soda,	0.10	
Trace of Potash.		
Oxide of Iron, -	9.75	0.7
Water, - -	24.00	15.5
	98.60	*
	Klaproth, Beit, b. iv s. 338.	*Bergmann*, Opusc. t. iv. p. 156.

Geognostic and Geographic Situations.

In England, it occurs in beds, sometimes below, sometimes above the chalk formation; at Rosswein, in Upper Saxony, under strata of greenstone-slate; and in different places in Austria, Bavaria, and Moravia, it is found immediately under the soil.

Uses.

This mineral was employed by the ancients for cleansing woollen, and also linen cloth, and they named it *Terra Fullonum*, and *Creta Fullonum*: hence the name *Fullers*

* Gehlen has discovered Chrome in fullers earth.

Fullers Earth. The *Morochtus* of Dioscorides, which he celebrates on account of its remarkable saponaceous properties, is conjectured to have been a variety of fullers earth. Some ancient writers describe it under the name *Galactites,* because it communicates to water a milk-white colour; also *Mellilites,* from the fancied sweet taste it communicates to water. The fullers earth of different countries varies in goodness: the most celebrated, and the best, is that found in Buckinghamshire and Surrey. Good fullers earth has a greenish-white or greenish-grey colour, falls into powder in water, appears to melt on the tongue like butter, communicates a milky colour to water, and deposites very little sand when mixed with boiling water. The remarkable detersive property of this substance depends on the alumina it contains; and it appears that the proportion of this should not be less than a fourth or fifth of the whole mass. It should not, however, be much more, for in that case the fullers earth would be too tenacious to diffuse itself through water *. Before the general use of soap, this substance was very universally employed for cleansing woollen cloth, but in consequence of the general substitution of soap, it is now much less used than formerly.

Observations.

1. Fullers earth, although nearly allied to Steatite, is distinguished from it by colour, fracture, opacity, and inferior specific gravity. Some varieties of steatite, particularly the greenish-grey, pass into fullers earth.

2. The Uneven Fullers Earth *(Unebene Walkerde)* of Karsten, which is considered as very nearly allied to common

* Kid's Mineralogy, vol. i. p. 176.

common fullers earth, is thus described by that celebrated mineralogist: "Its colour is brick-red, and is either pure, or spotted and veined with white and green. It occurs massive. Internally it is glistening, passing on the one side into glimmering, on the other into shining, and is resinous. The fracture is very coarse-grained uneven, passing into conchoidal. The fragments are rather sharp edged. It is translucent on the edges. Adheres feebly to the lips, but not at all to the tongue. It is soft, and very soft. It is sectile. It is rather light." According to Klaproth, it contains the following substances:—Silica, 48.50. Alumina, 13 50. Magnesia, 1.50. Iron, 6.50. Manganese, 0.50. Water, 25.50. A trace of Muriate of Soda. It occurs in the fissure of a vein of basalt which traverses granite on the Pringelberg, near Nümptsch in Silesia.

3. Werner is of opinion that the fullers earth of Rosswein in Saxony, is formed by the decomposition of greenstone-slate, as it is covered by it, and we can trace the gradation from the fully formed fullers earth to the fresh greenstone-slate. Steffens conjectures it to have been formed from previously existing strata, by a process analogous to that by which muscular fibre is converted into a kind of spermaceti: hence he says it is of newer formation than the bounding rocks.—May it not be an original deposition of greenstone, in a loose state of aggregation, resembling the disintegrated felspar in certain granites?

7. Steatite,

7. Steatite, or Soapstone.

Speckstein, *Werner.*

Creta Hispanica, *Wall.* t. i. p. 396.; Creta Briansonia, *Wall.* t. i. p. 390.—Speckstein, *Wid.* s. 451.—Semi-indurated Steatites, *Kirw.* vol. i. p. 151.—Speckstein, *Estner,* b. ii. s. 791. *Id. Emm.* b. i. s. 363.—Steatite compatta, *Nap.* p. 296.—Steatite, *Lam.* t. ii. p. 343.—La Steatite commune, *Broch.* t. i. p. 474.—Talc Steatite, *Hauy,* t. iii. p. 252.—Speckstein, *Reuss,* b. ii. s. 176. *Id. Lud.* b. i. s. 132. *Id. Suck.* 1r th. s. 544. *Id. Bert.* s. 141. *Id. Mohs,* b. i. s. 541. *Id. Hab.* s. 66. *Id. Leonhard,* Tabel. s. 27.—Talc Steatite, *Lucas,* s. 84.—Steatite commune, *Brong.* t. i. p. 496.—Talc Steatite, *Brard,* p. 198.—Dichter Speckstein, *Haus.* s. 100.—Speckstein, *Karst.* Tabel. s. 44.—Steatite, *Kid,* vol. i. p. 96. —Speckstein, *Steffens,* b. i. s. 233. *Id. Lenz,* b. ii. s. 644. *Id. Oken,* b. i. s. 380.

External Characters.

Its principal colour is white, of which it presents the following varieties: greyish, greenish, seldom yellowish, and reddish-white; the reddish-white borders on flesh-red; the greenish-white passes into mountain, oil, and lastly, siskin green. It is sometimes marked with spotted, and dendritic greyish-black delineations.

It occurs massive, disseminated in crusts; and also crystallised, in the following figures *.

1. Rectangular

---

* There is still a difference of opinion among mineralogists, in regard to the crystals of steatite, some considering them as true, others as false or supposititious. It would appear that those crystals which are implanted in the steatite, or interlaced with it, are of cotemporaneous formation with it, while the others appear to be supposititious.

1. Rectangular four-sided prism, acuminated with four planes.
2. Oblique four-sided prism.
3. Six-sided prism, acuminated with six planes, set on the lateral planes.
4. Rhomboidal dodecahedron.
5. Rhomb.
6. Six-sided pyramid.

The six-sided prism, rhomb, and six-sided pyramid, are considered to be supposititious crystals;—the prism originating from rock-crystal,—the rhomb from brown-spar,—and the six-sided pyramid from calcareous-spar.

The crystals are generally middle-sized and imbedded, and their surfaces are smooth and glistening.

Internally it is dull, seldom feebly glimmering.

The fragments are indeterminate angular, and blunt-edged

It is translucent on the edges, or faintly translucent.

It becomes shining in the streak.

It is soft, passing into very soft.

It is very sectile.

It is rather easily frangible.

It does not adhere to the tongue.

It feels very resinous.

Specific gravity, 2.382, *Karsten.* 2.608, *Brisson.*

Chemical Characters.

Before the blowpipe, it loses its colour, and becomes black, but is infusible without addition.

Constituent

Constituent Parts.

Steatite of Baireuth.		Steatite of Cornwall.		Steatite of Monte Ramuzo.	
Silica, -	59.50	Silica, -	45.00	Silica, -	44.00
Magnesia,	30.50	Magnesia,	24.75	Magnesia,	44.00
Alumina,		Alumina,	9.25	Alumina,	2.00
Iron, -	2.50	Iron, -	1.00	Iron, -	7.30
Potash, -		Potash, -	0.75	Manganese,	1.50
Water, -	5.50		18.00	Chrome,	2.00
				Trace of Lime, & Muriatic Acid.	
	98.00		98.75		100.80
Klaproth, Beit. b. ii. s. 179.		*Klaproth*, Beit. b. v. s. 24.		*Vauquelin.*	

Geognostic Situation.

It occurs frequently in small cotemporaneous veins, that traverse serpentine in all directions; and in angular and other shaped pieces in flœtz-trap rocks. It also occurs in metalliferous veins that traverse primitive rocks, accompanying different formations of galena, blende, copper and silver ores; in tinstone veins and beds: And veins in grey-wacke, filled with galena, sparry-ironstone, and other metalliferous substances, occasionally contain steatite.

Geographic Situation.

Europe.—It occurs in the serpentine of Portsoy and Zetland; in the limestone of Icolmkill; and in the trap rocks of Fifeshire, the Lothians, Arran, Skye, Canna, and other parts in Scotland. In England, in the serpentine of Cornwall. On the Continent of Europe, it is found in Norway, Sweden, Saxony, Bohemia, Baireuth, Salzburg, Switzerland, Spain, &c.

Asia—In different parts of Siberia.

Uses.

The Cornish steatite is used at Worcester, in the manufacture of porcelain. The Arabs use it in their baths, instead of soap, to soften the skin. Certain savage tribes eat it, either alone, or mix it with their food, to deceive hunger. M. Labillardiere informs us, that the inhabitants of New Caledonia eat considerable quantities of a soft steatite, in which Vauquelin found 0.37 magnesia, 0.36 silica, 0.17 oxide of iron, and which contains no nourishing ingredient. Humboldt assures us, that the Otomacks, a savage race on the banks of the Orinoco, live for nearly three months of the year principally on a kind of potters-clay. Mr Goldberry says, that the Negroes near the mouth of the Senegal mix their rice with a white steatite, and eat it without inconvenience; and it is well known that Negroes in general eat earthy substances with great avidity.

As steatite becomes hard in the fire, and does not alter its shape, it has been successfully employed in imitating engraved gems by M. Vilcot, an artist of Luettich, in the county of Liege. The subjects intended to be represented, are engraved on it with great ease: it is then exposed to a strong heat, when it acquires a considerable degree of hardness. It is afterwards polished, and may be coloured, by means of metallic solutions. A variety found in Arragon in Spain, is used by artists under the name *Spanish Chalk.* Like fullers-earth, it extracts grease from woollen stuffs; and when gently burnt, it is sometimes used as the basis of *rouge.*

Observations.

1. The yellowish-white variety approaches to Lithomarge, the flesh-red to Bole, and the siskin-green and greenish-grey to Fullers-Earth.

2. It

2. It is distinguished from *Talc*, by its wanting the slaty fracture: from *Serpentine*, by softness; and from *Chlorite*, by fracture.

3. Weiss and Steffens are of opinion, that steatite is not an original substance, but has been formed from other minerals, particularly felspar and mica, by a process somewhat resembling that which takes place with flesh, when it is converted into a fatty substance. This opinion will be fully considered in the geognostic part of the system.

4. The dendritic delineations that sometimes occur in steatite, according to Esper and Lenz, are not crystallizations of iron, but true fuci, or zosteræ *. Lenz affirms that he extracted plants from steatite, which were so fresh that they vegetated when put in water! All the specimens of these dendritic appearances I have had an opportunity of examining, do not differ from the common dendritic delineations observed in other minerals.

5. Karsten, in his Mineralogical Tables, divides this species into two subspecies, viz. Common Steatite, and Lamellar Steatite. Leonhard considers the lamellar steatite as a distinct mineral from common steatite, and describes it under the name Lamellar Talc (*Schaalentalk*). Werner is of opinion, that it is an intimate mixture of steatite and asbestus. Steffens agrees with Leonhard in considering it as a distinct species; and Hausmann arranges it with serpentine, under the name Lamellar Serpentine The follo ing is the description given of it by Leonhard: " Its colours are olive and mountain green, and ulphur-yellow. It occurs massive, disseminated, and in flakes. Externally it is splendent; internally

 shining,

* Vid. Annalen der Min. Soc. b. i. s. 315.; also Leonhard, Taschb. b. iv. s. 395.

shining, and resinous. The fracture is foliated, generally perfect curved foliated, seldom slightly inclining to fibrous. The fragments are rather blunt-edged. It sometimes exhibits indistinct large granular concretions; also thick lamellar, seldom thin prismatic concretions. It is translucent, or translucent on the edges. It is soft. It yields a pale greenish-grey coloured streak. It is rather brittle. Is rather easily frangible. Specific gravity, 2.6315, *Kopp*. It is infusible before the blowpipe. It is found near Zöplitz, in the Fichtelgebirge, partly in veins, partly in beds, in serpentine."

8. Figurestone, or Agalmatolite.

Bildstein, *Werner*.

Agalmatolith, *Klaproth*.

Steatites, particulis impalpabilibus, mollis, semi-pellucidus, lardites, colore flavescente, *Wall.* gen. 28. spec. 186. t. i. p. 399. —Indurated Steatites, *Kirw.* vol. i. p. 153.—La Pierre a Sculpture, *Broch.* t. i. p. 451.—Agalmatolit, *Lud.* b. ii. s. 151. *Id. Suck.* 1r th. s. 503. *Id. Bert.* s. 205. *Id. Leonhard,* Tabel. s. 27. *Id. Haus.* s. 86. *Id. Karst.* Tabel. s. 28. *Id. Kid,* vol. i. p. 181.—Talc graphique, *Hauy,* Tabl. p. 68.—Agalmatolith, *Steffens,* b. i. s. 240.—Bildstein, *Lenz,* b. ii. s. 594. *Id. Oken,* b. i. s. 379.

External Characters.

Its colours are greenish-grey, apple-green, pale yellowish-brown, flesh-red, and rose-red.

It occurs massive.

Internally it is glimmering and resinous.

The fracture in the large is slaty, in the small, splintery.

The

The fragments are sharp-edged, or slaty.
It is translucent, sometimes only in the edges.
It becomes resinous in the streak.
It is soft, approaching to very soft.
It is rather sectile.
It is easily frangible.
Specific gravity, 2.617, *Karsten*. 2.815, *Klaproth*.

Chemical Character.

It is infusible before the blowpipe.

Constituent Parts.

Chinese Figurestone.			Figurestone of Nagyag.
Silica,	35.00	54.50	55.00
Alumina,	29.00	34.00	33.00
Lime,	2.00		
Potash,	7.00	6.25	7.00
Iron, -	1.00	0.75	0.50
Water,	5.00	4.00	3.00
	99.00	99.50	98.50
	Vauquelin.	*Klaproth*, Beit. b. v. s. 21.	*Klaproth*, Id. s. 21.

Geographic Situation.

It occurs in China, and at Nagyag, in Transylvania, but the geognostic situations are unknown.

Uses.

This mineral, owing to its softness, can easily be fashioned into various shapes with the knife: hence, in China, where it frequently occurs, it is cut into figures, generally of men, also into pagodas, cups, snuff-boxes, &c. Baron Veltheim was of opinion, that the celebrated

Roman *Vasa murrhina*, brought from the most distant parts of India, were made of figurestone; whilst other antiquarians maintain, that they were of porcelain. Data are wanting for enabling us to decide in regard to these *vasa murrhina*.

Observations.

1. This substance was formerly confounded with Steatite, from which it is distinguished by lustre and fracture. It appears to be intermediate between Steatite and Nephrite.

2 Lenz, in the second volume of his Mineralogy, describes what he considers as a distinct subspecies of figurestone from Ochsenkopf, near Schneeberg in Saxony, where it occurs along with talc, corundum, and magnetic-ironstone. The following analysis of it has been published by Dr John: Silica, 51.50. Alumina, 32.50. Oxide of iron, 1.75 Oxide of manganese, 12.00. Potash, 6.00. Lime, 3.00. Water, 5.15.

XVI. TALC

XVI. TALC FAMILY.

THIS Family contains the following species: 1. Nephrite, 2. Serpentine, 3. Potstone, 4. Talc, 5. Nacrite, 6. Asbestus, 7. Picrolite.

1. Nephrite.

Nephrit, *Werner.*

This species is divided into two subspecies, viz. Common Nephrite, and Axestone.

First Subspecies.

Common Nephrite.

Gemeiner Nephrit, *Werner.*

Fetter Nephrit, *Saussure.*

Jade, *Kirw.* vol. i. p. 171.—Le Nephrite commune, *Broch.* t. i. p. 467.—Gemeiner Nephrit, *Reuss*, b. ii. 2. s. 187. *Id. Lud.* b. i. s. 131. *Id. Suck.* 1r th. s. 551. *Id. Bert.* s. 144. *Id. Mohs,* b. i. s. 335.—Jade nephrite, *Lucas,* p. 197. *Id. Brong.* t. i. p. 347. *Id. Brard,* p. 410.—Gemeiner Nephrit, *Leonhard,* Tabel. s. 28.—Nephrit, *Haus.* s. 100. *Id. Karsten,* Tabel. s. 44.—Nephrite, *Kid,* vol. i. p. 113.—Jade nephretique, *Hauy,* Tabl. p. 61.—Gemeiner Nephrit, *Steffens,* b. i. s. 266. *Id, Lenz,* b. ii. s. 507. *Id. Oken,* b. i. s. 331.

External Characters.

Its colour is leek-green, of various degrees of intensity, and sometimes passes into mountain-green, greenish-grey, and greenish-white.

It occurs massive, in blunt-edged pieces, and rolled pieces.

Internally dull or glimmering, owing to intermixed talc and asbestus.

The fracture is coarse-splintery, and the splinters are greenish-white.

The fragments are rather sharp-edged.

It is strongly translucent.

It scratches glass, but is scratched by rock-crystal.

It is difficultly frangible.

It feels rather greasy.

It is rather brittle.

Specific gravity, 2.962, Oriental, according to *Karsten*. 3.020, Mexican, *Karsten*. 2.970, 3.071, *Saussure* the Father. 2.957, *Saussure* the Son.

Chemical Characters.

Before the blowpipe, it melts into a white enamel.

Constituent Parts.

Silica,	50.50
Magnesia,	31.00
Alumina,	10.00
Iron,	5.50
Chrome,	0.05
Water,	2.75

Kastner.

Geognostic and Geographic Situations.

Europe.—In Switzerland, nephrite occurs in granite and gneiss; in the Hartz, in veins that traverse primitive greenstone.

Asia.—The most beautiful varieties of this mineral are brought

brought from Persia and Egypt, and from the mines of Seminowski, near Kolyvan in Siberia *.

America.—It is found on the banks of the river Oronooko, and near Tlascala in Mexico.

Uses.

It is principally prized as an ornamental stone. The Turks cut it into handles for sabres and daggers. Artists sometimes engrave figures of different kinds on it; and it is said to be highly esteemed as a talisman by the savage tribes of the countries where it is found. Although it takes a good polish, it has always an oily and muddy aspect. It was formerly believed to be useful in alleviating or preventing nephritic complaints: hence it has been called *Nephritic Stone.*

Observations.

1. This mineral is characterised by colour, coarse-splintery fracture, white-coloured splinters, resinous aspect, and considerable hardness and weight.

2. It is nearly allied, both to Steatite and Serpentine, as was observed by Pott, Baumer, Lehman, Vogel, and other older mineralogists; and its external characters shew that it is a different substance from Saussurite, (*Magerer Nephrit* of some mineralogists), and Lamellar Serpentine, (*Schaalentalk*).

Second

* Some authors mention China as one of the localities of nephrite, but the Chinese nephrite, or jade, as it is sometimes called, is prehnite.

Second Subspecies.

Axestone.

Beilstein, *Werner.*

Panamustein, *Blumenbach.*

Beilstein, *Estner,* b. ii s. 851. *Id. Emm.* b. iii. s. 351.—La Pierre de Hache, *Broch.* t. i. p. 470.—Beilstein, *Reuss,* b. ii. 2. s. 120. *Id. Leonhard,* Tabel. s. 28.—Jade axinien, *Brong.* t. i. p. 349.—Neuseelandischer Nephrit, *Oken,* b. i. s. 331.

External Characters.

Its colour is intermediate between mountain-green and leek-green, and passes into dark grass-green, oil-green, and greenish-grey.

It occurs massive.

Internally its lustre is strongly glimmering, inclining to glistening.

The fracture is slaty in the great, and more or less distinctly splintery in the small.

The fragments are tabular.

It is translucent.

It is semi-hard, approaching to hard.

It is rather easily frangible.

Specific gravity, 3.008, 3 000, *Karsten.* 3.007, *Lichtenberg.*

Geographic Situation.

It occurs in Tavaipunamu, one of the New Zeland group of islands; and it is said also to occur in China.

Uses.

It is used by the natives of New Zeland, and other islands in the South Sea, for hatchets and ear-pendants.

Observations.

Observations.

1. It is so nearly allied to Common Nephrit, that Werner has placed it in the system as a subspecies of that mineral.

2. It was first brought to Europe by Captain Cook, and was communicated to the mineralogists of Germany by Dr Forster, who accompanied that illustrious commander in his second voyage round the world.

2. Serpentine.

Serpentin, *Werner.*

This species is divided into two subspecies, viz. Common Serpentine, and Precious Serpentine.

First Subspecies.

Common Serpentine.

Gemeiner Serpentin, *Werner.*

Steatites serpentinus, *Wall.* t. i. p. 156.—Serpentin, *Wid.* s. 462. *Id. Kirwan,* vol. i. p. 156. *Id. Emm.* b. i. s. 384. *Id. Estner,* b. ii. 855.—La Serpentine, *Broch.* t i. p. 481.—Roche serpentineuse, *Hauy,* t. iv. p. 436.—Gemeiner Serpentin, *Reuss,* b. ii. 2. s. 210. *Id Lud* b. i. s. 133 *Id. Suck.* 1r th s. 561. *Id. Bert.* s. 146. *Id. Mohs,* b i. s. 551. *Id. Hab.* s. 58. *Id. Leonhard,* Tabel. s. 28 —Serpentine commune, *Brong.* t. i. p. 486.—Gemeiner Serpentin, *Haus.* s. 100. *Id. Karsten,* Tabel. s. 42.—Serpentine, *Kid,* vol. i. p. 93.—Gemeiner Serpentin, *Steffens,* b. i. s. 268. *Id. Lenz,* b. ii. s. 651. *Id. Oken,* b. i. s. 378.

External Characters.

Its principal colour is green, of which it presents the following

following varieties: leek, oil, and olive green; from oil-green it passes into mountain-green and greenish-grey; from leek-green it passes into greenish-black; from greenish-black into blackish-green; sometimes it occurs straw-yellow, and rarely yellowish-brown, and liver-brown: further, red; of which the following varieties occur; blood-red, brownish-red, peach-blossom-red, and scarlet-red. The peach-blossom and scarlet-red are the rarest. The colour is either uniform, or veined, spotted, dotted, and clouded; and frequently several of these delineations occur together.

It occurs massive.

Internally it is dull, or glimmering, owing to intermixed foreign parts.

The fracture is small-grained uneven, frequently splintery, but also even, sometimes inclining to conchoidal.

The fragments are rather sharp-edged.

It is translucent on the edges.

It is soft. It does not yield to the nail, but is scratched by calcareous-spar.

It is rather sectile.

It is rather difficultly frangible.

It feels somewhat greasy.

Specific gravity, 2.348, *Karsten.* 2.587, *Brisson.* 2.561, 2.574, *Kirwan.*

Physical Characters.

Some varieties of serpentine not only move the magnetic-needle, but even possess magnetic poles.

Chemical Characters.

It is infusible before the blowpipe, but on exposure to a higher temperature, it melts with difficulty into an enamel.

Constituent

Constituent Parts.

Silica, - -	31.50	28.00	Silica, -	32.00
Magnesia, -	47.25	34.50	Magnesia,	37.24
Alumina, -	3.00	23.00	Alumina, -	0.50
Lime, - -	0.50	0.50	Lime, -	10.60
Iron, - -	5.50	4.50	Iron, -	0.66
Oxide of Manganese,	1.50		Volatile matter, and	
Water, -	10.50	10.50	Carbonic Acid,	14.16
	John, Chem. Untersuch, ii. s. 94.	*Rose.*	*Hisinger*, Afhandlingar i Fysik, iii. p. 303.	

Richter and Rose discovered a small portion of Chrome in the serpentine of Saxony.

Geognostic Situation.

Serpentine occurs in primitive, transition, and flœtz rocks. In primitive mountains, it occurs in beds, often of great thickness, in gneiss, mica-slate, and clay-slate: in transition rocks, it is associated with clay-slate; and in flœtz rocks, it is imbedded in greenstone, into which it seems to pass. These beds, particularly those that occur in primitive mountains, contain many of the minerals of the talc and steatite families, and not unfrequently ores, particularly of magnetic ironstone, and veins of native copper.

Geographic Situation.

Europe.—In Scotland, it occurs in the islands of Unst and Fetlar, in Zetland; Isle of Glass in the Hebrides *; at Portsoy in Banffshire; at the bridge of Cortachie in Forfarshire; between Ballantrae and Girvan, in Ayrshire; and near Burntisland in Fifeshire. It abounds in some districts in Cornwall in England; and it occurs at Cloghan Lee, on the west coast of Ireland, in the county of Donnegal †. On the Continent of Europe, it occurs in

* Neill.

† Greenough.

in Saxony, Bohemia, Silesia, Bavaria, Salzburg, Tyrol, Austria, Switzerland, Savoy, Italy, and the island of Corsica.

Asia.—It is found in different districts in Siberia; and in New Holland.

America.—Island of Cuba.

Uses.

As it is soft and sectile, and takes a good polish, it is cut and turned into vessels and ornaments of various kinds. At Zöblitz in Upper Saxony, many people are employed in quarrying, cutting, turning, and polishing the serpentine which occurs in that neighbourhood; and the various articles into which it is there manufactured, are carried all over Germany. At Portsoy in Banffshire, the serpentine is also turned into a variety of elegant ornamental articles, which, on account of the beauty of the stone, are sold at a high price. The serpentine of Portsoy much exceeds that of Zöblitz in beauty and variety of colour, and hence is deservedly more esteemed. Those varieties which have an intermixture of blood-red, peach-blossom-red, and scarlet-red, and yellowish-green, are the most highly prized: indeed, in Saxony, they are in such estimation, as to be arranged with the precious stones, and claimed as the property of the State. In ancient times, serpentine was an article of the materia medica: it was prescribed with wine as a remedy for the stone, recommended as a certain cure for the bite of serpents, and was considered as possessing talismanic powers in lethargy, small-pox, poisoning, and madness. Boetius de Boot gravely remarks, that serpentine has a repulsion for poison of every kind, so that the moment the poisonous liquid is poured into a vessel of this mineral, it begins to foam, and is expelled from it.

Observations.

Observations.

1. It is distinguished from *Precious Serpentine* by its numerous colours, its uneven or splintery fracture, inferior translucency, and inferior hardness.

2. It passes into Steatite, and from thence into Talc, Asbestus, and Amianthus.

3. Some of the varieties are marked in the manner of the skins of serpents: hence the name *Serpentine* applied to the species. The greenish black, with white or red veins, is named *Verde di Prato*; the green with white veins *Verde di Susa.*

4. It is worthy of remark, that Common Serpentine passes on the one hand into Greenstone, and on the other into Asbestus, which passes into Actynolite or Hornblende.

Second Subspecies.

Precious Serpentine.

Edler Serpentin, *Werner.*

Le Serpentine noble, *Broch.* t. i. p. 484.—Edler Serpentin, *Reuss,* b. ii. 2. s. 210. *Id. Lud.* b. i. s. 134. *Id. Suck.* 1r th. s. 563. *Id. Bert.* s. 147. *Id. Leonhard,* Tabel. s. 28.—Serpentine Noble, *Brong.* t. i. p. 485.—Edler Serpentin, *Haus.* s. 100. *Id. Karst.* Tabel. s. 42.—Noble Serpentine, *Kid,* vol. i. p. 94.—Edler Serpentin, *Steffens,* b. i. s. 271. *Id. Lenz,* b. ii. s. 656. *Id. Oken,* b. i. s. 331.

This subspecies is divided into two kinds, viz. Splintery Precious Serpentine, and Conchoidal Precious Serpentine.

First

First Kind.

Splintery Precious Serpentine.

Edler splittriger Serpentin, *Werner.*

External Characters.

Its colour is dark leek-green.
It occurs massive.
Internally it is feebly glimmering.
The fracture is splintery.
The fragments are rather sharp-edged.
It is translucent.
It is soft, passing into semi-hard.
Specific gravity, 2.173, *Karsten.*
In other characters it agrees with Common Serpentine.

Second Kind.

Conchoidal Precious Serpentine.

Edler muschlicher Serpentin, *Werner.*

External Characters.

Its colour is leek-green, which sometimes passes into pistachio-green.

It occurs massive.
Its lustre is glistening, passing into glimmering.
The fracture is flat conchoidal.
The fragments are sharp-edged.
It is translucent.
It is intermediate between soft and semi-hard.
In other characters agrees with the foregoing.

Constituent

Constituent Parts.

Silica, - - -	42.50
Magnesia, - -	38.63
Lime, - - -	0.25
Alumina, - -	1.00
Oxide of Iron, -	1.50
Oxide of Manganese, -	0.62
Oxide of Chrome, -	0.25
Water, - - -	15.20

John, Chem. Untersuchungen, ii. s. 218.

Geognostic Situation.

It is almost always accompanied with granular foliated limestone, with which it is intermixed. and forms beds in gneiss and mica-slate. In these beds, it is associated with arsenical-pyrites, magnetic-pyrites, and galena or leadglance. These beds are neither so large nor so abundant as those of common serpentine.

Geographic Situation.

It is found in Holyhead Island *. It occurs in Italy, where the compound of serpentine and limestone is named *Marmore verde antico ;* also at Reichenbach in Silesia : it occurs also in Sweden, Bohemia, and the Tyrol.

Observations.

1. The simple colours, greater degree of transparency, and superior hardness, distinguish Precious from *Common Serpentine.*

* Greenough.

2. In ancient authors, a mineral is described under the name *Ophites*, from *οφιτης*, and this from *οφις*, *a serpent;* and in later writings, it is named *Egyptian Ophite*, *Green Egyptian Marble*, and *Egyptian Green*. Some mineralogists consider it as a variety of common serpentine: others, as Dr John, with more accuracy, describe it as a mixture of reddish-brown common serpentine, leek and pistachio green precious serpentine, white granular foliated limestone, and small portions of diallage.

3. Potstone, or Lapis ollaris.

Topfstein, *Werner*.

Potstone, *Kirw.* vol. i. p. 155.—Topfstein, *Reuss*, b. ii. 2. s. 236. *Id. Lud.* b. i. s. 115. *Id. Suck.* 1r th. s. 576. *Id. Hab.* s. 30. *Id. Leonhard*, Tabel. s. 26.—Serpentine ollaire, *Brong.* t. i. p. 486.—Topfstein, *Karsten*, Tabel. s. 42.—Talc ollaire, *Hauy*, Tabl. p. 56.—Topfstein, *Steffens*, b. i. s. 231. *Id. Lenz*, b. ii. s. 598. *Id. Oken*, b. i. s. 381.

External Characters.

Its colour is greenish-grey, of different degrees of intensity; the darker varieties incline to leek-green.

It occurs massive.

Internally it is glistening, inclining to shining, and is pearly.

The fracture is curved and imperfect foliated, which passes into slaty.

The fragments are indeterminate angular, or slaty.

It occurs in imperfect coarse granular concretions.

It is translucent on the edges.

It affords a white-coloured streak.

It

It is very soft.
It is perfectly sectile.
It feels greasy.
It is rather difficultly frangible.
Specific gravity, 2.880, *Saussure & Karsten*.

Chemical Characters.

It is infusible before the blowpipe.

Constituent Parts.

Silica, - -	39
Magnesia, -	16
Oxide of Iron, -	10
Carbonic Acid, -	20
Water, - -	10

Tromsdorf.

Geognostic Situation.

It occurs in beds, in primitive clay-slate.

Geographic Situation.

Europe.—It occurs abundantly on the shores of the Lake Como in Lombardy, and at Chiavenna in the Valteline. It occurs in different parts in Norway; and it is mentioned as a production of Finland *.

Africa.—It is said to occur in Upper Egypt.

America.—It is found in the country around Hudson's Bay; and in Greenland.

Uses.

When newly extracted from the quarry, it joins to a considerable degree of softness a kind of tenacity, so that

* Steffens and Von Buch.

that it can be turned and cut with great ease: hence it is frequently fashioned into various kinds of culinary vessels, which harden in drying, and are very refractory in the fire. These vessels do not communicate any taste to the food boiled in them, and have been used for culinary purposes for ages. Pliny mentions them, and describes the mode of making them, and the changes they experience by using. In those times, potstone was named *Lapis comensis*, and *Lapis Siphnius*, from the island of Siphnus, (the present Siphanto), where it was found. In Upper Egypt, this mineral is named *Pierre de Baram*, and is used for culinary vessels. Quarries of potstone were worked on the banks of the Lake Como, from the beginning of the Christian era to the 25th of August 1618, when they fell in and destroyed the neighbouring town of Pleurs. It was there used for culinary vessels and oven-soles, both of which were uncommonly durable. In proof of this, it is mentioned, that an oven at Liddus in the Valais, stood unimpaired for several hundred years. The town of Pleurs drew annually from those quarries, stone to the value of 60,000 ducats. In Greenland and Hudson's Bay, culinary vessels and lamps are made of potstone; and in Norway and Sweden, it is used for lining stoves, ovens, and furnaces.

Observations.

1. It is very nearly allied to Indurated Talc, from which it is distinguished by its deeper grey colour, higher lustre, kind of fracture, distinct concretions, and white streak.

2. It is so nearly allied to Mica, that Werner has placed it in the system beside it: we have, for obvious reasons, preferred arranging it with the talcaceous minerals.

4. Talc.

4. Talc.

Talk, *Werner*.

This species is divided into three subspecies, viz. Common Talc, Indurated Talc, and Columnar Talc.

First Subspecies.

Common Talc.

Gemeiner Talk, *Werner*.

Talcum albicans, lamellis subpellucidis flexis, *Waller*. gen. 27. spec. 180.—Gemeiner Talk, *Wid*. p. 441.—Common Talc, or Venetian Talc, *Kirw*. vol. i. p. 150.—Gemeiner Talk, *Estner*, b. ii. s. 824. *Id. Emm.* b. i. s. 391.—Talco compatto, *Nap.* p. 293.—Talc cailleux, *Lam.* t. ii. p. 342.—Talc laminaire, *Hauy*, t. iii. p. 252.—Le Talc commun, *Broch.* t. i. p. 487.—Gemeiner Talk, *Reuss*, b. ii. 2. s. 229. *Id. Lud.* b. i. s. 136. *Id. Suck.* 1r th. s. 571. *Id. Bert.* s. 139. *Id. Hab.* s. 64.—Talc laminaire, *Lucas*, p. 83.—Gemeiner Talk, *Leonhard*, Tabel. s. 29.—Talc laminaire, *Brong.* t. i. p. 503. *Id. Brard*, p. 197.—Blättricher Talk, *Haus.* s. 91.—Gemeiner Talk, *Karst.* Tabel. s. 42.—Laminated and Venetian Talk, *Kid*, vol. i. p. 107. & 108.—Talc hexagonal, laminaire, ecailleux, *Hauy*, Tabl. p. 56.—Gemeiner Talk, *Steffens*, b. i. s. 228. *Id. Lenz*, b. ii. s. 665. *Id. Oken*, b. i. s. 389.

External Characters.

Its colours are silver-white, greenish-white, apple-green, asparagus-green, and leek-green, which latter colour passes into duck-blue.

It occurs massive, disseminated; and crystallised in small, delicate, regular six-sided tables *.

It is generally splendent, or shining, and is pearly, or semi-metallic, sometimes passing into metallic.

The fracture is perfect and curved foliated, with a single cleavage; sometimes also broad and small diverging radiated.

The fragments are wedge-shaped, seldom splintery.

It occasionally occurs in large, small, and fine granular distinct concretions; the radiated varieties in wedge-shaped concretions.

It is translucent; in thin folia transparent.

It is flexible, but not elastic.

It is very soft.

It is perfectly sectile.

It feels very greasy.

Specific gravity, 2.695, 2.795, *Kirwan.* 2.770, *Karsten.*

Chemical Characters.

It becomes white before the blowpipe, and at length, with difficulty, affords a small globule of enamel.

Constituent Parts.

Silica, -	62.00	Silica, -	61.75
Magnesia,	27.00	Magnesia,	30.50
Alumina, -	1.50	Potash, -	2.75
Oxide of Iron,	3.50	Oxide of Iron,	2.50
Water, -	6.00	Water, -	0.25
		Loss, -	2.25
Vauquelin, Jour. d. Min. N. 88. p. 243.		*Klaproth*, Karst. Tab. s. 43.	

Geognostic

* According to Hauy, the primitive form is a right rhomboidal prism, in which the angles of the base are 120° and 60°.

Geognostic Situation.

It occurs in beds, in granular limestone and dolomite; also in cotemporaneous veins, in beds of indurated talc, serpentine, and porphyry.

Geographic Situation.

Europe.—It is found in Aberdeenshire and Banffshire; and on the Continent of Europe, in Norway, Sweden, Saxony, Bohemia, Switzerland, the Tyrol, and Salzburg.

Asia.—Persia, China, India.

Uses.

It enters into the composition of the cosmetic named *rouge*. This substance is prepared by rubbing together in a warm mortar, generally of serpentine, certain proportions of carmine and finely powdered talc, with a small portion of oil of benzoin. This cosmetic communicates a remarkable degree of softness to the skin, and is not pernicious. The Romans prepared a beautiful blue or purple colour, by combining this substance with the colouring fluid of particular kinds of testaceous animals *; and the flesh polish is given to gypsum figures, by rubbing them with talc. The Persians, according to Tavernier, whiten the walls of their houses and gardens by means of lime-water, and then powder them with silver-white coloured talc, which is said to give them a beautiful appearance. The Chinese burn talc, mix it with wine, and use it internally, as a cordial for curing diseases, and procuring long life: even European physicians, at one period, prescribed the powder of talc in dysenteric and hæmorrhoidal affections.

* The Buccinum reticulatum and Buccinum lapillus, that abound on the coasts of the Mediterranean.

Observations.

Common Talc is often confounded with *Mica*, but is distinguished from it by its greasy feel, inferior hardness, want of elasticity, and colour. It connects the Talc species with Chlorite and Mica.

Second Subspecies.

Indurated Talc.

Verhärteter Talk, *Werner*.

Verhærteter Talk, *Estner*, b. ii. s. 828. *Id. Emm.* b. iii. s. 280. —Le Talc endurcie, *Broch.* t. i. p. 489.—Verhärteter Talk, *Reuss*, b. ii. 2. s. 233. *Id. Lud.* b. i. s. 136. *Id. Suck.* 1r th. s. 573. *Id. Bert.* s. 140. *Id. Mohs*, b. i. s, 565. *Id. Leonhard*, Tabel. s. 29.—Talç endurcie, *Brong.* t. i. p. 504. —Verhärteter Talc, *Karsten*, Tabel. s. 42.—Indurated Talc, *Kid*, vol. i. p. 109.—Verhärteter Talc, *Steffens*, b. i. s. 230, *Id. Lenz*, b. ii. s, 669. *Id. Oken*, b. i. s. 390.

External Characters.

Its colour is greenish-grey, of various degrees of intensity.

It occurs massive.

Its lustre is shining, passing to glistening, and is pearly.

The fracture is intermediate between imperfect foliated and curved slaty: some varieties even pass into fibrous and small radiated.

The fragments are tabular.

It is strongly translucent on the edges.

It is soft.

It is rather sectile.

It

It is rather easily frangible.
It feels somewhat greasy.
Specific gravity, 2.982, *Wiedeman.*

Geognostic Situation.

It occurs in primitive mountains, where it forms beds in clay-slate and serpentine, and is associated with amianthus, chlorite, rhomb-spar, garnet, actynolite, quartz, &c.

Geographic Situation.

It occurs in Perthshire, Banffshire, the Zetland islands; and on the Continent of Europe, in Sweden, Saxony, Silesia, the Tyrol, Austria, and Switzerland.

Uses.

It is employed for drawing lines by carpenters, tailors, hat-makers, and glaziers. The lines are not so easily effaced as those made by chalk, and besides remain unaltered under water. Dr Kid remarks: "If lines be traced on glass by means of a piece of indurated talc, they remain invisible, or are scarcely perceptible by the naked eye, till breathed on. I have not met with an explanation of the effect produced in this instance; but it may perhaps in part depend on the comparative softness of the substance with which the impression is made: the condensation of the breath taking place more readily on the glass, than on the talc covering the glass, and the impression of the talc becoming more apparent by the simple contrast *." It is sometimes made into culinary vessels; and when reduced to powder, may be employed for the purpose of removing stains occasioned by grease, from silk.

Observations.

* Kid's Mineralogy, vol. i. p. 109.

Observations.

It connects the Talc species with Potstone, a mineral with which it has been frequently confounded.

Third Subspecies.

Columnar Talc.

Stänglicher Talk, *Karsten.*

Stänglicher Talk, *Karsten,* Tabel. s. 42.—Stänglicher Speckstein, *Haus.* s. 100.—Stänglicher Talk, *Steffens,* b. i. s. 231. *Id. Lenz,* b. ii. s. 669.

External Characters.

Its colour varies from pale apple-green to greenish-grey.

It occurs massive.

Internally it is glimmering and resinous.

The longitudinal fracture is coarse-fibrous; the cross fracture is splintery.

The fragments are rather sharp-edged.

It occurs in thin columnar or prismatic concretions.

It is opaque.

It is rather light.

5. Nacrite.

5. Nacrite.

Nacrite, *Brongniart*.

Erdiger Talk, *Werner*.

Erdiger Talk, *Wid.* s. 439.—Talcite, *Kirw.* vol. i. p. 149.—Erdiger Talk, *Estner*, b. ii. s. 821. *Id. Emm.* b. i. s. 389.—Talco terroso, *Nap.* p. 293.—Le Talc terreux, *Broch.* t. i. p. 486.—Erdiger Talk, *Reuss*, b. ii. s. 227. *Id. Lud.* b. i. s. 135. *Id. Suck.* 1r th. s. 570. *Id. Bert.* s. 299. *Id. Mohs*, b. i. s. 560. *Id. Leonhard*, Tabel. s. 29.—Nacrite, *Brong.* t. i. p. 505.—Schuppiger Thon, *Karst.* Tabel. s. 28. *Id. Haus.* s. 85.—Talc granuleux, *Hauy*, Tabl. p. 67.—Schuppiger Thon, *Steffens*, b. i. s. 202.

External Characters.

Its colours are cream-yellow, greenish-white, and greenish-grey.

It consists of scaly parts, which are more or less compacted; the most compact varieties have a thick or curved slaty fracture.

It is strongly glimmering, and is pearly, inclining to resinous.

It is friable.

It feels rather greasy.

It soils; and when we breath on it, it gives out a clayey smell.

It is light.

Chemical Characters.

It melts easily before the blowpipe.

Constituent

Constituent Parts.

Silica, - -	50.0	Silica, -	60.20
Alumina, -	26.0	Alumina,	30.83
Potash, -	17.0	Oxide of Iron,	3.55
A little Iron, Lime, and Muriatic acid.		Water, -	5.00
	Vauquelin.		*John.*

Geognostic and Geographic Situations.

It occurs in scales or masses in the cavities of primitive rocks, particularly of quartz. It is found at Meronitz in Silesia; Bohemia; Freyberg in Saxony; Gieren, in Silesia; and Sylva in Piedmont.

Observations.

This mineral is by Werner arranged as a subspecies of Talc; but its chemical composition and external characters induce me, with Brongniart, to place it in the system as a distinct species, under the name *Nacrite.*

6. Asbestus.

Asbest, *Werner.*

This species is divided into four subspecies, viz. Rock-Cork, Amianthus, Common Asbestus, and Rock-Wood.

Fourth

First Subspecies.

Rock-Cork.

Berg Kork, *Werner.*

Aluta montana, *Wall.* t. i. p. 414.; Suber montanum, Id. p. 415. —Berk Kork, *Wid.* s. 469.—Suber montanum, Corium montanum, *Kirw.* vol. i. p. 163.—Berg Kork, *Estner,* b. ii. s. 864. *Id. Emm.* b. i. s. 399.—Sughero montano, *Nap.* p. 319.—Varieté d'Amianthe, *Lam.* p. 367.—Le Siege de montagne, *Broch.* t. i. p. 492.—Asbeste tresse, *Hauy,* t. iii. p. 248.—Holz Asbest, *Reuss,* b. ii. 2. s. 253.—Kork Asbest, *Lud.* b. i. s. 137.—Berg Kork, *Suck.* 1r th. s. 263. *Id. Bert.* s. 148. *Id. Mohs,* b. i. s. 567.—Asbeste tresse, *Lucas,* p. 81. —Berg Kork, *Leonhard,* Tabel. s. 29.—Asbeste suberiforme, *Brong.* t. i. p. 479.—Asbeste tresse, *Brard,* p. 194.—Schwimmender Asbest, *Karst.* Tabel. s. 42. *Id. Haus.* s. 99.—Compact spongy Amianthus, *Kid,* vol. i. p. 103.—Asbeste tresse, *Hauy,* Tabl. p. 55.—Berg Kork, *Steffens,* b. i. s. 278. *Id. Lenz,* b. ii. s. 670.—Korkichter Asbest, *Oken,* b. i. s. 326.

External Characters.

Its colours are yellowish and greyish white, and yellowish and ash grey.

It occurs massive, in plates, that vary in thickness *, and with impressions.

Internally it is feebly glimmering.

The fracture at first sight appears fine grained uneven; but on more accurate inspection, it is found to be promiscuous fibrous.

The

---

* The variety in plates has received the following names: *Mountain Flesh;* Bergfleish; Caro montana; Chair de montagne; Chair fossile: *Mountain Paper;* Papier fossile: Bergpapier; *Mountain Leather;* Bergleder; Corium montanum; Cuir de montagne.

The fragments are blunt-edged.

It is opaque; very rarely feebly translucent on the edges.

It is very soft.

It is sectile.

It is slightly elastic-flexible.

It is difficultly frangible.

It emits a grating sound when we handle it.

It feels meagre.

It is so light as to swim on water.

Specific gravity, 0.679, 0.991, *Brisson*. 0.991, *Hauy*.

Chemical Characters.

It melts with great difficulty before the blowpipe into a milk-white nearly translucent glass.

Constituent Parts.

Silica, - -	56.2
Magnesia, - -	26.1
Alumina, - -	2.0
Lime, - - -	12.7
Oxide of Iron, -	3.0
	100

Bergmann, Opusc. t. iv.

Geognostic Situation.

It occurs in cotemporaneous veins in serpentine; also in metalliferous veins in primitive and transition rocks; and occasionally in mineral beds.

Geographic Situation.

It occurs in veins in the serpentine of Portsoy; and in plates in the lead-veins at Lead Hills and Wanlockhead in Lanarkshire. At Sala in Sweden, it is in a metalliferous

ferous bed, along with asbestus, steatite, calcareous-spar, rhomb-spar, and brown-spar: in veins along with ores of silver, calcareous-spar, and heavy-spar, at Kongsberg in Norway; in the silver mines of Johanngeorgenstadt in Saxony; at Falleyas in Spain, in beds along with meerschaum and talc; and in primitive rocks in Carinthia, Idria, France, &c.

Second Subspecies.

Amianthus, or Flexible Asbestus *.

Amiant, *Werner.*

Asbestinum Græcorum, *Plinius*, 19. 1.—Asbestus maturus, *Wall.* t. i. p. 410.; Amianthus, Id. p. 408.—Amianth, *Wid.* s. 464.—Amianthus, *Kirw.* vol. i. p. 161.—Amianth, *Estner*, b. ii. s. 368. *Id. Emm.* b. i. s. 402.—Amiantho, *Nap.* p. 316. —L'Amianth, *Lam.* t. ii. p. 365. *Id. Broch.* t. i. p. 494.—Asbest flexible, *Hauy*, t. iii. p. 245.—Biegsamer Asbest, *Reuss*, b. ii. 2. s. 243.—Amianth, *Lud.* b. i. s. 137.—Amianth-asbest, *Suck.* 1r th. s. 265.—Amianth, *Bert.* s. 149. *Id. Mohs*, b. i. s. 569.—Amianth-asbest, *Hab.* s. 64.—Asbest flexible, *Lucas*, p. 81.—Biegsamer Asbest, *Leonhard*, Tabel. s. 30.—Asbest amianthe, *Brong.* t. i. p. 478.—Asbest flexible, *Brard*, p. 194.—Biegsamer Asbest, *Karst.* Tabel. s. 42.—Amianth, *Haus.* s. 99.—Loosely fibrous and flexible Amianthus, *Kid*, vol. i. p. 101.—Asbest flexible, *Hauy*, Tabl. p. 55.—Amiant, *Steffens*, b. i. s. 276. *Id. Lenz*, b. ii. s. 672.—Biegsamer Asbest, *Oken*, b. i. s. 325.

External Characters.

Its most common colour is greenish-white, of different degrees

* Amianthus, from αμιαντος, *unstained*, *unsoiled*, which refers to the property this substance possesses, of remaining unsoiled in the fire. It is also named Rock-flax and Rock-wool.

degrees of intensity, which passes into greenish-grey, and rarely into light olive-green. It is sometimes blood-red, particularly when it occurs in veins in serpentine.

It occurs massive, and in small veins.

Internally its lustre is shining and pearly, occasionally approaching to semi-metallic.

The fracture is parallel fibrous.

The fragments are generally loose splintery.

It is generally translucent on the edges.

It is very soft.

It is sectile.

It is intermediate between common and elastic flexible.

It splits easily.

Specific gravity, 2.444, *Muschenbröck*. According to *Brisson*, it varies considerably in specific gravity: he found the long silky amianthus to vary from 0.9088 to 2.3134, before it had absorbed water; from 1.5662 to 2.3803, after it had absorbed water.

Chemical Characters.

Before the blowpipe, it phosphoresces, and melts with difficulty into a whitish or greenish slag.

Constituent Parts.

	Asbestus of Swartioick, in Sweden.	Asbestus of Tarentaise, in Savoy.	Asbestus of Torias, in Spain.	
Silica, - -	64.0	64.0	72.00	59.00
Carbonate of Magnesia, - -	17.2	18.6	12.19	25.00
Alumina, -	2.7	3.3	3.03	3.00
Carbonate of Lime,	13.9	6.9	10.05	9.05
Barytes, -		6.0		
Oxide of Iron, -	2.2	1.2	2.02	2.25
	Bergmann, Opusc. t. iv.	*Bergmann*, Id.	*Bergmann*, Id.	*Chenevix*.

Geognostic

Geognostic Situation.

It occurs frequently, along with common asbestus, in cotemporaneous veins in serpentine; in similar veins in primitive and flœtz greenstone; in gneiss, and in mica-slate; and it occasionally forms one of the constituent parts of metalliferous beds.

Geographic Situation.

Europe.—It occurs in serpentine in the islands of Unst and Fetlar in Zetland; and in the same rock at Portsoy; in veins in mica-slate, in Glenelg in Inverness-shire; in different parts of Aberdeenshire, and Argyleshire: in flœtz greenstone in the middle division of Scotland, as in Fifeshire, particularly in Inchcolm, and other quarters. In England, it occurs in veins in serpentine, at St Kevern's in Cornwall *. On the Continent, it occurs in the Hartz, in veins in primitive greenstone; in Bohemia, in metalliferous beds, along with magnetic-ironstone; in Upper Saxony, in veins in serpentine; and in a similar situation in Silesia and Switzerland. In Dauphiny, and in St Gothard, it is found in cotemporaneous veins in gneiss and mica-slate, along with felspar, earthy and common chlorite, and rock-crystal. Uncommonly beautiful varieties are met with in the Tarentaise mountains in Savoy, and in the island of Corsica.

Asia.—It abounds in the serpentine rocks in the Uralian and Altain mountains.

Uses.

This mineral, on account of its flexibility, and its resisting the action of considerable degrees of heat, was

* Greenough.

woven into those incombustible cloths in which the ancients sometimes wrapped the bodies of persons of distinction, before they were placed on the funeral-pile, that their ashes might be collected free from admixture *. After the body was consumed, the cloth was withdrawn from the fire, the ashes taken out of it, washed with milk and wine, and sprinkled with consecrated water, and inclosed in an urn, either with or without the fossile-cloth in which the body had been consumed. The goodness of the amianthus for this purpose, depends on the length of its fibres, which vary from an inch to a foot in length, its whiteness and flexibility. In preparing the cloth, the amianthus is previously well washed, to free it of all impurities, then combed straight, and woven with flax. The cloth is placed on glowing coals, by which the flax and oil used in the operation of weaving are consumed, and the cloth is deprived of its stains †. In this manner are manufactured, not only large pieces of cloth, but also gloves, purses, belts, and napkins. All these articles have a shining appearance, and white colour; but various tints may be communicated to them by artificial means. At Nerwinski in Siberia, gloves, caps and purses are made of amianthus; and it is worked into girdles, ribbons, and other articles, in the Pyrenees. The finest girdles are made by weaving the most beautiful varieties of amianthus with silver-wire: they are much prized by the women, not only on account of their beauty, which is certainly very considerable, but from certain mysterious

* Dioscorides says: "Amianthus lapis in Cypro nascitur, scissili alumini similis, quo elaborato utpote flexili, telas spectaculi gratia texunt, quæ ignibus injectæ ardent quidem sed flammis invectæ splendidiores exeunt."

† We are told that the Emperor Charlemagne had a table-cloth of amianthus, which he used to throw into the fire after dinner, that it might burn clean, by way of amusing his guests.

rious properties they are supposed to possess. When a number of fibres are placed together, we can form of them a wick for lamps. It is said the Romans made use of wicks of this kind in the lamps placed in their temples and cemeteries: hence, it has been alleged that these lamps never required to be renewed. It is well known, however, that the duration of amianthus wick is not considerable; for Rozier found that they did not continue for more than twenty hours *. Paper has been made of this mineral, but it is too hard for use. It has been proposed to preserve valuable documents from fire, by writing them on paper of amianthus. Such a plan might deserve consideration, if we possessed fire proof ink. Dolomieu informs us, that it is used by the Corsicans in the composition of a kind of pottery, which is thereby rendered very light, and less liable to be broken by sudden alternations of temperature, or even by falling, than other kinds of pottery. The Chinese pound and knead it with gum tragacanth, and form it into a kind of furnace, which they affirm to be very durable. Ancient physicians prescribed it for different diseases. Thus, in the state of salve, it was considered as very useful in restoring vigour to enfeebled limbs: the itch was said to yield readily to its drying powers; and in affections of the stomach, it was not to be disregarded, as it restored the appetite when entirely lost.

Observations.

1. It is distinguised from *Common Asbestus* by its higher lustre, its fibres being more easily separated, and its flexibility.

2. It is said to be the *Lapis carystius* of Strabo.

Third

* It is said that the natives of Greenland make use of amianthus for the wicks of their lamps.

Third Subspecies.

Common Asbestus *.

Gemeiner Asbest, *Werner*.

Asbestus immaturus, *Wall.* t. i. p. 411.—Gemeiner Asbest, *Wid.* s. 471.—Asbestus, *Kirwan*, vol. i. p. 159.—Gemeiner Asbest, *Estner*, b. ii. s. 872. *Id. Emm.* b. i. s. 406.—Asbesto commune, *Nap.* p. 314.—Asbeste, *Lam.* t. ii. p. 369.—Asbeste dur, *Hauy*, t. iii. p. 247.—L'Asbeste commune, *Broch.* t. i. p. 497.—Gemeiner Asbest, *Reuss*, b. ii. 2. s. 248. *Id. Lud.* b. i. s. 138. *Id. Suck.* 1r th. s. 267. *Id. Bert.* s. 150. *Id. Mohs*, b. i. s. 571. *Id. Hab.* s. 63.—Asbeste dur, *Lucas*, p. 81.—Gemeiner Asbest, *Leonhard*, Tabel. s. 30.—Asbest dur, *Brong.* t i. p. 479. *Id. Brard*, p. 194.—Gemeiner Asbest, *Karsten*, Tabel. s. 42. *Id. Haus.* s. 99.—Asbeste dur, *Hauy*, Tabl. p. 55.—Gemeiner Asbest, *Steffens*, b. i. s. 274. *Id. Lenz*, b. ii. s. 679.—Steifer Asbest, *Oken*, b. i. s. 325.

External Characters.

Its colours are dark leek-green, and mountain-green; also greenish-grey and yellowish grey.

It occurs massive, and very rarely in capillary crystals, that appear to be rhomboidal prisms †.

Internally it is glistening and resinous, passing into pearly.

The

---

* The literal signification of this term is *unextinguishable;* but as the verb ἀποσβέννυμι is metaphorically used in the sense of *aboleo*, or *perdo*, it may be rendered *imperishable;* this explanation being more appropriate than the former to the peculiar character of this substance.—*Kid.*

† Count de Bournon found them to be tetrahedral rhomboidal prisms. —*Cat. Min.* p. 123.

The fracture is parallel, and slightly curved, rather coarse-fibrous.

The fragments are splintery.

It is translucent, or only translucent on the edges.

It is soft, approaching to very soft.

It is rather brittle.

It is difficultly frangible.

It feels rather greasy.

Specific gravity, 2.000, *Karsten.* 2.542, *Kirwan.*

Chemical Characters.

It melts before the blowpipe into a blackish glass.

Constituent Parts.

According to Mr Chenevix, it contains nearly the same constituent parts as amianthus. Gehlen discovered chrome in the leek-green asbestus of Zöblitz, and manganese in a variety from Siberia.

Geognostic Situation.

Like amianthus, it occurs in veins in serpentine, and in primitive greenstone: it also occurs in metalliferous beds, along with magnetic ironstone, iron-pyrites, magnetic pyrites, calcareous-spar, garnet, and indurated talc, and sometimes along with ores of copper, viz. copper-pyrites, copper-glance, and grey copper ore.

Geographic Situation.

Europe.—It occurs in the serpentine of Zetland, Portsoy, and Cornwall; and on the Continent of Europe, it is found in all the serpentine districts, and in metalliferous beds in the Saxon Erzgebirge, Salzburg, &c.

Asia.—It is found at Sisertskoi and Sawod, and other parts in Siberia.

Observations.

Observations.

Common Asbestus connects the Asbestus species with Indurated Talc, and Actynolite. Some varieties of common asbestus are so like asbestous actynolite, that Count de Bournon, Cordier, and other mineralogists, have been led to suppose, that all the subspecies of asbestus are but modifications of Hornblende,—an opinion which is plausible and ingenious.

Fourth Subspecies.

Rock-Wood, or Ligneous Asbestus.

Bergholz, *Werner.*

Bergholz, *Wid.* s. 473.—Ligniform Asbestus, *Kirw.* vol. i. p. 161. —Bergholz, *Estner*, b. ii. s. 877. *Id. Emm.* b. i. s. 410.— Ligno Montano, *Nap.* p. 321.—Asbeste ligniforme, *Hauy*, t. iii. p. 240.—Le Bois de Montagne, *Broch.* t. i. p. 499.— Asbest ligniforme, *Brong.* t. i. p. 480.—Holzasbest, *Reuss*, b. ii. 2. s. 253. *Id. Leonhard*, Tabel. s. 30.—Asbeste ligniforme, *Lucas*, p. 81. *Id. Brong.* t. i. p. 48. *Id. Brard*, p. 195. —Holzasbest, *Karst.* Tabel. s. 42.—Holzartiger Asbest, *Haus.* s. 99.—Ligniform Asbestus, *Kid*, vol. i. p. 105.—Asbeste ligniforme, *Hauy*, Tabl. p. 55.—Bergholz, *Steffens*, b. i. s. 280. *Id. Lenz*, b. ii. s. 680.—Holzicher Asbest, *Oken*, b. i. s. 326.

External Characters.

Its colour is wood-brown, of various degrees of intensity.

It occurs massive, and in plates.

Internally its lustre is glimmering.

The fracture in the large is curved slaty; in the small, delicate and promiscuous fibrous.

The

The fragments are tabular.
It becomes shining in the streak.
It is soft, passing into very soft.
It is opaque.
It is sectile.
It is rather difficultly frangible.
It is slightly elastic-flexible.
It feels meagre.
It creaks when we handle it.
Specific gravity, 2.051, *Wiedeman.*

Chemical Character.

It is infusible before the blowpipe.

Geognostic and Geographic Situations.

It occurs at Sterzing in the Tyrol, along with many different fossils, as common asbestus, actynolite, quartz, garnet, blende, iron-pyrites, galena, and calamine; and its repository, as Mohs remarks, appears to be a bed, as it is accompanied with minerals that often occur in such situations. It is also found in Dauphiny, and in Stiria; and Steffens conjectures, from the descriptions of Georgi, that it occurs in different places in the mountains of Archangel and Olocnezk.

Observations.

It is distinguished from *Rock-Cork*, by its wood-brown colour, higher lustre, and double fracture, being curved slaty in the large, and delicate fibrous in the small.

7. Picrolite.

7. Picrolite.

Pikrolith, *Hausmann.*

Pikrolith, *Hausmann,* Von Moll's Ephem. 4. 3. s. 401. *Id. Steffens,* b. i. s. 273. *Id. Lenz,* b. ii. s. 650. *Id. Oken,* b. i. s. 386.

External Characters.

Its colours are muddy leek-green, mountain-green, and straw-yellow.

It occurs massive.

Internally it is dull, or glimmering and pearly.

The fracture is long splintery, which passes on the one side through fine splintery into even and flat conchoidal; on the other into delicate and concentric fibrous.

The fragments are long splintery.

Sometimes it occurs in concretions, which are conical, or undulating lamellar, which latter passes into fortification-lamellar.

It is translucent on the edges.

It is semi-hard.

It is rather brittle.

It yields a dull white streak.

It is uncommonly difficultly frangible.

It feels meagre.

Specific gravity, 2.5380.

Chemical Characters.

It becomes white before the blowpipe, but is infusible.

Constituent

Constituent Parts.

According to Hausmann, it is principally composed of Carbonate of Magnesia.

Observations.

The above is all the information I have been able to collect in regard to the appearance of this species. It was first described by Hausmann, and he proposes to place it in the system beside Serpentine. It is here placed at the end of the Talc Family, where it can remain until its true place be ascertained.

END OF VOLUME FIRST.

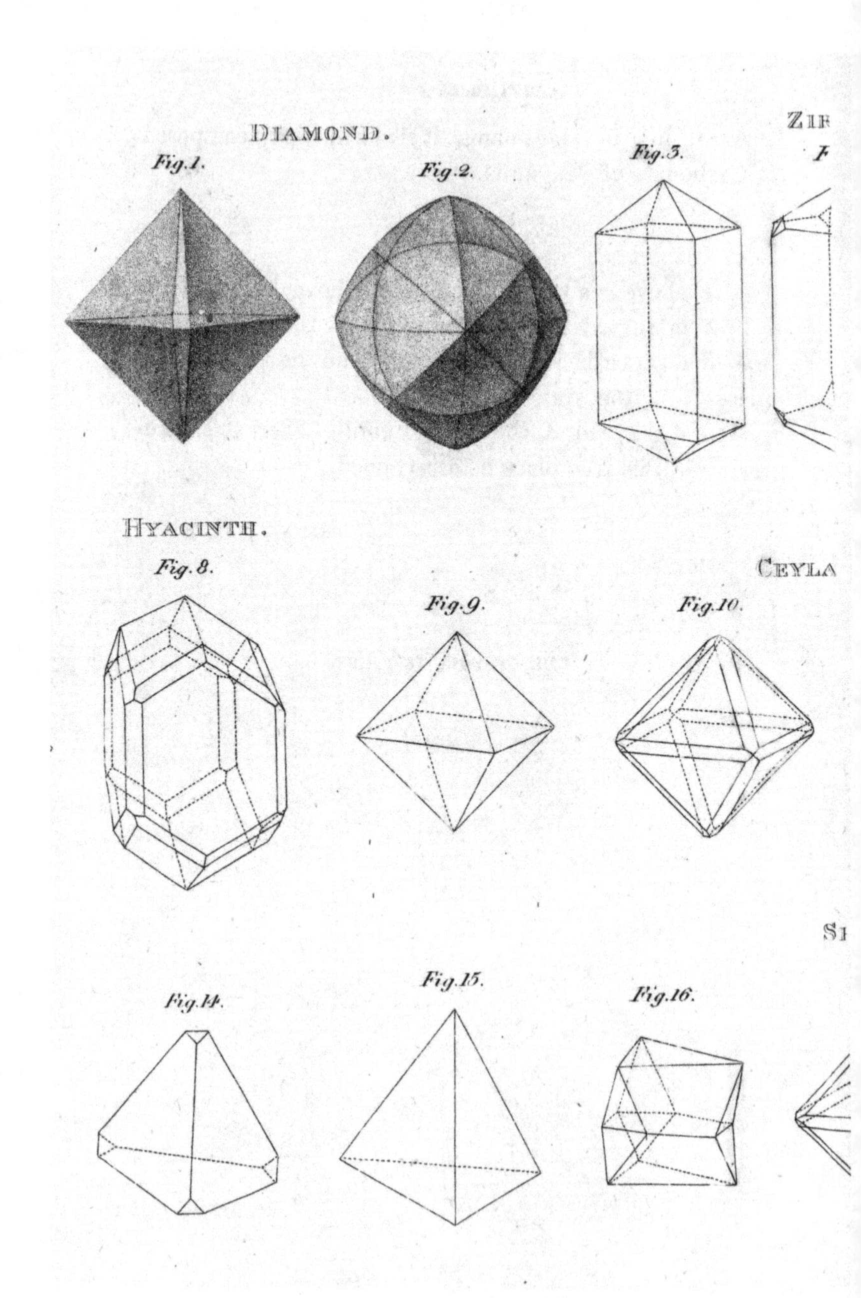
DIAMOND.
Fig.1.
Fig.2.
Fig.3.
ZIR
HYACINTH.
Fig.8.
Fig.9.
Fig.10.
CEYLA
Fig.14.
Fig.15.
Fig.16.
S

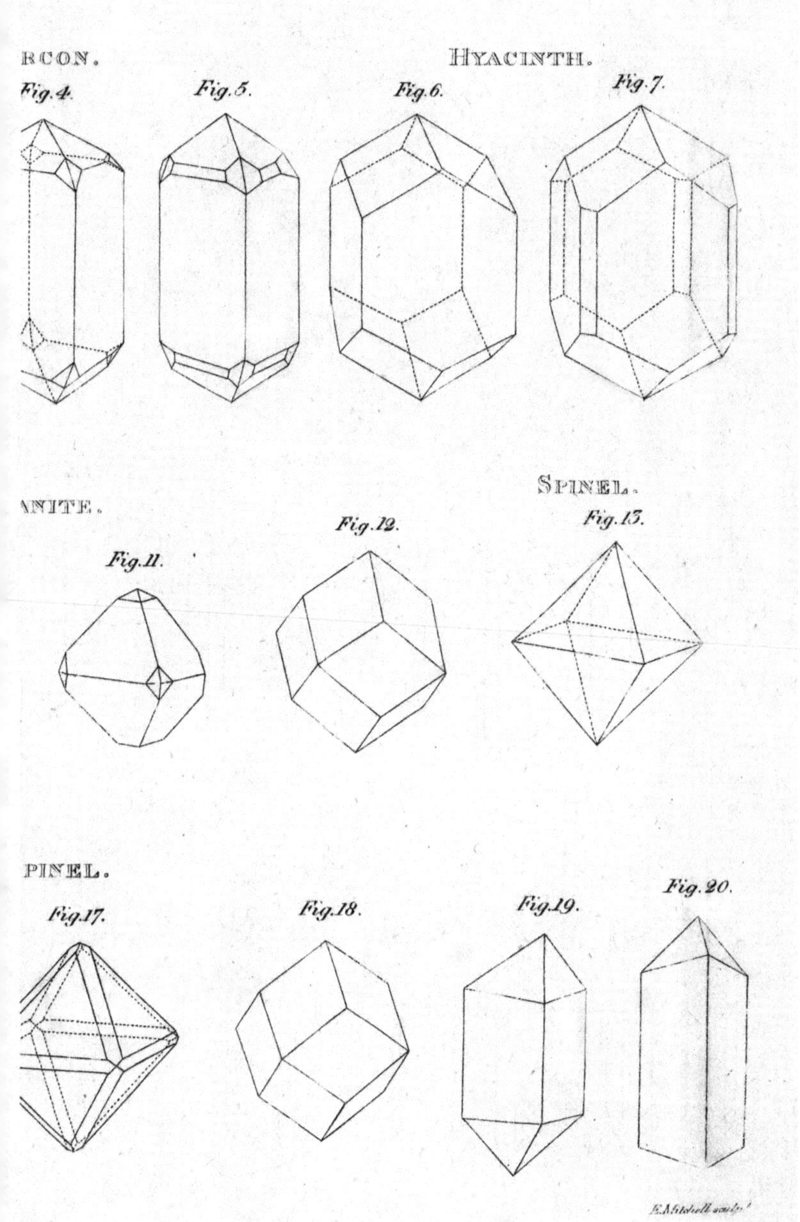
PLATE I
RCON.
HYACINTH.
Fig. 4.
Fig. 5.
Fig. 6.
Fig. 7.
ANITE.
SPINEL.
Fig. 11.
Fig. 12.
Fig. 13.
PINEL.
Fig. 17.
Fig. 18.
Fig. 19.
Fig. 20.
E. Mitchell sculpt

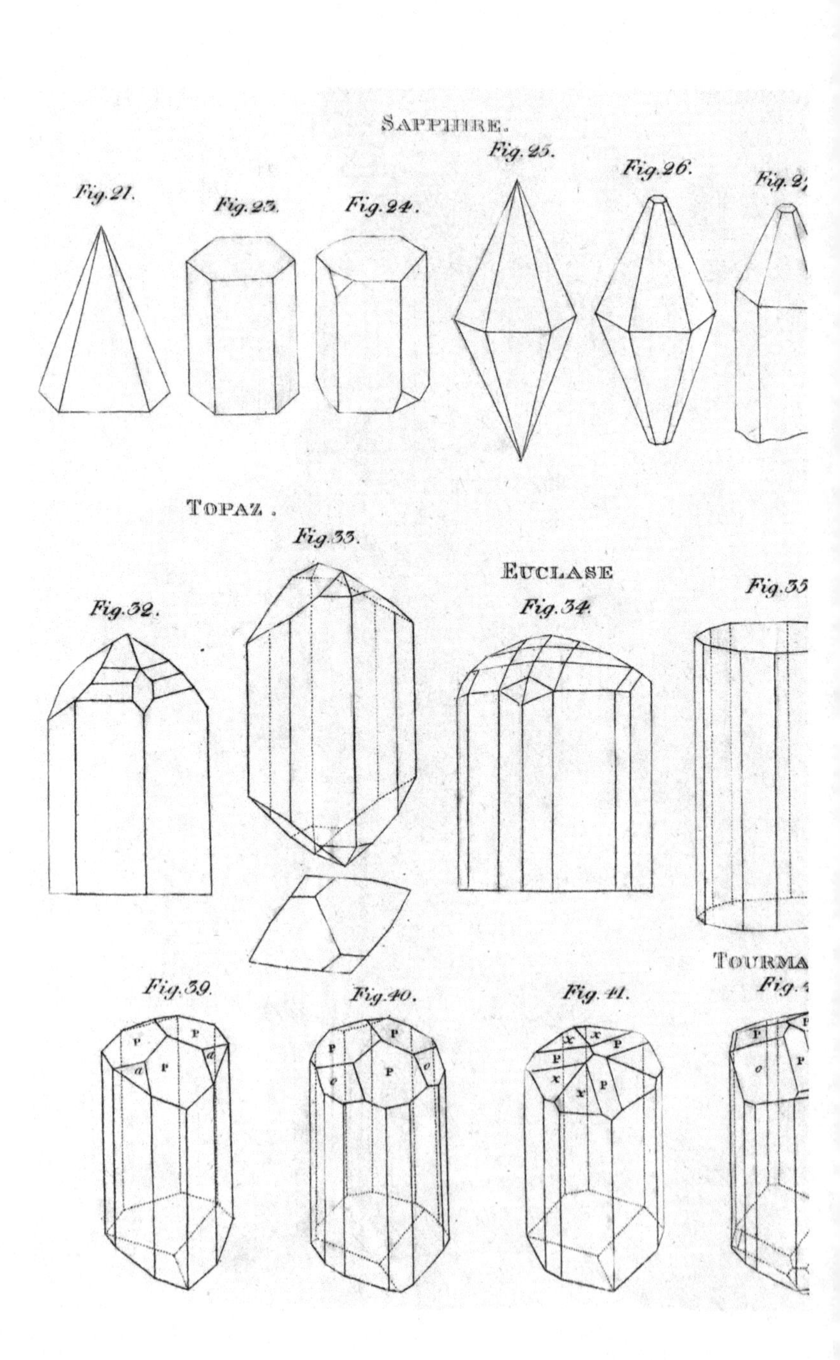
SAPPHIRE.
Fig. 21.
Fig. 23.
Fig. 24.
Fig. 25.
Fig. 26.
TOPAZ.
Fig. 32.
Fig. 33.
EUCLASE
Fig. 34.
Fig. 39.
Fig. 40.
Fig. 41.
TOURMA

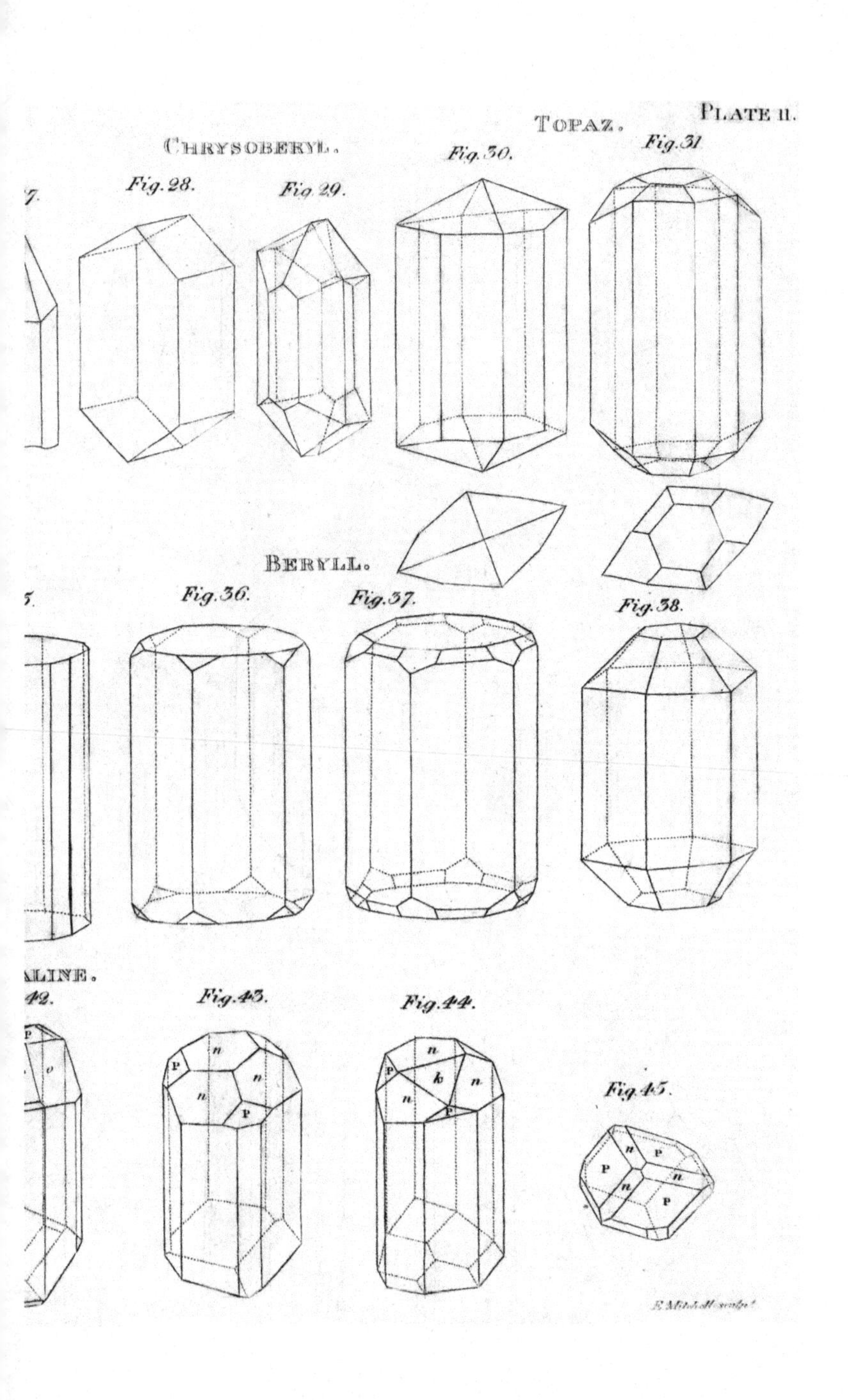
PLATE II.
TOPAZ.
CHRYSOBERYL.
Fig. 28.
Fig. 29.
Fig. 30.
Fig. 31
BERYLL.
Fig. 36.
Fig. 37.
Fig. 38.
Fig. 43.
Fig. 44.
Fig. 45.
E. Mitchell sculpt

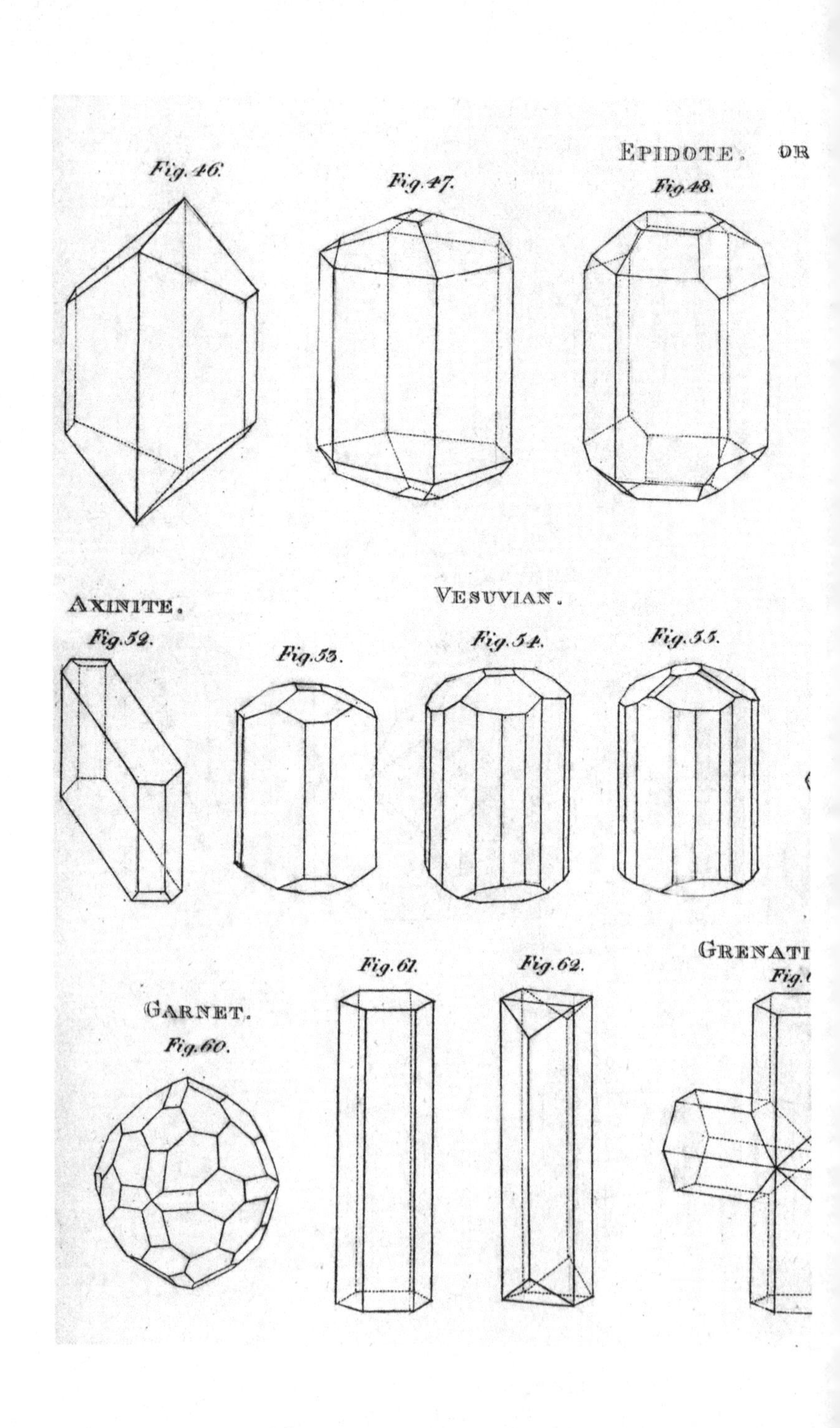

EPIDOTE. OR
Fig. 46.
Fig. 47.
Fig. 48.
AXINITE.
VESUVIAN.
Fig. 52.
Fig. 53.
Fig. 54.
Fig. 55.
GRENATI
GARNET.
Fig. 60.
Fig. 61.
Fig. 62.

PLATE III

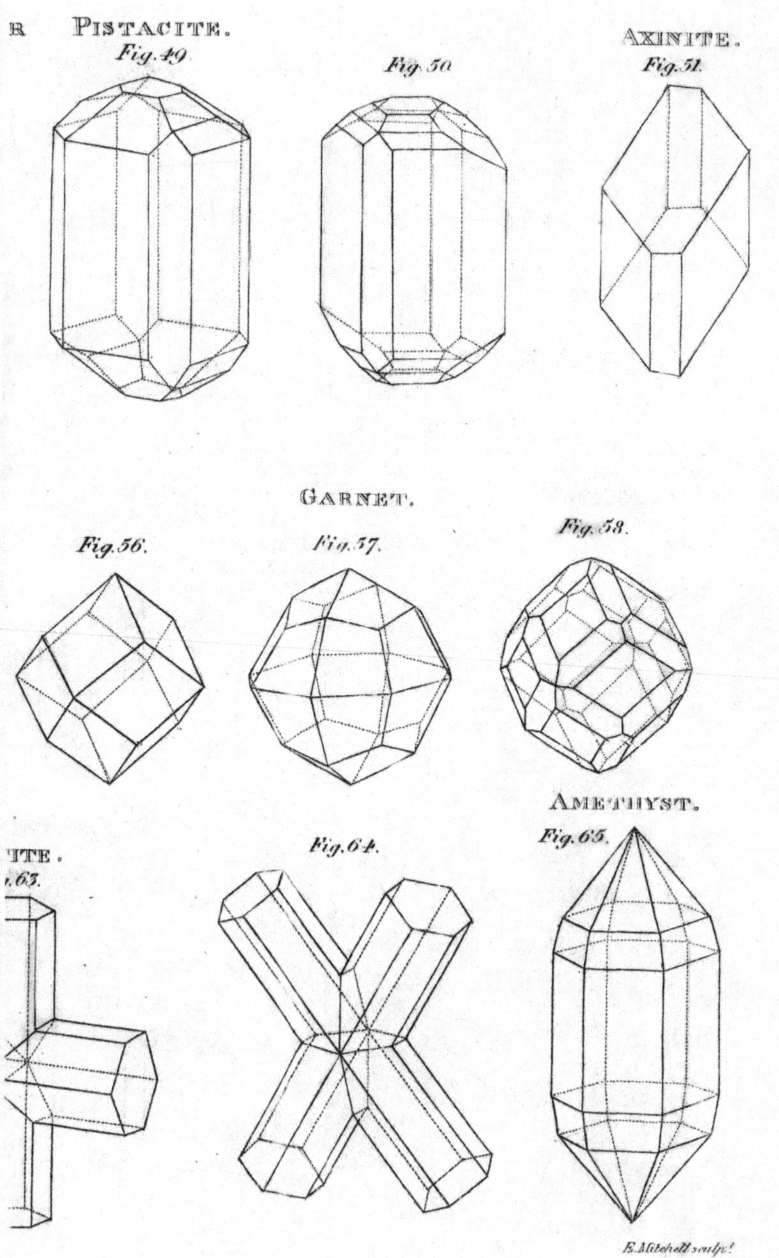
R
PISTACITE.
Fig. 49
Fig. 50
AXINITE.
Fig. 51.
GARNET.
Fig. 56.
Fig. 57.
Fig. 58.
AMETHYST.
Fig. 65.
ITE.
.63.
Fig. 64.
E. Mitchell sculp.t

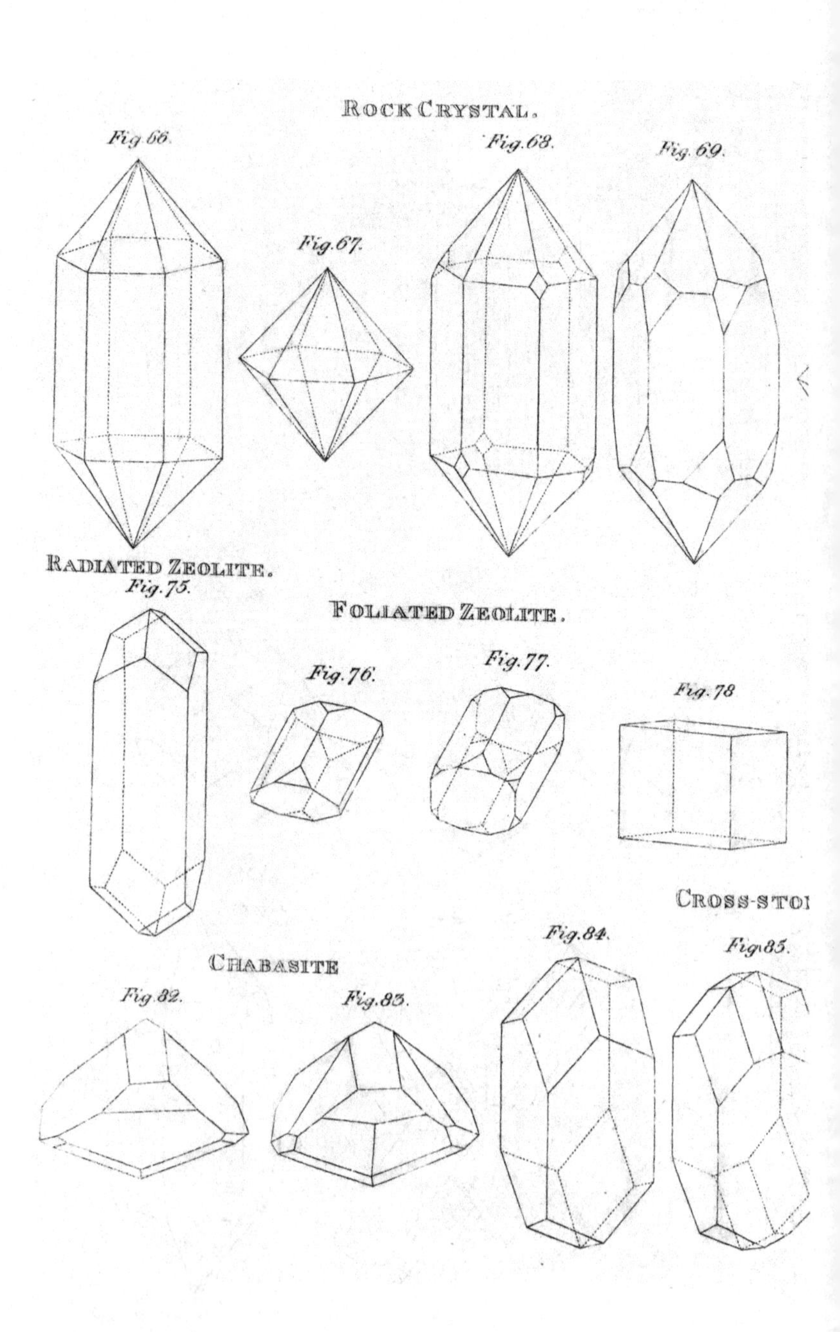

ROCK CRYSTAL.
Fig. 66.
Fig. 67.
Fig. 68.
Fig. 69.
RADIATED ZEOLITE.
Fig. 75.
FOLIATED ZEOLITE.
Fig. 76.
Fig. 77.
Fig. 78
CROSS-STO
Fig. 84.
Fig. 85.
CHABASITE
Fig. 82.
Fig. 83.

PLATE IV.

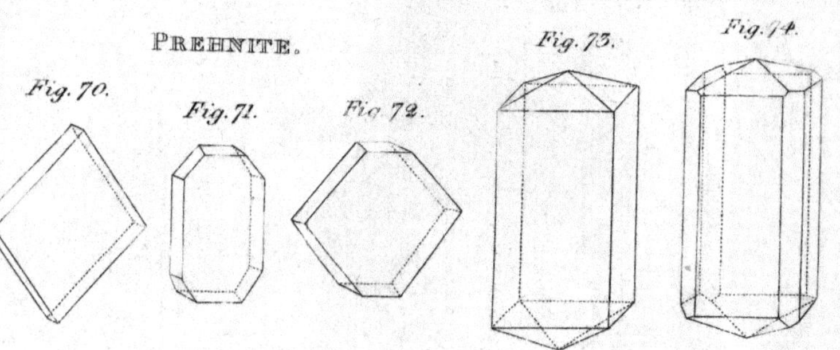
NEEDLE ZEOLITE.
PREHNITE.
Fig. 70.
Fig. 71.
Fig. 72.
Fig. 73.
Fig. 74.

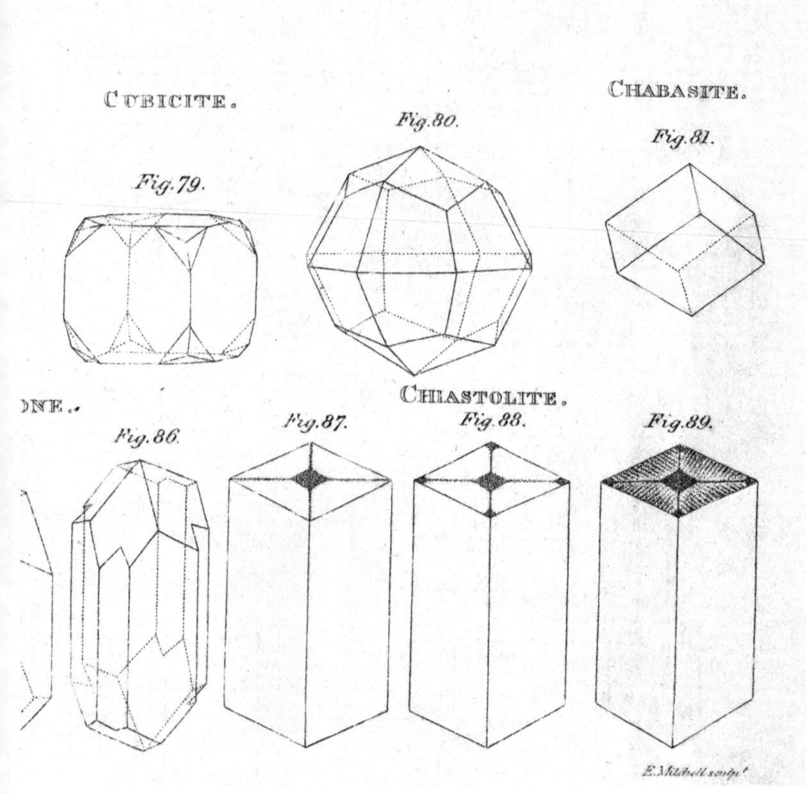
CUBICITE.
CHABASITE.
Fig. 80.
Fig. 81.
Fig. 79.
CHIASTOLITE.
Fig. 86.
Fig. 87.
Fig. 88.
Fig. 89.
E. Mitchell sculpt

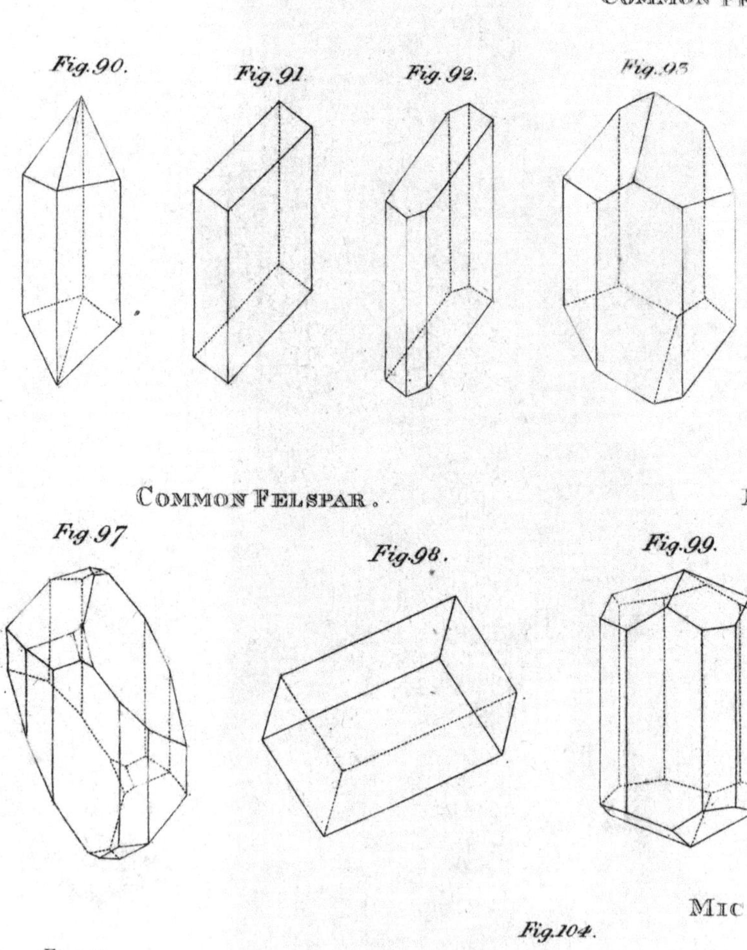
Fig.90.
Fig.91
Fig.92.
Fig.93
Common Felspar.
Fig.97
Fig.98.
Fig.99.
Fig.104.
Fig.103.a
Fig.103.b

PLATE V.

ELSPAR.

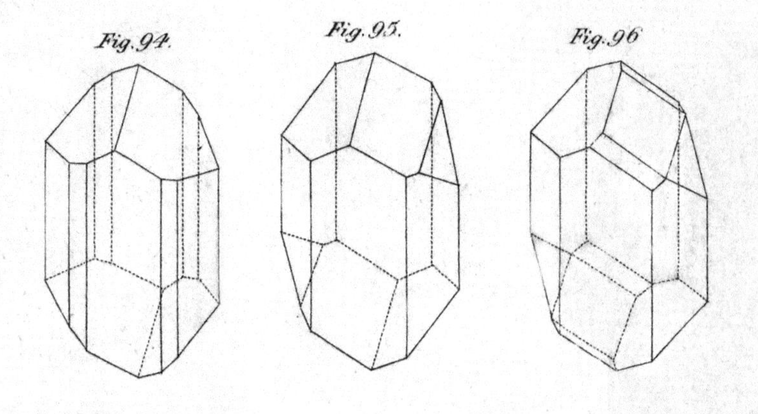

Fig. 94. Fig. 95. Fig. 96.

MEIONITE.

Fig. 100.

NEPHELINE.

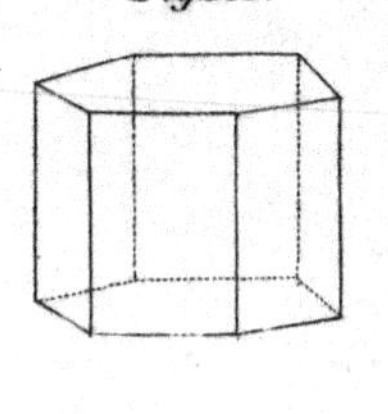

Fig. 101.

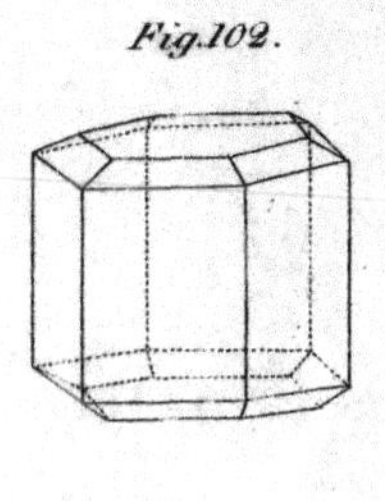

Fig. 102.

CA.

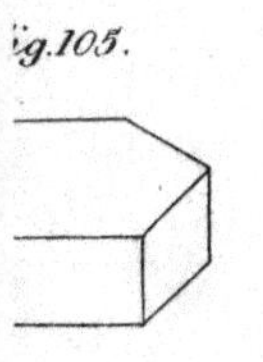

Fig. 105.

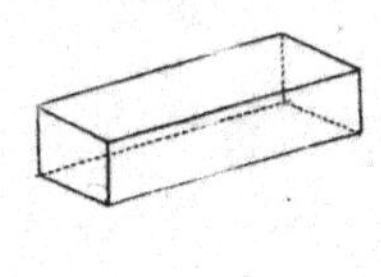

Fig. 106.

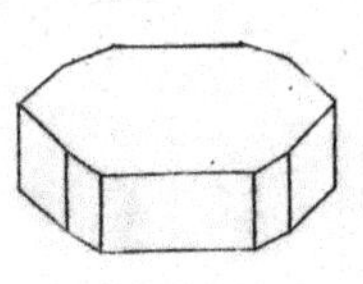

Fig. 107.

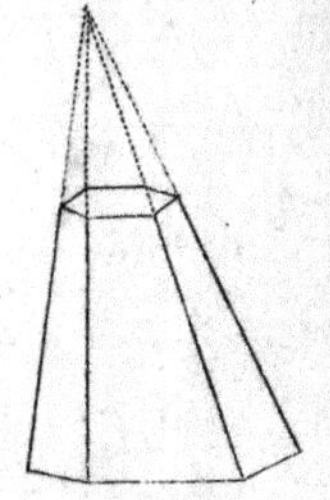

Fig. 108.

Printed in the United States
By Bookmasters